Urban Rail Transportation: Planning and Management

Urban Rail Transportation: Planning and Management

Edited by Callen Lucas

CLANRYE INTERNATIONAL
www.clanryeinternational.com

Clanrye International,
750 Third Avenue, 9th Floor,
New York, NY 10017, USA

ISBN: 978-1-63240-862-4

Cataloging-in-Publication Data

Urban rail transportation : planning and management / edited by Callen Lucas.
p. cm.
Includes bibliographical references and index.
ISBN 978-1-63240-862-4
1. Railroads. 2. Urban transportation. 3. Urban transportation--Planning.
4. Urban transportation--Management. I. Lucas, Callen.
HE1031 .U73 2019
385--dc23

For information on all Clanrye International publications
visit our website at www.clanryeinternational.com

CLANRYE INTERNATIONAL

Printed in China.

Contents

Preface

The main aim of this book is to educate learners and enhance their research focus by presenting diverse topics covering this vast field. This is an advanced book which compiles significant studies by distinguished experts in the area of analysis. This book addresses successive solutions to the challenges arising in the area of application, along with it; the book provides scope for future developments.

Rail transport refers to the transfer of goods and passengers from one place to another on wheeled vehicles, which run on tracks or rails. Urban rail transit includes providing transit services to passengers and goods in and around urban and sub-urban areas. The planning and management of urban rail transit includes the maintenance, safety and financial planning of the rail transport. Regular inspections done via track geometry cars and walking inspections are a part of rail maintenance. The socio-economic benefits and energy efficiency are some of the primary concerns while planning an urban rail system. This book discusses the fundamentals as well as modern approaches of urban rail transportation. It explores all the important aspects of urban rail transportation in the present day scenario. The book is appropriate for students seeking detailed information in this area as well as for experts.

It was a great honour to edit this book, though there were challenges, as it involved a lot of communication and networking between me and the editorial team. However, the end result was this all-inclusive book covering diverse themes in the field.

Finally, it is important to acknowledge the efforts of the contributors for their excellent chapters, through which a wide variety of issues have been addressed. I would also like to thank my colleagues for their valuable feedback during the making of this book.

Editor

1

A Time-Domain Model for the Study of High-Frequency Wheelset–Track Interaction

Wenshan Fang[1] · José Martinez-Casas[2] · Stefano Bruni[1]

Abstract A mathematical model of dynamic wheelset–track interaction is proposed in this paper. The model is defined in the time domain in order to introduce and correctly evaluate non-linear and time-variant phenomena related to the contact model and boundary conditions which play a very important role in rail surface degradation phenomena. The complete model can be divided into three main components: the model of the wheelset, the model of the track and the model of wheel–rail contact forces. In the paper, the wheelset is described as a rotating flexible body, and the gyroscopic and inertial effects associated with wheelset rotation are introduced to this model using an 'Eulerian' finite element approach based on 3D quadratic solid elements. The discrete supported track is modelled using finite Timoshenko beam element, which takes into account both the vertical and the lateral rail vibration valid up to 1500 Hz. The wheelset and the track are coupled by means of a contact model based on the nonlinear Hertz and Kalker theories. The flexible components of the interaction model make it possible to describe the train–track dynamics in a relatively high-frequency range, which allows the investigation of specific aspects such as rail corrugation. Some numerical results are presented in terms of contact forces and rail–wheel vibration speed in the paper. The effect of wheelset and track flexibility in specific frequency range on train–track interaction dynamics is briefly discussed.

✉ Wenshan Fang
wenshan.fang@polimi.it

[1] Mechanical Engineering Department, Politecnico di Milano, 20156 Milan, Italy

[2] Centro de Investigación en Ingeniería Mecánica, Universitat Politècnica de València, Camino de Vera s/n, 46022 Valencia, Spain

Editor: Xuesong Zhou

Keywords Flexible wheelset · Flexible track · Train–track interaction · High-frequency dynamics

1 Introduction

Train–track interaction has been extensively studied in the last 40 years at least, leading to modelling approaches that can deal satisfactorily with many dynamic problems arising at the wheel–rail interface. Some important dynamic phenomena, for instance, rolling noise [1] and vehicle–track coupled vibration response [2] caused by the roughness of the wheel and rail running surfaces have been investigated with frequency-domain wheel–rail interaction models, where the dynamic response characteristics at wheel–rail contact are derived by combining the wheelset and track frequency response functions with the assumed rail roughness [3]. However, frequency-domain method gets its limit when comes to nonlinear components in wheel–rail contact and vehicle suspension. To take consideration of the effect of nonlinearities in the track and in wheel–rail contact, train–track interaction models in time domain are widely studied today: in [4–14], the train–track interaction is defined in time domain with the vehicle and/or the track model defined using the finite element method.

In state-of-the-art papers, different configurations of the vehicle model and the track model are used to investigate the dynamic influences of a train running along a track under specific targets. To investigate the track dynamics, detailed track model coupled with relatively simple wheelset model was used. Zhai et al. [9] investigated the influence of elastic track structures on the lateral hunting stability of different railway vehicles. Baeza et al. [10] developed a cyclic boundary track model based on modal substructuring. Di Gialleonardo et al. [13] investigated the effect of different levels of track flexibility on running behaviours of both tangent and curved tracks. To investigate the wheelset dynamics, efforts have been made to develop detailed rotating flexible wheelset model. Arnold et al. [5] and Kaiser et al. [7, 13], proposed the use of a semi-analytical solution of a rotational symmetric structure in two dimensions, using Navier's equations. Baeza et al. [8, 11], on the other hand, modelled the rotating wheelset based on Brown and Shabana's formulation [15, 16] of kinematics of flexible rotating structure using a Lagrangian coordinate. However, open problems still exist related to wheel–rail interaction in the high-frequency range and its relationship with wheelset flexibility. Furthermore, the effect of wheelset and track flexibility at different levels still needs to be investigated in full.

The aim of this paper is to define a wheelset–track interaction model valid in a frequency range up to 1500 Hz and hence suitable for the investigation of typical problems related to high-frequency train–track interaction such as short-pitch rail corrugation. To this aim, a new coupling of a full rotating flexible wheelset model together with a discrete supported Timoshenko finite element track model is developed. The wheelset model is developed using a Eulerian approach [14]. The equations of wheelset motion are deducted using virtual work principal, and the convergence of the model is guaranteed using integral by parts, which has not been treated in previous work. The track model describes displacement and rotation of both vertical and lateral directions. The results of this new model are presented by comparing the running behaviours under different model configurations.

2 Mathematic Model of the Wheelset–Track Interaction

The vehicle model used in this paper considers one single wheelset and the primary suspension only. The second suspension system is ignored, with the bogie frame and the car body masses represented by two static forces applied through the primary suspension on the two sides. The reason of this simplification is due to the mechanical filter effect introduced by the suspensions: the dynamics of the sprung masses of vehicle (bogie frame and car body) occur in low-frequency area under 20 Hz and thus are well isolated from the high-frequency vibrations of wheelsets and can consequently be simplified as lumped masses [10, 11, 14, 17] when modelling the high-frequency vehicle–track interaction.

2.1 Model of the Rotating Flexible Wheelset

When modelling the rotating flexible wheelset, the interested contact points are always applied to the interface with the rail. The external forces of the wheelset are applied in the direction and displacement of the contact points, which is a fixed spatial point. Thus, a Eulerian approach originally from [14] is introduced to describe the kinematic motion of the wheelset. The position vector $\mathbf{r}$ is defined as:

$$\mathbf{r} = \mathbf{u} + \mathbf{w}(\mathbf{u}, t), \tag{1}$$

where $\mathbf{u}$ is the position of an undeformed spatial point and $\mathbf{w}(\mathbf{u}, t)$ is the displacement associated with wheelset flexibility.

The wheelset is set to rotate with spin velocity Ω in the second axle, and the angular velocity tensor verifies:

$$\tilde{\mathbf{\Omega}} = \begin{bmatrix} 0 & 0 & \Omega \\ 0 & 0 & 0 \\ -\Omega & 0 & 0 \end{bmatrix} = \Omega\mathbf{J}; \quad \text{and} \quad \mathbf{J}\cdot\mathbf{J} = -\mathbf{E}; \tag{2}$$

where J and E are defined as:

$$\mathbf{J} = \begin{bmatrix} 0 & 0 & 1 \\ 0 & 0 & 0 \\ -1 & 0 & 0 \end{bmatrix}; \quad \mathbf{E} = \begin{bmatrix} 1 & 0 & 0 \\ 0 & 0 & 0 \\ 0 & 0 & 1 \end{bmatrix}. \tag{3}$$

The spin velocity of the wheelset is:

$$\mathbf{V} = (\, v_1 \quad v_2 \quad v_3 \,)^{\mathrm{T}} = \tilde{\mathbf{\Omega}}\mathbf{u} = \Omega\mathbf{J}\mathbf{u} = \Omega\tilde{\mathbf{u}} \tag{4}$$

where $\tilde{\mathbf{u}} = (\, u_3 \quad 0 \quad -u_1 \,)^{\mathrm{T}}$.

The equation of wheelset motion is formulated by virtual work principle. The velocity and acceleration of the particle are computed through the material derivatives as follows:

$$\begin{aligned} \dot{\mathbf{r}} &= \frac{D\mathbf{r}}{Dt} = \frac{D\mathbf{u}}{Dt} + \frac{D\mathbf{w}}{Dt} = \sum_{i=1}^{3} v_i \frac{\partial \mathbf{u}}{\partial u_i} + \dot{\mathbf{w}} + \sum_{i=1}^{3} v_i \frac{\partial \mathbf{w}}{\partial u_i} \\ &= \Omega\mathbf{J}\mathbf{u} + \dot{\mathbf{w}} + \Omega \sum_{i=1}^{3} \tilde{\mathbf{u}}_{\mathbf{i}} \frac{\partial \mathbf{w}}{\partial u_i}; \end{aligned} \tag{5}$$

$$\begin{aligned} \ddot{\mathbf{r}} &= \frac{D^2\mathbf{r}}{Dt^2} \\ &= \Omega\mathbf{J}\frac{\partial u}{\partial t} + \frac{\partial \dot{\mathbf{w}}}{\partial t} + \sum_{i=1}^{3} v_{\mathbf{i}} \frac{\partial \dot{\mathbf{w}}}{\partial u_i} + \Omega \sum_{i=1}^{3} \frac{\partial \tilde{u}_i}{\partial t}\frac{\partial \mathbf{w}}{\partial u_i} \\ &\quad + \Omega \sum_{i=1}^{3} \tilde{u}_i \frac{\partial^2 \mathbf{w}}{\partial u_i^2}\frac{\partial u_i}{\partial t} + \Omega \sum_{i=1}^{3} \tilde{u}_i \frac{\partial \dot{\mathbf{w}}}{\partial u_i} \end{aligned} \tag{6}$$

The virtual work associated with the inertial forces is:

$$\begin{aligned} \delta W &= \int_{\mathrm{Vol}} \rho\delta\mathbf{w}^{\mathrm{T}}\ddot{r}\mathrm{d}v = \int_{\mathrm{Vol}} \rho\delta\mathbf{w}^{\mathrm{T}}\ddot{\mathbf{w}}\mathrm{d}v + 2\Omega \int_{\mathrm{Vol}} \rho\delta\mathbf{w}^{\mathrm{T}} \sum_{i=1}^{3} \tilde{u}_i \frac{\partial \dot{\mathbf{w}}}{\partial u_i}\mathrm{d}v \\ &\quad + \Omega^2\Bigg(- \int_{\mathrm{Vol}} \rho\delta\mathbf{w}^{\mathrm{T}} \sum_{i=1}^{3} \mathbf{E}u_i \frac{\partial \mathbf{w}}{\partial u_i}\mathrm{d}v \\ &\quad + \int_{\mathrm{Vol}} \rho\delta\mathbf{w}^{\mathrm{T}} \sum_{i=1}^{3}\sum_{j=1}^{3} \tilde{u}_i\tilde{u}_j \frac{\partial^2 \mathbf{w}}{\partial u_i \partial u_j}\mathrm{d}v - \int_{\mathrm{Vol}} \rho\delta\mathbf{w}^{\mathrm{T}} \sum_{i=1}^{3} \mathbf{E}u_i\mathrm{d}v \Bigg) \end{aligned} \tag{7}$$

The virtual work equation is then solved by finite element integration. In this step, the integrals in the volume of the solid are computed as the sum of integrals on each volume of the finite element elements:

$$\int_{\text{Vol}} \cdot\, \mathrm{d}v = \sum_{\text{elements}} \int_{V^e} \cdot \mathrm{d}v \tag{8}$$

To reduce the computational cost, a modal approach and nodal interpolation are adopted after the finite element integration. The displacement of the eth element in volume V^e is:

$$\mathbf{w}(\mathbf{u},t) = \mathbf{N}^e(\mathbf{u})\mathbf{w}^e(t) = \mathbf{N}^e(\mathbf{u})\Phi^e(\mathbf{u})\mathbf{q}(t) \quad \mathbf{u} \in V^e \tag{9}$$

where $\mathbf{N}^e(\mathbf{u})$ is the shape function matrix of the eth element, $\mathbf{w}^e(t)$ is the nodal displacement of the eth element, $\mathbf{\Phi}^e(\mathbf{u})$ is the mode shapes of the eth element computed from the finite element model of the wheelset, and $\mathbf{q}(t)$ is the modal coordinate vector. Introduce Eq. (9) into Eq. (7), one can obtain:

$$\delta W = \sum_{e=1}^{\text{N elements}} \delta \mathbf{q}^{e\mathrm{T}} \Phi_{FE}^{e\mathrm{T}} \times \begin{pmatrix} \int_{V^e} \rho \mathbf{N}^{e\mathrm{T}}\mathbf{N}^e \mathrm{d}v \ddot{\mathbf{q}}^e + 2\Omega \int_{V^e} \rho \mathbf{N}^{e\mathrm{T}} \sum_{i=1}^{3} \tilde{u}_i \frac{\partial \mathbf{N}^e}{\partial u_i} \mathrm{d}v \dot{\mathbf{q}}^e \\ + \Omega^2 (- \int_{V^e} \rho \mathbf{N}^{e\mathrm{T}} \sum_{i=1}^{3} \mathbf{E} u_i \frac{\partial \mathbf{N}^e}{\partial u_i} \mathrm{d}v \\ + \int_{\text{Vol}} \rho \mathbf{N}^{e\mathrm{T}} \sum_{i=1}^{3}\sum_{j=1}^{3} \tilde{u}_i \tilde{u}_j \frac{\partial^2 \mathbf{N}^e}{\partial u_i \partial u_j} \mathrm{d}v \\ - \int_{\text{Vol}} \rho \mathbf{N}^{e\mathrm{T}} \sum_{i=1}^{3} E u_i \mathrm{d}v) \mathbf{q}^e \end{pmatrix} \Phi_{FE}^{e} \tag{10}$$

Define matrices $\mathbf{M}^e$, $\mathbf{V}^e$, $\mathbf{A}^e$, $\mathbf{C}^e$, $\mathbf{L}^e$ as follows:

$$\begin{aligned} \mathbf{M}^e &= \int_{V^e} \rho \mathbf{N}^{e\mathrm{T}} \mathbf{N}^e \mathrm{d}v; \\ \mathbf{V}^e &= \int_{V^e} \rho \mathbf{N}^{e\mathrm{T}} \sum_{i=1}^{3} \tilde{u}_i \frac{\partial \mathbf{N}^e}{\partial u_i} \mathrm{d}v; \\ \mathbf{C}^e &= \int_{V^e} \rho \mathbf{N}^{e\mathrm{T}} \sum_{i=1}^{3} \mathbf{E} u_i \frac{\partial \mathbf{N}^e}{\partial u_i} \mathrm{d}v; \\ \mathbf{A}^e &= \int_{V^e} \rho \mathbf{N}^{e\mathrm{T}} \sum_{i=1}^{3}\sum_{j=1}^{3} \tilde{u}_i \tilde{u}_j \frac{\partial^2 \mathbf{N}^e}{\partial u_i \partial u_j} \mathrm{d}v; \\ \mathbf{L}^e &= \int_{V^e} \rho \mathbf{N}^{e\mathrm{T}} \sum_{i=1}^{3} \mathbf{E} u_i \mathrm{d}v; \end{aligned} \tag{11}$$

where matrices $\mathbf{V}^e$, $\mathbf{A}^e$, $\mathbf{C}^e$, $\mathbf{L}^e$ are related to the gyroscopic and inertial effects associated with wheelset rotation.

With 3D solid element, the polynomial shape functions are C^0 continuous. Thus, in Eq. (11), the convergence of the integral of matrix $\mathbf{A}^e$ cannot be guaranteed due to the second derivative of the shape function. This problem is solved by integrating $\mathbf{A}^e$ by parts. The integration is taken over all of the three directions in space, which leads to:

$$\begin{aligned} \mathbf{A}^e &= \int_{V^e} \rho \mathbf{N}^{e\mathrm{T}} \sum_{i=1}^{3}\sum_{j=1}^{3} \tilde{u}_i \tilde{u}_j \frac{\partial^2 \mathbf{N}^e}{\partial u_i \partial u_j} \mathrm{d}v \\ &= \int_{S^e} \rho \mathbf{N}^{e\mathrm{T}} \sum_{i=1}^{3} \tilde{u}_i \tilde{u}_1 \frac{\partial \mathbf{N}^e}{\partial u_i} \mathrm{d}u_2 \mathrm{d}u_3 + \int_{S^e} \rho \mathbf{N}^{e\mathrm{T}} \sum_{i=1}^{3} \tilde{u}_i \tilde{u}_2 \frac{\partial \mathbf{N}^e}{\partial u_i} \mathrm{d}u_3 \mathrm{d}u_1 \\ &+ \int_{S^e} \rho \mathbf{N}^{e\mathrm{T}} \sum_{i=1}^{3} \tilde{u}_i \tilde{u}_3 \frac{\partial \mathbf{N}^e}{\partial u_i} \mathrm{d}u_1 \mathrm{d}u_2 \\ &- \int_{V^e} \rho \left(\sum_{i=1}^{3}\sum_{j=1}^{3} \tilde{u}_i \tilde{u}_j \frac{\partial \mathbf{N}^{e\mathrm{T}}}{\partial u_i} \frac{\partial \mathbf{N}^e}{\partial u_j} - \mathbf{N}^{e\mathrm{T}} \sum_{i=1,3} u_i \frac{\partial \mathbf{N}^e}{\partial u_i} \right) \mathrm{d}v \end{aligned} \tag{12}$$

With Eqs. (8), (10), (11) and (12), the matrices of each element can be assembled into global matrices using a standard finite element method. Consequently, the final motion of equation of the wheelset is:

$$\begin{aligned} &\ddot{\mathbf{q}} + 2\Omega \Phi_{FE}^{\mathrm{T}} \mathbf{V} \Phi_{FE} \dot{\mathbf{q}} + (\mathbf{K} + \Omega^2 \Phi_{FE}^{\mathrm{T}} (\mathbf{A} - \mathbf{C}) \Phi_{FE}) q \\ &\quad = \Omega^2 \Phi_{FE}^{\mathrm{T}} \mathbf{L} + \mathbf{Q} \end{aligned} \tag{13}$$

where $\mathbf{Q}$ is the generalized forces of the wheelset and $\mathbf{K}$ is the modal stiffness matrix of the finite element wheelset model.

In present work, an ETR500 wheelset profile is modelled (Fig. 1). 3D solid element with 20 nodes is used in the mesh, which can describe the bending motion better than linear solid element. Fifty modes of the wheelset are adopted in the simulation, covering a frequency range up to 1500 Hz approximately. Some of the mode shapes of the flexible wheelset and the corresponding natural frequencies are reported in Fig. 2 and Table 1. The modal frequencies are validated by experimental data (from [17]) with an acceptable error, see Table 1.

In Fig. 2a, d are mode shapes with multiplicity 1: the first torsional mode and the first umbrella mode. These

Fig. 1 Finite element model of the wheelset

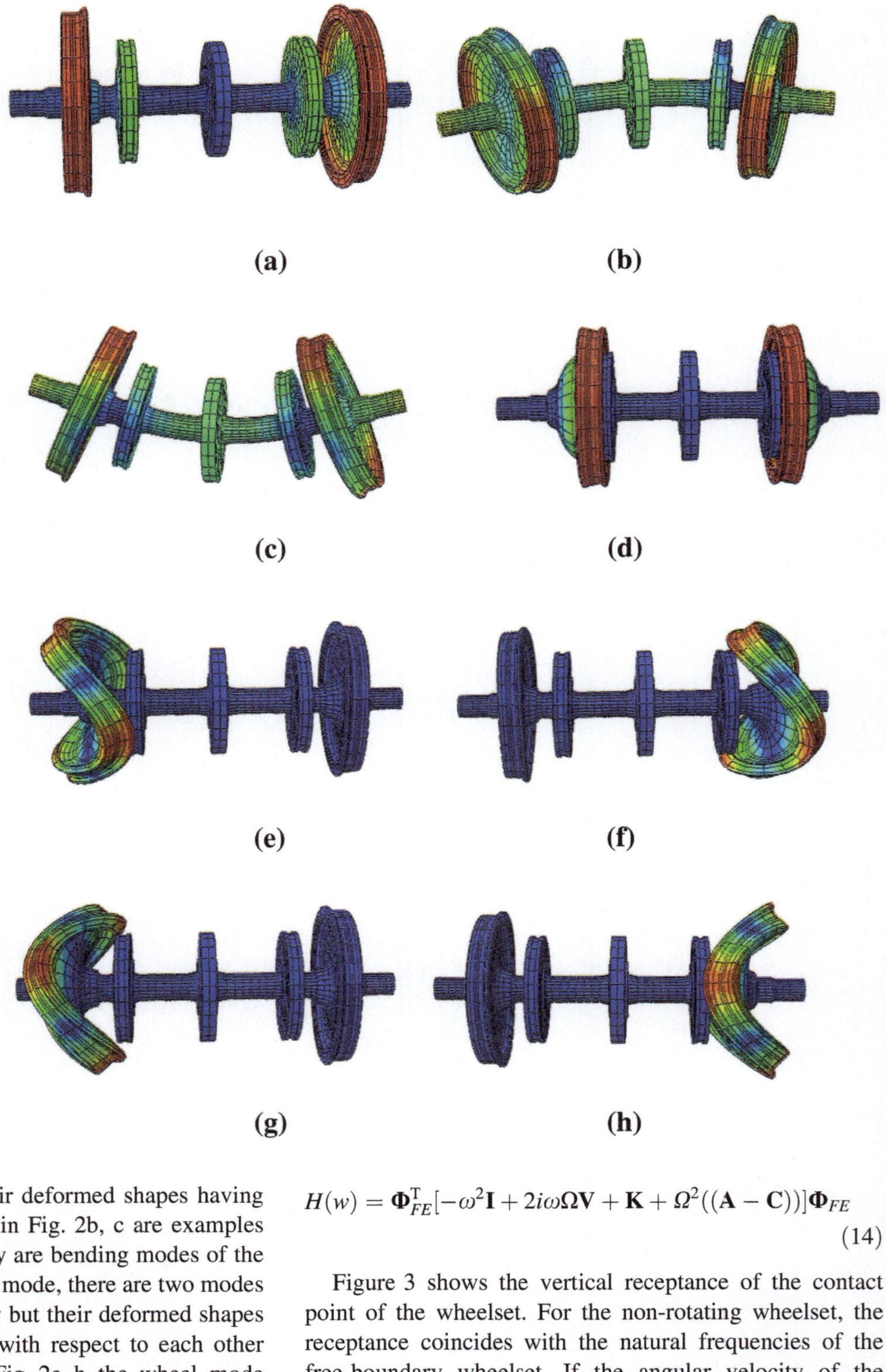

Fig. 2 Mode shapes of the flexible wheelset: **a** first torsional mode; **b**, **c** first bending mode; **d** first umbrella mode; **e–h** first wheelset bending mode

modes are characterized by their deformed shapes having axial symmetry. Modes shown in Fig. 2b, c are examples of modes of multiplicity 2. They are bending modes of the wheelset axis. For each bending mode, there are two modes with the same natural frequency but their deformed shapes are rotated by a certain angle with respect to each other around the wheelset axis. In Fig. 2e–h the wheel mode shapes with two nodal diameters and no nodal circle are shown. Although the natural frequency corresponding to these mode shapes is the same (341.3 Hz), they must be regarded as two different modes with multiplicity 2.

The receptance expression is obtained from Eq. (13):

$$H(w) = \mathbf{\Phi}_{FE}^{\mathrm{T}}[-\omega^2\mathbf{I} + 2i\omega\Omega\mathbf{V} + \mathbf{K} + \Omega^2((\mathbf{A} - \mathbf{C}))]\mathbf{\Phi}_{FE} \tag{14}$$

Figure 3 shows the vertical receptance of the contact point of the wheelset. For the non-rotating wheelset, the receptance coincides with the natural frequencies of the free-boundary wheelset. If the angular velocity of the wheelset is nonzero, the peaks of the FRF associated with modes with multiplicity 2 (bending modes) produce two peaks that can be interpreted as the forward and backward modes. For instance, when the vehicle velocity is 100 km/h, the backward mode and the forward mode of the first

Table 1 Vibration modes of the wheelset

Vibration mode	Frequency by MATLAB (Hz)	Experimental frequency (Hz) (from [18])
First torsional mode	71.9	–
First vertical and horizontal bending modes (2)	75.0	77
Second vertical and horizontal bending modes (2)	121.9	122
Third vertical and horizontal bending modes (2)	176.4	184
First umbrella mode	203.3	203
Second umbrella mode	247.3	254
Fourth vertical and horizontal bending modes (2)	317.9	315
First bending modes of wheels (2 + 2)	341.3	335
Fifth vertical and horizontal bending modes (2)	570.7	548
Second bending modes of the wheels (2 + 2)	936.8	922

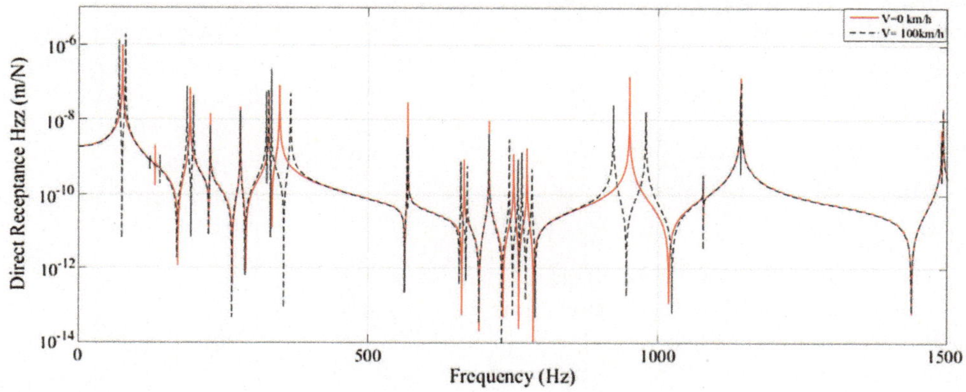

Fig. 3 Direct wheelset receptance at the contact point Hz for different vehicle velocities

bending mode are 65 and 85 Hz, respectively. The peaks of the receptance associated with modes with multiplicity 1 (torsional modes and umbrella modes) remain the same.

A Campbell diagram for the wheelset shows the frequency intersection points of the model. Several intersection points of 'forward' and 'backward' modes of wheelset modes with multiplicity 2 are shown in Fig. 4, which indicates wheelset resonances are affected by the rotation of the wheelset.

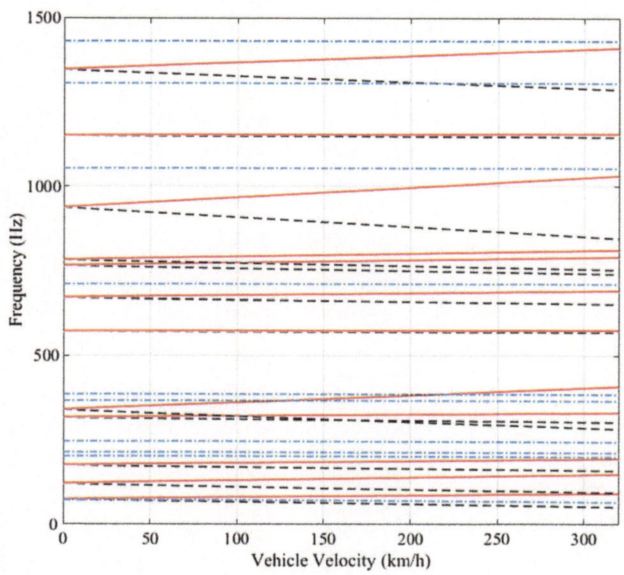

Fig. 4 Campbell diagram for the wheelset

2.2 Model of the Track

The rail is modelled using Timoshenko beam elements on discrete supports with four elements per sleeper bay. The deformation of the cross section of the rail is neglected in current study since the frequency range interested in the current work is under 1500 Hz [1]. Half the track only is considered due to the symmetric structure. The half sleepers are modelled as equally spaced masses, while railpads and ballast are modelled as two linear viscoelastic layers with a simple spring–damper model (Fig. 5). A rail with UIC60 section is considered in the track model. The railpad and track bed data come from the EUROBALT project. Parameters are listed in Table 2.

Only horizontal and vertical displacements and angular deflections around horizontal and vertical axes are considered, and motions related to axial and torsional rail

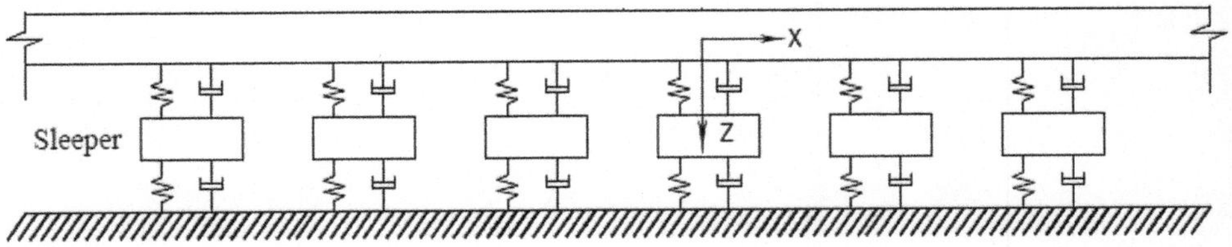

Fig. 5 Finite element track model

deformation are not included in the analysis. The displacement vector $\{\mathbf{u}_{ij}\}$ for a general element of the model linking nodes i and j is then

$$\{\mathbf{u}_{ij}\} = \{\, v_i \quad \theta_i \quad u_i \quad \varphi_i \quad v_j \quad \theta_j \quad u_j \quad \varphi_j \,\}' \tag{15}$$

The mass and stiffness matrices of the beam elements are determined using Timoshenko beam elements including shear deformation and rotatory inertial effects. The homogeneous Euler–Lagrangian equations for a Timoshenko beam are [18]:

$$\begin{cases} kAG(v'' - \theta') = 0; \\ kAG(u'' + \varphi') = 0; \\ kAG\theta + EI_y\theta'' - kAGv' = 0; \\ kAG\varphi + EI_z\varphi'' + kAGu' = 0; \\ kG(I_y + I_z)\varphi'' = 0. \end{cases} \tag{16}$$

The shape functions of the beam element are formulated by solving the homogeneous Euler–Lagrangian equations with corresponding boundary conditions of the beam element. The stiffness matrix of the rail element is therefore given by:where the coefficients are defined as:

Table 2 Simulation parameters

Description	Value
Wheelset type	ETR500
Wheelset mass	1375 kg
Axle load	110 KN
Rail type	UIC 60
Sleeper spacing	0.6 m
Track length	42 m
Second moment of area of rail	30.55×10^{-6} m^4
Second moment of area of rail	51.92×10^{-7} m^4
Half mass of sleeper	170 kg
Railpad vertical stiffness	3×10^8 N/m
Railpad vertical damping	3×10^4 Ns/m
Railpad lateral stiffness	2×10^7 N/m
Railpad lateral damping	8×10^3 Ns/m
Ballast vertical stiffness	8×10^7 N/m
Ballast vertical damping	8×10^6 N/m
Ballast lateral stiffness	1×10^5 Ns/m
Ballast lateral damping	8×10^3 Ns/m

$$\mathbf{K}_r = \begin{bmatrix} 3\beta_z & 3a\beta_z & 0 & 0 & -3\beta_z & 3a\beta_z & 0 & 0 \\ & a^2\beta_z(4+3\alpha_z) & 0 & 0 & -3a\beta_z & a^2\beta_z(2-3\alpha_z) & 0 & 0 \\ & & 3\beta_y & -3a\beta_y & 0 & 0 & -3\beta_y & -3a\beta_y \\ & & & a^2\beta_y(4+3\alpha_y) & 0 & 0 & 3a\beta_y & a^2\beta_y(2-3\alpha_y) \\ & & & & 3\beta_z & -3a\beta_z & 0 & 0 \\ & & & & & a^2\beta_z(4+3\alpha_z) & 0 & 0 \\ & & & & & & 3\beta_y & 3a\beta_y \\ & & & & & & & a^2\beta_y(4+3\alpha_y) \end{bmatrix} \tag{17}$$

$$\alpha_z = \frac{EI_z}{kAGa^2}, \quad \beta_z = \frac{EI_z}{2a^3(1+3\alpha_z)}, \quad \alpha_y = \frac{EI_y}{kAGa^2}, \quad \beta_y = \frac{EI_y}{2a^3(1+3\alpha_y)} \tag{18}$$

being E the Young's modulus, I the second moment of area, A the cross-sectional area, k the Timoshenko shear coefficient, G the shear modulus, and $a = 0.075$ m is half the element length.

A Rayleigh damping is used for the rail, with a corresponding damping ratio 0.6%. Mass, stiffness and damping matrix for the railpad element (including the mass of the sleeper) and ballast element are set as two-layer parallel

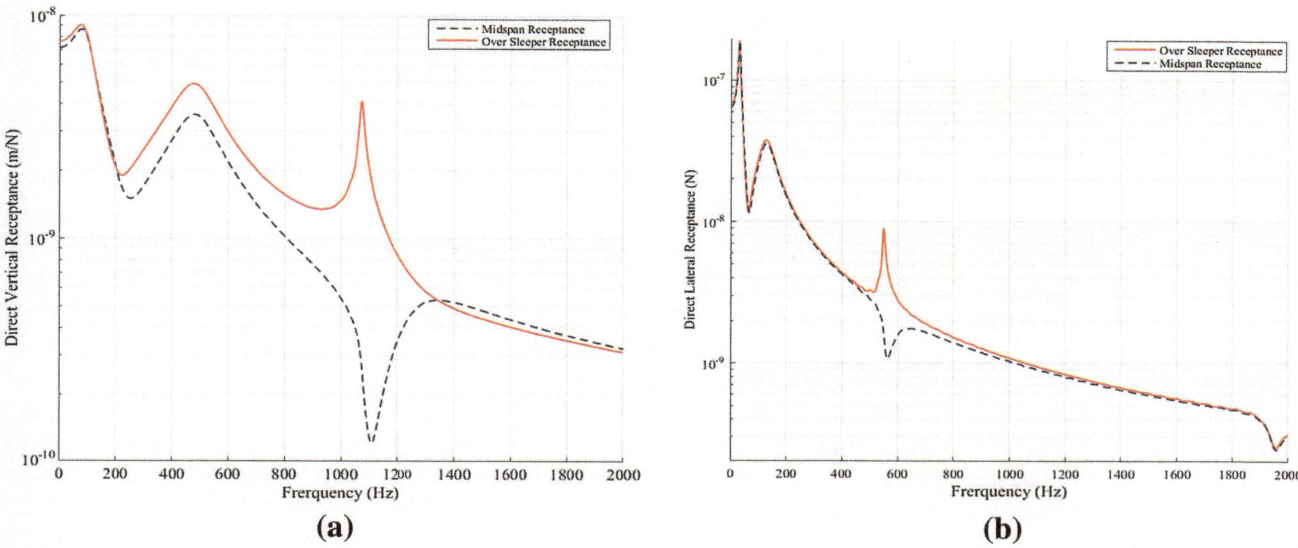

Fig. 6 Track receptance: **a** direct receptance of the vertical direction; **b** direct receptance of the lateral direction

spring–dashpot elements with the ground nodes being clamped constrained. A set of equations of motion for the track is therefore written in a matrix form:

$$\mathbf{M}_t\ddot{\mathbf{u}}_t + \mathbf{C}_t\dot{\mathbf{u}}_t + \mathbf{K}_t\mathbf{u}_t = \mathbf{F}_{ct} \tag{19}$$

where $\mathbf{M}_t$, $\mathbf{C}_t$ and $\mathbf{K}_t$ are the mass, damping and stiffness matrix of the track model, respectively. $\mathbf{F}_{ct}$ is the generalized force applied on the contact point of the rail surface.

The direct receptance of the track in both vertical and lateral direction is obtained (cf. Fig. 6), and the results match with that obtained in previous study [4], which is proved to be valid for dynamic study of frequency up to 1500 Hz.

2.3 Model of Wheel–Rail Contact Force

The nonlinear normal force between wheel and rail surface is solved using Hertz's theory [19]:

$$F_n = C_h\delta^{1.5} \tag{20}$$

where the constant C_h depends from the contact surfaces bending and from material characteristics and δ is the value of the elastic penetration between the two bodies.

Tangential and lateral forces have been calculated according to Kalker's linear theory [20]:

$$\varepsilon_t = \frac{(\dot{x}_{cw} - \dot{x}_{ct})}{V}, \quad \varepsilon_l = \frac{(\dot{y}_{cw} - \dot{y}_{ct})}{V}; \quad F_t = f_{11}\cdot\varepsilon_t, \quad F_l = f_{22}\cdot\varepsilon_l \tag{21}$$

where V represents the speed of the relative reference with respect to an absolute observer (i.e. the wheelset speed) and f_{11} and f_{22} are constants dependent on dimensions of the contact area and the elastic material properties. Since a motion of the wheelset on a straight path at consistent speed has been taken into account in the present model, the creepages are very little, which meets the hypothesis of Kalker's linear theory. The ODE45 routine based on an explicit Runge–Kutta formula in MATLAB is used to solve simultaneously the rail and the wheel equations.

3 Results and Discussion

This section presents the results of simulations under different rail irregularities: a random rail corrugation and several single-harmonic corrugations. The simulation is to investigate the influence of different track and wheelset modelling options on wheel–rail contact forces, which are important to wear and damage phenomena in the rolling surfaces.

3.1 Response to Random Rail Corrugation

In order to investigate how the corrugation wavelength may influence the results from the track model and the wheelset models, simulations are carried out when the rail roughness is described by a white-noise band function. Roughness is defined as an additional component of the relative displacement between the wheel and the rail. The amplitude of the random corrugation is 0.1 μm, and the wavelengths were chosen between 2.7 and 67.7 cm, covering the frequency range from 60 to 1500 Hz if the vehicle velocity is 40 m/s.

Figure 7 shows the contact force of three directions for a flexible track and an infinitely rigid track, respectively. A rotating flexible wheelset model was used for both

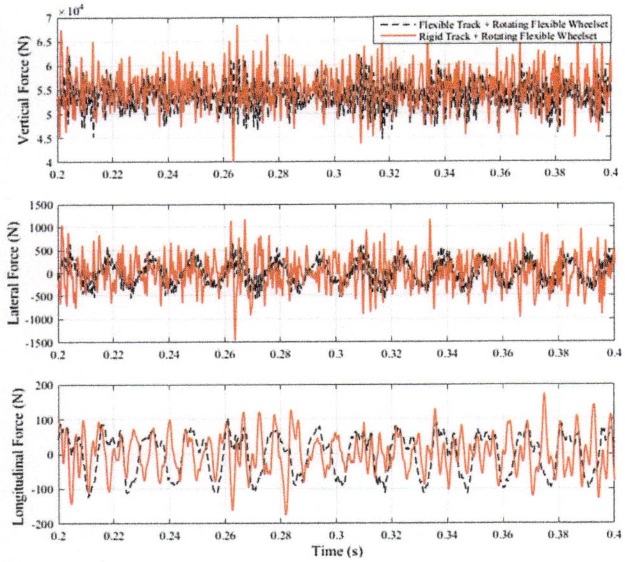

Fig. 7 Contact forces for different track options

interaction models. With a flexible track, a periodic vibration component corresponding to the sleeper bay could be seen from the three directions. To investigate the difference caused by track flexibility, a discrete Fourier transform modulus of the contact forces was taken (Fig. 8). Four peaks can be observed at frequency 67, 134, 201 and 268 Hz for the flexible track model, which is due to the sleeper bay and the length of each beam element of the track. In the frequency range below 800 Hz, the spectra of the contact forces are on average higher for the model considering a rigid track, compared to the model considering a flexible track. This is due to the reduction in impedance produced in this frequency range by track flexibility. For frequency higher than 1000 Hz, the flexibility of the track leads to slightly higher amplitude of both vertical and longitudinal forces.

Figure 9 shows the contact force of three directions with a non-rotating flexible wheelset and a rotating flexible wheelset. A flexible track model was used for both interaction models. The time histories of the vertical and lateral contact forces of the two different wheelset models are almost the same, while the longitudinal force of the rotating flexible wheelset shows a higher amplitude with the non-rotating one. Actually, the longitudinal component of the non-rotating flexible wheelset is very small, with a max value around 25 N, i.e. 200 times smaller than the amplitude of the dynamic fluctuation of the vertical force. Considering wheelset rotation instead, a small longitudinal force is generated, due to the effect of gyroscopic forces on the modes of the rotating wheelset involving axle bending in the horizontal plane. Despite the fact the force amplitude is still small compared to the vertical and lateral force components, it shows a big difference from the case of the non-rotating wheelset.

The Fourier transform modulus of the vertical and lateral forces (Fig. 10) shows slight difference with non-rotating and rotating flexible wheelset. For longitudinal force, the rotating flexible wheelset shows a higher component than the non-rotating flexible wheelset under 800 Hz. The peak frequencies of rotating flexible wheelset can be found to correspond with the forward and backward mode frequencies at 40 m/s. For instance, the first two peak frequencies of rotating flexible wheelset appear at 67 and 135 Hz. At 67 Hz, the peak meets the backward first bending mode and the frequency due to the sleeper bay. At

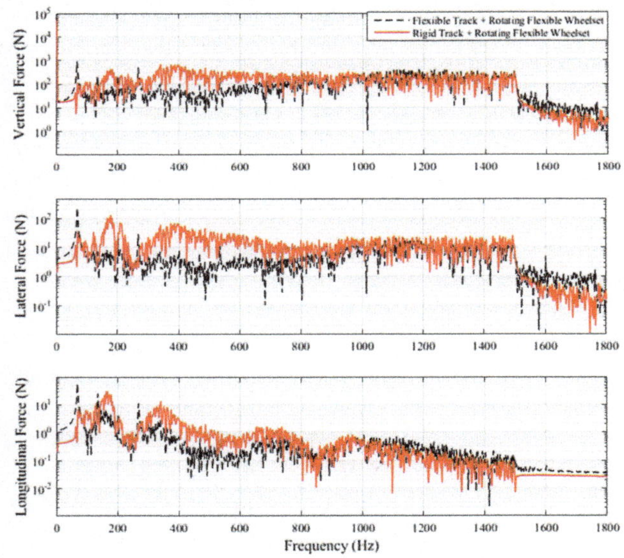

Fig. 8 Spectra of the contact forces for different track options

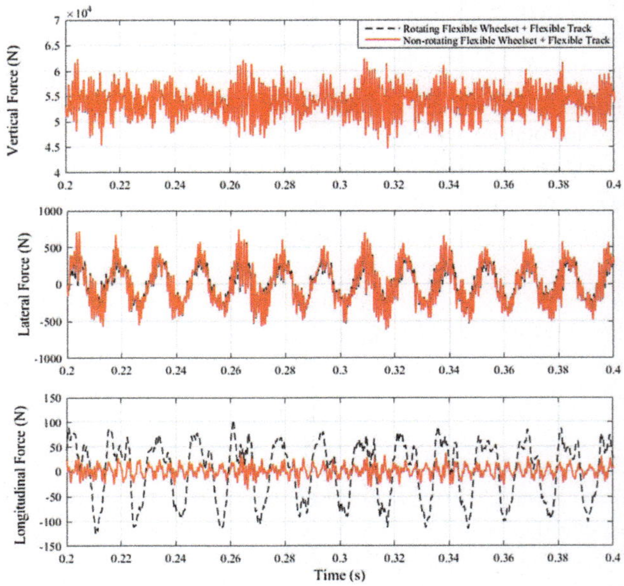

Fig. 9 Contact forces of the wheelset running on randomly worn rails

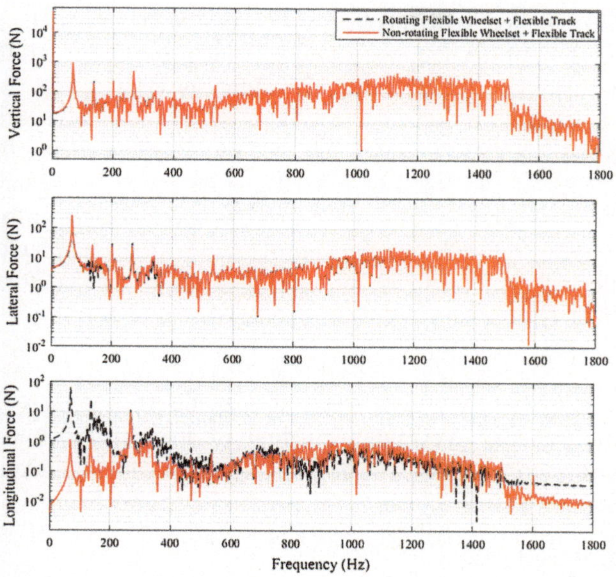

Fig. 10 Spectra of the contact forces for different wheelset options

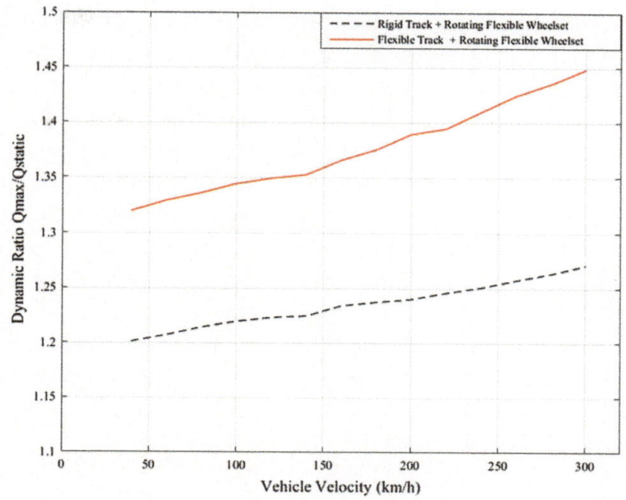

Fig. 11 Dynamic ratio of different track models for an excitation at pinned–pinned frequency

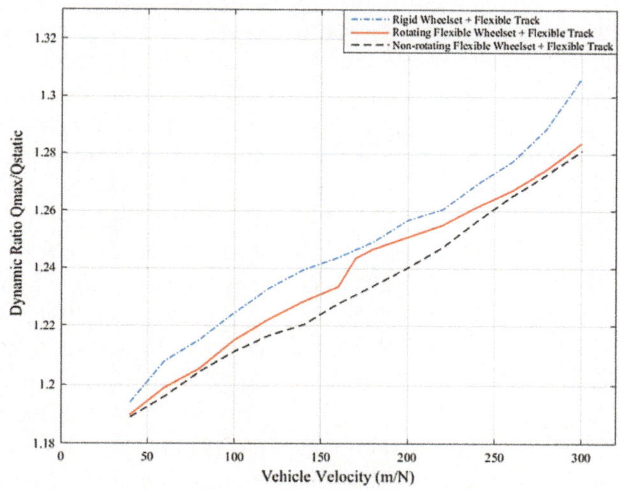

Fig. 12 Dynamic ratio of different track models for an excitation at the fifth backward frequency of the rotating wheelset

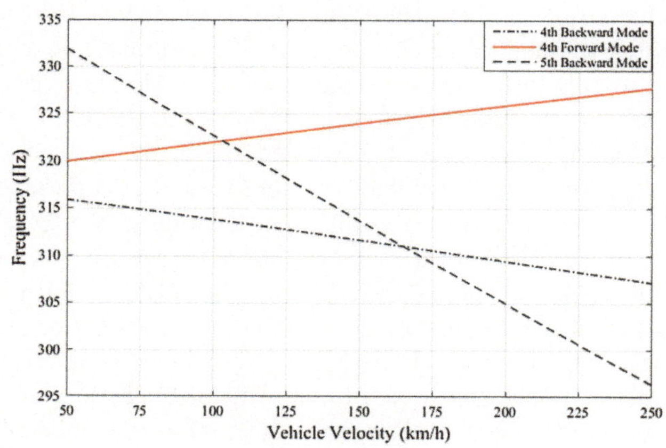

Fig. 13 Intersection points for the fifth bending backward mode of the rotating wheelset

about 135 Hz, the peak appears due to the forward second bending mode of the wheelset and the frequency due to the mid-sleeper bay.

3.2 Response to Single-Harmonic Rail Corrugation

In order to quantify the relevance of dynamic effects affecting the wheel–rail contact forces, the peak values for the vertical component of the contact forces normalized with respect to the static wheel load are reported with increasing vehicle speed under several single-harmonic rail corrugations which may excite the resonances of the wheelset and the track. Two kinds of harmonic excitation are chosen: the first one has an excitation frequency equal to the pinned–pinned frequency of the finite element track, and the second one coincides with the fifth backward mode of the rotating wheelset.

Figure 11 shows the dynamic ratio of the vertical contact force peak value for an excitation at pinned–pinned frequency with different track models. For both models, the dynamic ratio increases with increasing velocity. For the flexible track model, the ratio is 10–15% higher than the rigid track and shows a greater increasing trend compared to the rigid one, confirming the influence of the flexible track model.

Figure 12 shows the ratio of peak value of the vertical contact force to the static wheel load with different wheelset models. A rigid wheelset is added to this analysis. The fifth backward mode of the rotating wheelset is chosen as the excitation frequency to investigate influence of the resonance frequencies of the rotating wheelset, which is shown as intersection points at 322 Hz with a velocity of

103 km/h and at 311 Hz with a velocity of 165 km/h (Fig. 13). It is observed that the average dynamic ratio is higher for the model considering a rigid wheelset than for the model accounting for wheelset flexibility. This is due to a compensation of the wheelset flexibility to the contact forces. For the flexible wheelset model, the dynamic ratio of the rotating flexible wheelset shows a slightly higher amplitude with the non-rotating flexible wheelset. A peak ratio at around 165 km/h for the rotating wheelset can be found, which is the resonance frequency related to the intersection point of the fifth backward and the fourth backward bending modes, while the intersection point at 103 km/h does not show significant amplification of the dynamic contact force.

4 Conclusion

This paper proposed a vehicle–track interaction model in time domain, which is valid for dynamic simulation up to 1500 Hz. The effect of wheelset and track flexibility on dynamics of wheelset–track interaction was investigated.

Results of the contact forces under different structure flexibilities are presented for two kinds of rail corrugation: a randomly corrugated track and several single-harmonic corrugated tracks. For the random corrugation, the Timoshenko beam track model shows obvious difference compared with a rigid track model. Force peaks on the three directions could be found in the discrete support flexible track at each sleeper bay position. In the frequency range below 800 Hz, the flexible track model shows lower contact forces than the rigid model, while for frequency higher than 1000 Hz, the contact forces of the flexible track model are slightly higher than for the rigid track model. The rotating flexibility of the wheelset produces a higher longitudinal force when compared with a non-rotating flexible wheelset, while the vertical and lateral forces show little difference between the two wheelset models. For the single-harmonic corrugation case, the numerical results show that the wheel–rail contact force amplitude is generally increasing with growing velocities and shows a peak when the frequency of excitation meets the resonance frequencies of the flexible structure. However, it must be highlighted that the effect of velocity is more remarkable.

Overall, the results presented above show that simulation results are highly sensitive to the track model and wheelset model adopted. Neglecting wheelset rotating flexibility and track flexibility may lead to an overestimation or underestimation of the dynamic contact forces generated by the vehicle in response to track imperfections, depending on the type of excitation and on the frequency range considered.

The model in this paper is valid up to 1500 Hz in wheel–rail interaction analysis, which can be used for investigation of specific interaction problems, e.g. rail short-pitch corrugation. However, for problems related to higher-frequency range, e.g. rolling noise problem, the model of the track still needs to be improved. Future developments of this work will be addressed to further expanding the frequency range of validity of the train–track interaction model, by introducing a more detailed track model based on solid FEM in a moving reference.

Acknowledgements This work has been financially supported by the China Scholarship Council.

References

1. Thompson D (2008) Railway noise and vibration: mechanisms, modelling and means of control. Elsevier, London
2. Sun W, Zhou J, Thompson D, Gong D (2014) Vertical random vibration analysis of vehicle–track coupled system using Green's function method. Veh Syst Dyn 52(3):362–389
3. Thompson D, Jones C (2000) A review of the modelling of wheel/rail noise generation. J Sound Vib 231(3):519–536
4. Knothe K, Grassie S (1993) Modelling of railway track and vehicle/track interaction at high frequencies. Veh Syst Dyn 22(3–4):209–262
5. Arnold J, Kaiser I, Schupp G (2005) Simulation of a railway vehicle's running behaviour: how elastic wheelsets influence the simulation results. Veh Syst Dyn 41:242–251
6. Gómez J, Vadillo E, Santamaría J (2006) A comprehensive track model for the improvement of corrugation models. J Sound Vib 293(3):522–534
7. Kaiser I, Popp K (2006) Interaction of elastic wheelsets and elastic rails: modelling and simulation. Veh Syst Dyn 44(sup1):932–939
8. Baeza L, Fayos J, Roda A, Insa R (2008) High frequency railway vehicle–track dynamics through flexible rotating wheelsets. Veh Syst Dyn 46(7):647–659
9. Zhai W, Wang K (2010) Lateral hunting stability of railway vehicles running on elastic track structures. J Comput Nonlinear Dyn 5(4):1–9
10. Baeza L, Ouyang H (2011) A railway track dynamics model based on modal substructuring and a cyclic boundary condition. J Sound Vib 330(1):75–86
11. Baeza L, Vila P, Xie G, Iwnicki SD (2011) Prediction of rail corrugation using a rotating flexible wheelset coupled with a flexible track model and a non-Hertzian/non-steady contact model. J Sound Vib 330(18):4493–4507
12. Di Gialleonardo E, Braghin F, Bruni S (2012) The influence of track modelling options on the simulation of rail vehicle dynamics. J Sound Vib 331(19):4246–4258
13. Kaiser I (2012) Refining the modelling of vehicle–track interaction. Veh Syst Dyn 50(sup1):229–243
14. Martínez-Casas J, Mazzola L, Baeza L, Bruni S (2013) Numerical estimation of stresses in railway axles using a train–track interaction model. Int J Fatigue 47:18–30
15. Shabana AA (2013) Dynamics of multibody systems. Cambridge University Press, Cambridge

16. Brown M, Shabana A (1997) Application of multibody methodology to rotating shaft problems. J Sound Vib 204(3):439–458
17. Francesco R, Roberto A, Bruni S (2007) A time domain model for the study of high frequency train–track interaction. Paper presented at the 7th international conference on railway bogies and running gears, Budapest
18. Petyt M (2010) Introduction to finite element vibration analysis. Cambridge University Press, Cambridge
19. Johnson KL, Johnson KL (1987) Contact mechanics. Cambridge University Press, Cambridge
20. Kalker JJ (1990) Three-dimensional elastic bodies in rolling contact, vol 2. Kluwer, Amsterdam

2

Public Perception of and Attitude to Driverless Train: A Case Study of Sydney, Australia

Anna Fraszczyk[1] · **Corinne Mulley**[2]

Abstract There are 32 cities around the world with automated metro systems in operation. The majority of these are located in Europe (13) and Asia (14) but none so far in Australia. However, the picture is changing in 2019 when the first stage of the 'Sydney Metro' starts its operation on the North West link in Sydney, New South Wales, Australia. The automated train is planned to be used safely without human interaction, thus reducing significantly the labour input in the provision of service. Although the proposal for a fully automated system came late in the planning phase of the Sydney Metro project, it appears that, from a technical point of view, the project is clear and well planned. However, providing information to Sydneysiders and understanding their attitudes towards automation has received no consideration. Moreover, how the public perceive these changes and their attitudes to aspects of the new system, including driverless trains, may well be crucial in properly positioning and marketing the new services to ensure the expected patronage. This paper aims to fill this gap by investigating public opinion of the new Sydney Metro service by undertaking a research on people's perceptions of a driverless train as well as their attitudes to the new transport system and public transport more generally

Keywords Driverless train · Public attitude · Sydney Metro

✉ Anna Fraszczyk
anna.fraszczyk@gmail.com

[1] Mahidol University, Nakhon Pathom, Thailand

[2] The University of Sydney, Sydney, Australia

Editors: Jiaqi Ma, Yihui Wang, Xiaolu Rao, Shane Wu and Alireza Khani.

1 Introduction

A person's travel choice is a complex issue, which has been attracting the attention of social and transport scientists over the last decades across the world. Researchers looked at the ways in which the role of attitudes affects travel choices (e.g. [1, 2] in the USA) and studied the subject from various perspectives, including e.g. youth travel (e.g. [3] in Australia) or emerging technologies (e.g. [4] in Finland).

This paper is motivated by the need to understand public attitudes to driverless trains, Grade 4 of Automation [5], as a prerequisite for ensuring that operators and/or authorities have the information to correctly market their benefits and to allay any fears that citizens may have.

A case study approach is used to seek respondents to a questionnaire undertaken in Sydney, the capital of New South Wales, Australia, where a proposed driverless train extension to the heavy rail network is being built. Respondents to the survey are used to draw conclusions as to the public perception of driverless trains, to identify recommendations for Sydney Trains in how to manage public perceptions and to provide lessons for other cities considering the implementation of such a system in the future.

The paper is structured as follows. First, the background to the topic is presented including a contextual review of the literature on public attitudes to autonomous cars and to various aspects of autonomous trains. This leads to the establishment of the research gap to be filled by this paper. The following section describes the Sydney case study. This is followed by a description of the survey methodology for this study and the results of the survey before turning to the discussion section and conclusions with pointers to future directions for research in this area.

2 Background

A rapid development of automation of transport over the last century is a fact, with personal vehicles as well as public transport experiencing improvements in terms of technologies to support or even replace the driver. Vehicle manufacturers are investing in new research of technologies and development of new automated vehicles, for both private (cars) and public transport (buses, trains) use. The benefits of transport automation are many, as explained in [5] or [6], including an overall decrease in cost, and improvements in terms of system's safety and sustainability [7, 8] due to machines taking control over vehicles and optimising their performance, according to parameters programmed by humans.

2.1 Public Attitudes to Autonomous Cars

The last decade brought significant technological progress with the development of an autonomous car, with large companies such as Tesla, Audi or Google investing in research and testing phases of this new transport concept. These developments in technology are followed by a large amount of research in the area of public attitudes to a driverless car. The expanding literature on the topic of driverless cars tends to focus on public opinions, attitudes, acceptability, views and interest in new technologies with research conducted at a local (e.g. [9]) as well as an international (e.g. [10]) scale. Kyriakidis et al. [10], for example, investigated user acceptance and willingness to buy automated vehicles (AV), with a distinction between three levels of partly, highly and fully automated driving. In their online survey, they collected responses from 109 countries with an original sample size of 5000 people. The results presented show that respondents perceive manual driving as most enjoyable, but they are also fascinated by an option of a fully automated driving. Interestingly, the respondents were not keen on entirely removing the steering wheel from a vehicle, even in a fully automated mode. Software hacking, legal issues and safety were some of the top concerns cited by respondents in the potential use of AV [10].

On a country-scale research, Payre et al. [7] focused on the acceptability of fully automated driving (FAD) across France. They collected 421 online responses and analysed the acceptance of and intention to use FAD vehicles with the conclusion that there is a strong positive correlation between attitudes to FAD and intentions to actually use vehicles in a FAD mode.

On a more local-scale, Piao et al. [11] studied views of La Rochelle residents (France) on implementation of automated vehicles in urban areas, including buses, taxis and cars. The authors used an online survey and telephone interviews as tools for data collection, and they found that attitudes of the public were positive towards operation of automated buses in urban areas. However, Piao et al. [11] admitted that their trials, which were part of a greater project, were in a setting where traffic volume was reduced therefore not 100% real.

Another local perspective is presented by Bansal et al. [9], who surveyed 347 people in Austin, USA. The authors investigated public's interest in new technologies, including connected and autonomous vehicles (CAVs) at four different automation levels, via an online survey. They found that the majority of respondents perceive improved safety (fewer crashes) as the primary benefit of CAVs. However, as in the previous research, a potential equipment failure is the respondent's main concern. Also, based on people's socio-economics (e.g. gender, income, location) the authors identified groups of respondents more interested in new technologies than others (e.g. higher-income males living in urban areas who experienced more crashes).

2.2 Public Attitudes to Various Aspects of Trains

In terms of public transport research, there is a large volume of publications on public attitudes to buses, trains or public transport in general. In this review, the focus is on rail-related literature.

Harvey et al. [12] focused on heavy rail and studied the burning issue of public attitudes to high-speed rail (HSR) in the UK, where the HSR2 line is under live public discussion and potential development. The authors questioned 1799 people using an online questionnaire and tested seven hypotheses. Analysis of results revealed that e.g. there are socio-economic differences in attitudes (e.g. security concerns, HSR prestige, importance of comfort) between gender, age and occupational groups. Also, they confirmed that there is a relationship between respondents' previous travel behaviour and their attitudes and perceptions of HSR and long-distance travel.

Again in a rail context, a pro-environmental approach to attitudes is presented by Hess et al. [13]. The authors studied UK's rail travellers' attitudes to reductions in greenhouse gas pollution in a wider context of general attitudes towards the environment. Their analyses revealed that there are respondents with certain socio-economic characteristics, e.g. females or older respondents or respondents with a university degree, whose attitudes are more environmentally friendly than others.

A local metro case study is presented by Karvonen et al. [4] who, based on plans for an introduction of a driverless metro operation in Helsinki (metro plans were later postponed due to technical issues [14]), investigated roles of a driver on a train. The authors used various data collection techniques, including interviews, observations and workshops, and concluded that driver's role is much more complex than driving a train only. They identified four main and obvious duties of metro drivers, such as: operating the train, taking care of passengers, observing events outside the train and acting in exceptional situations. In addition, within the four main tasks they distinguished 19 'hidden' roles or subtasks, e.g. making announcements to the passengers or fixing small faults in exceptions, when necessary, which help to provide a 'human link' between passengers and the system. In their conclusions, Karvonen et al. [4] suggest that in advance of implementing a new driverless metro system a careful thought needs to be given to the ways all hidden roles of a driver will be replaced by a driverless system as a whole, so that no subtask is missed or overestimated.

Fraszczyk et al. [6] looked at public perception of driverless trains using a questionnaire as a tool for data collection. A paper-based questionnaire was distributed amongst European students and professors attending a rail summer school in the UK and provided a sample size of 50. The analysis of results revealed that for example 93% of females and 72% of males within the sample would like to see a 'fake' driver room on a driverless train. Interestingly, in the context of their own use of driverless trains, the respondents were not much worried about system maintenance (28% of males vs. 14% of females) and much more worried about a 'human error' (e.g. staff communication issues; 36% of males vs. 43% of females) and potential technical failures (50% of males vs. 43% of females).

Wang et al. [15] reviewed a number of studies focusing on driverless train operation. They list and describe the various opportunities and challenges driverless train operation brings. In the benefits section, the authors list seven main opportunities: lower operation costs, increased capacity and reliability, increased flexibility, energy efficiency, high levels of safety and security. In terms of challenges, the concerning issues relate to: safety, train control technology, communication systems, platform screen doors, terminal designs and detection and management of emergency situations. Overall, they highlight the need for high standards and intelligent solutions to achieve a safe and reliable driverless system.

2.3 Research Gap

In contrast to research on public attitudes to driverless car or to trains in general, the phenomenon of attitudes to automation of trains, increasingly present in Europe and Asia since early 1980s (e.g. Lille, France; Osaka, Japan), still has not attracted much attention by researchers. With the exception of the work of Fraszczyk et al. [6], where research on the public's perception of driverless trains was undertaken with a sample collected in Newcastle, UK, the majority of knowledge on public attitudes to and perception of driverless trains comes from news channels and workers unions' publications [16–19] rather than from evidence-based rigorous academic research. The lack of social research into the role of attitudes in the area of train automation presents a significant research gap, which this study aims to address.

2.4 Sydney Case Study

Unattended train operation (UTO) is gaining more popularity worldwide due to the undisputable benefits of cost efficiency and train capacity. Given the longevity of investment, this substitute of capital investment at the start of the project in return for lower ongoing labour costs motivates the choice of technology. Currently there are 48 lines worldwide with UTO systems in operation with 10+ more lines to be opened by 2025 [5]. From a technical perspective, the debate on the advantages over the disadvantages of such systems is well documented in the literature and there is now little room for debate. However, as identified above, the acceptability of automated trains by citizens is not well documented.

Currently under construction, one of the new driverless systems is planned to open in Sydney, Australia, in 2019. This Sydney Metro project will be the first UTO system in Australia and is planned to be used safely without human interaction, thus reducing significantly the labour input in the provision of service [20–23].

From a technical point of view, the Sydney Metro project is clear and well planned. However, little information on how it will operate has been provided to Sydneysiders and little research has been undertaken to understand attitudes towards automation. The issue of how Sydneysiders perceive driverless trains is crucial in properly positioning the new service and to ensure that marketing and other aspects positively combine to guarantee that the expected patronage emerges for this new infrastructure when it opens.

As few studies have considered public perception of driverless trains, the Sydney Metro case study offers a unique opportunity to investigate citizen attitudes to a new train line including the perceptions of fully automated metro services, attitudes to new driverless transport systems and linking this to attitudes to public transport more generally.

3 Methodology

3.1 Questionnaire Design

The questionnaire used in the study was inspired by a survey completed in 2014, which focused on public perception of driverless trains [6]. It reflects many of the issues raised by Wang et al. [15]. The final version of the questionnaire was enriched by a Sydney context and additional questions related to a general attitude towards public and private transport as well as socio-economic questions.

The questionnaire used was designed as part of an MSc project which focused on public perception of driverless trains. Divided into three parts, the questionnaire included questions about attitudes to and perception of driverless trains, personal information and an additional section for comments.

Most of the questions included in the questionnaire were closed questions, and the responents were expected to simply tick one answer only. Overall, the questionnaire involved eight personal questions only with the majority of the survey dedicated to attitudes to and perception of driverless trains. An example question is presented in Fig. 1.

3.2 Data Collection

Data collection for the study was conducted online in June 2016 by the University of Sydney using the panel of GMI Lightspeed. The panel sought to collect 300 responses, and following this data were cleaned and exported to IBM SPSS software where they were coded and prepared for statistical analysis.

4 Data Analysis

4.1 Sample Size and Age

The questionnaire was answered by 300 people from Greater Sydney area in Australia. The age range included in the sample was between under 18s and over 65s. Initial data collection provided too high a proportion of older people and for analysis, a random sample of the older people was made to ensure that the final proportion of older people was proportional to the number of persons over 65, as shown by the census data [25]. This was done to better represent the population by being more representative of people in each age category. The final sample size was 219 respondents with a gender split of 52.97% females, 46.58% males and 0.45% of respondents (1 person) who preferred not to answer the gender question. A breakdown of respondents by age and sex showed roughly equal male and female respondents in all age groups except 40–45 years old where there were more females than males and in the 50–55 and 60–65 age groups where there were more males than females, especially in the 60–65 age group.

Q1 Over the last 12 months how often have you used the following methods of travel? Please tick the relevant box for each of the methods listed.

	5 or more days per week (1)	3-4 days per week (2)	1-2 days per week (3)	1-3 times per month (4)	Less than once per month (5)	Not used in the last twelve months (6)
Airplane (1)	○	○	○	○	○	○
Train (2)	○	○	○	○	○	○
Bus (3)	○	○	○	○	○	○
Metro (4)	○	○	○	○	○	○
Light Rail (12)	○	○	○	○	○	○
Ferry (13)	○	○	○	○	○	○
Car as driver (5)	○	○	○	○	○	○
Car as passenger (6)	○	○	○	○	○	○
Motorcycle (7)	○	○	○	○	○	○
Taxi (8)	○	○	○	○	○	○
Cycling (9)	○	○	○	○	○	○
Walking (10)	○	○	○	○	○	○

Fig. 1 Example of a question used in the survey [24]

4.2 Descriptive Statistics in Relation to Attitudes to Train Travel and Driverless Trains

4.2.1 Rail Vehicles Usage

Respondents who use rail vehicles (train, metro or light rail) at least once a month (this combined answers to four options: 5+ days, 3–4 days, 1–2 days and 1–3 days per month; see Fig. 1 for details) totals up to 120 unique users, which is 55% of the sample. Of the 117 respondents who stated that they use a train (metro or light rail respondents not included) at least once a month, 96 were Sydney Trains users, 18 the country services of NSW TrainLink users and 3 respondents identifying that they were users of trains outside of Sydney. Sydney Trains users therefore comprise 44% of the sample.

4.2.2 Satisfaction and Priorities on Sydney Trains

Overall satisfaction with Sydney Trains, amongst the sample of 96 respondents using them at least once a month, is mainly positive with 24% of people very satisfied and 52% somewhat satisfied. Only 9% of the Sydney Trains users within the sample are unsatisfied with the service and 15% of respondents is neutral.

Sydney Train users were asked about their priorities when using the service with an option for answering on a 6-point Likert scale, with 1 being most important and 6 being least important.

The results displayed in Fig. 2 show that safety and price stand out as the two most important priorities for the majority of the sample of Sydney Trains users with answer options of "1" or "2" or "3" ticked by approx. 70% for each of the two priorities. The results also show that over 40% of respondents are not much bothered about their journeys' comfort (option "4" ticked), with accessibility

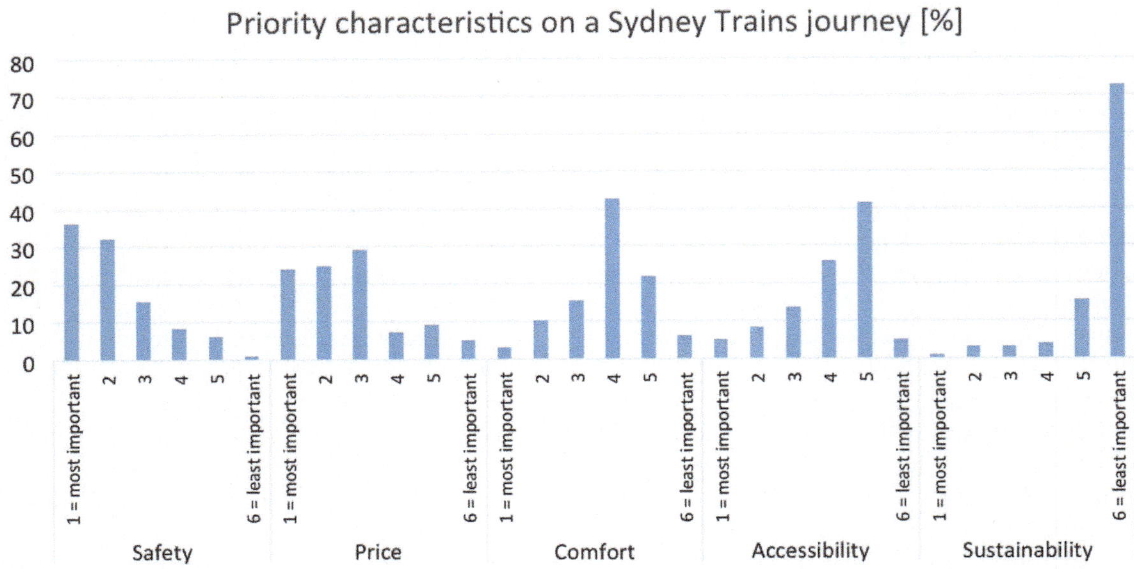

Fig. 2 Priority characteristics on a Sydney Trains journey (%)

and the sustainability (environmental issues) not being priorities for the majority at all (mainly "5" and "6" ticked).

4.2.3 Sydney Trains Users and Non-users Versus General Attitude Towards a Driverless Train

Respondents were asked a number of questions that elicited attitudes towards a driverless train. These are illustrated in Figs. 3, 4, 5, 6 and 7, showing the difference between users and non-users of Sydney Trains.

Figures 3 and 4 show how much more positive users of Sydney Trains are towards driverless trains. For the general question, Fig. 3 shows that users either strongly support (10%) or support (29%) driverless trains. In Fig. 4, whilst the majority of Sydney Trains users and non-users rated the importance of a driver on a train as either 'very important' (43 vs. 71%) or 'important' (28 vs. 13%), it is still the case that users appear more positive on the absence of a driver on a driverless train. This shows the fact that although UTO is designed to be fully functional without a driver on board the respondents are still valuing the importance of a driver, even when from a technical point of view they are no longer needed. This is in line with findings presented by Karvonen et al. [4] on very many hidden roles of the train driver.

4.2.4 Sydney Trains Users and Non-users Versus Driver's Cab

In some existing situations UTOs have retained the driver's cab as part of the train design, partly to prepare citizens for a movement to full UTO (e.g. Shanghai Metro Line 10 operated with a driver in a cab; Budapest Metro Line 4 with a train attendant for the first year of operation). Here the user and non-user respondents were much closer in their views, as shown by Fig. 5, and over 60% of respondents in each of the groups agreed that the driver cab should still be included on a driverless train. However, 14% of Sydney Trains users said "No" to the driver cab followed by 32% of the users being "Not sure".

4.2.5 Sydney Trains Users and Non-users and Safety Issues

A first question asked respondents a general view of safety, and this was followed up by a more specific question on the impact of driverless trains on events as a result of human error. Importantly 50% of users and over half of non-users (67%) responded that they would be worried about safety. This is an important factor for Sydney Trains to take account of: for users it needs strategies to ensure that users are convinced about safety and for non-users the safety concerns need to be addressed if modal shift towards rail is to be achieved. Interestingly, 43% of users responded that driverless trains were likely to make no difference to the occurrence of a human error which suggests another area where user education would be wise.

4.2.6 Benefits of Driverless Trains

Unsurprisingly, the main driver of UTO is cost savings but the literature identifies other benefits, as highlighted in the earlier sections. Respondents were asked about a number of

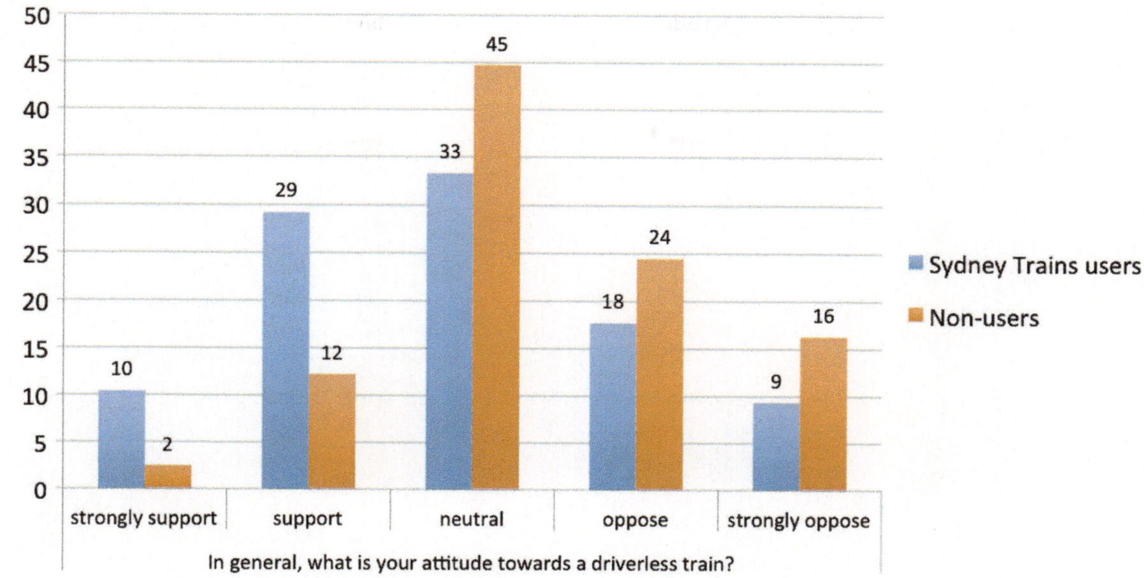

Fig. 3 Attitude towards a driverless train (%)

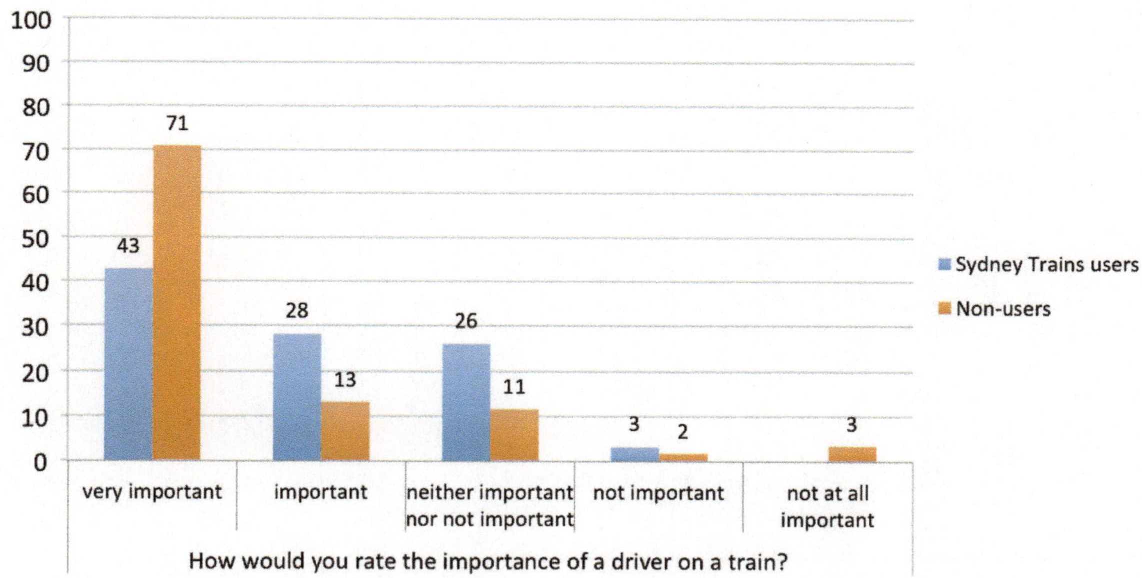

Fig. 4 Importance of a driver on a train (%)

potential benefits resulting from moving to driverless operation. Figure 6 shows how users and non-users of Sydney Trains responded to four questions relating to benefits. This shows how fares ("reduced ticket price") and frequency ("increased train frequency") are most important to users. Interestingly, lower fares are very important to non-users, too. The non-users scored the areas of reduced risk and greater sustainability more highly than users, but still with the maximum of 30% of non-users being concerned only. In summary, users are understandably more interested in the practical side of train use whereas non-users have the luxury of being able to be more philosophical.

4.2.7 Using a Driverless Train

As identified above, Sydney Trains will be introducing the first driverless train in 2019 on its new Sydney Metro service. Respondents were asked whether they would use this new service. Half of existing users (50%) responded "Yes" with a significant minority (35%) being unsure. The non-users were more emphatic with 40% answering "No" to this question

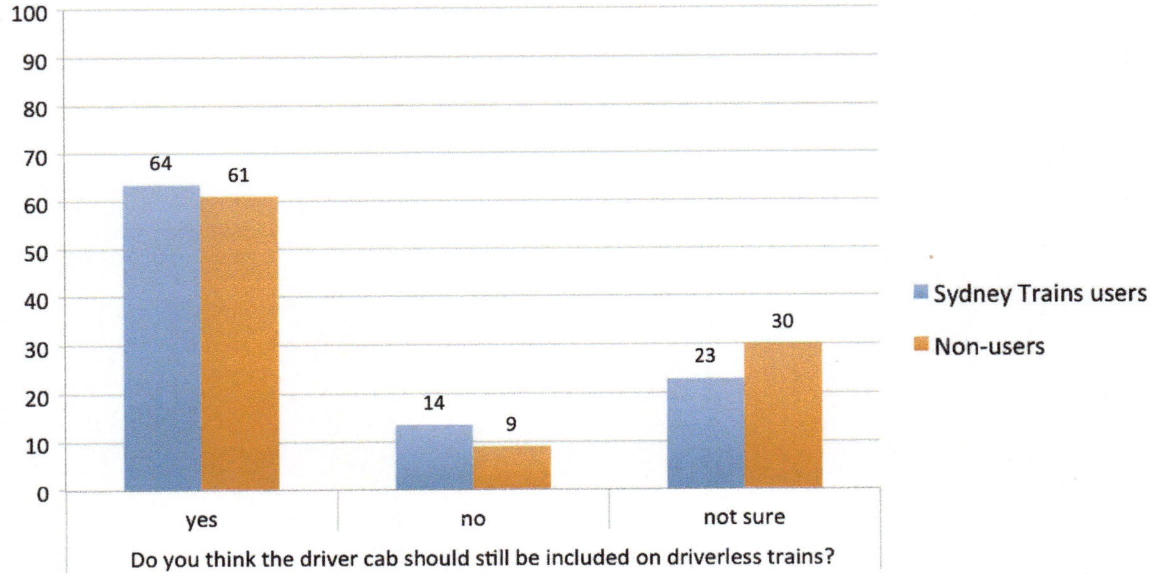

Fig. 5 Inclusion of the driver cab on a driverless train (%)

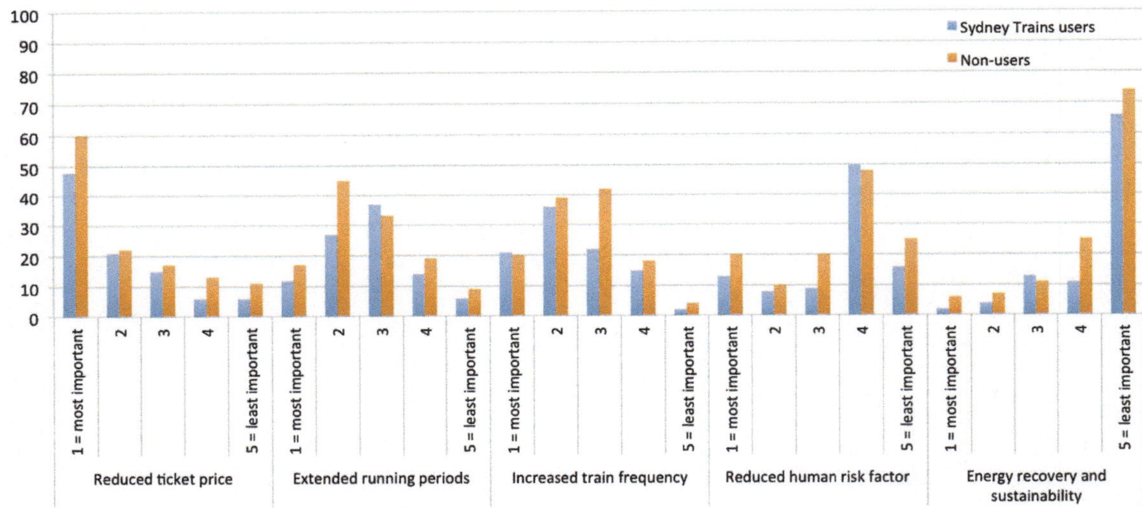

Fig. 6 Potential benefits resulting from moving to driverless train operation (%)

with 46% unsure. As there was no a follow-up question asking for reasons of potential use of Sydney Metro from 2019 onwards, the explanation of this decision is difficult as it could be to the lack of willingness to change travel patterns or scepticism towards UTO or other unknown issues.

4.3 The Role of Attitudes to Travelling on Attitudes to Driverless Trains

The analysis above shows a number of trends. First, there seems to be quite a difference between the attitudes of users and non-users of Sydney Trains towards different aspects of driverless trains. Second, there is a significant proportion of all respondents who oppose or strongly oppose the use of driverless trains. This is problematic in Sydney where driverless trains are coming soon and where there is an imperative to create some modal shift from car to public transport to increase sustainability and to reduce congestion, for drivers who lose time but also for the city as a whole which suffers from losses in productivity and environmental degradation as a result of car-based congestion.

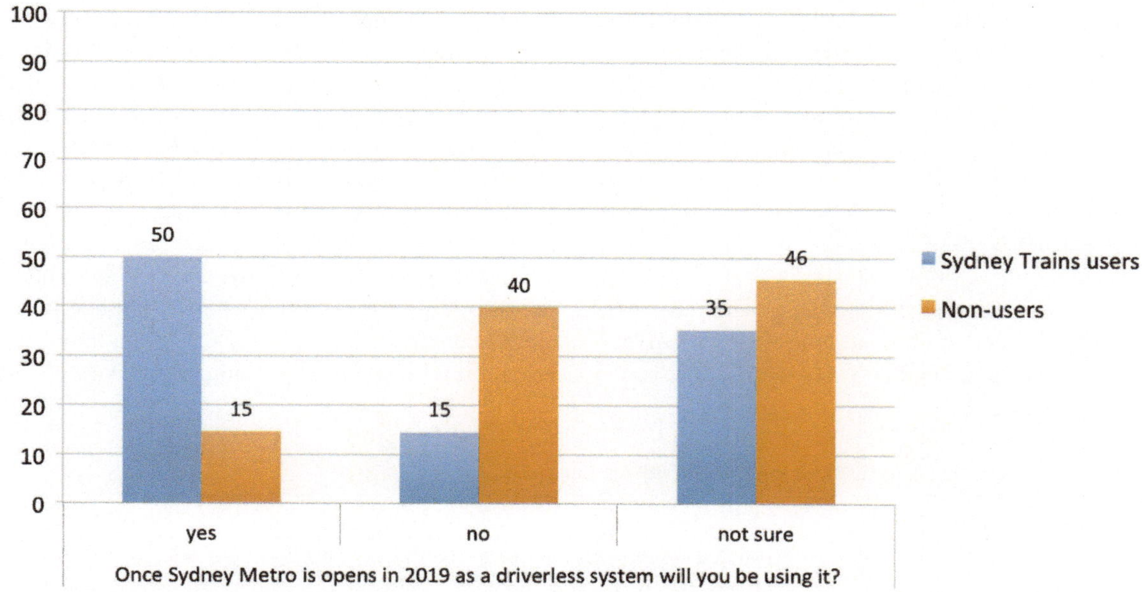

Fig. 7 Willingness to use driverless Sydney Metro when it opens in 2019 (%)

4.3.1 Factor Analysis to Identify Latent Variables

Travel behaviour research shows that general attitudes to transport do influence the mode of travel. Hence, the questionnaire included 21 travel attitude questions, broadly based on [1] and [26]. These were synthesised into latent variables using common factor analysis, following the identification of the number of factors using parallel analysis. Analysis using these factors to discriminate between respondents can form the basis of a targeted marketing and information campaign to improve the understanding and thereby creating a more positive response to the introduction of driverless trains.

Parallel analysis was chosen as the most accurate way of identifying the number of factors over the alternatives of the eigenvalue test (or Kaiser's criterion) or the scree test (Catel's scree test) because the latter have a tendency to over-estimate the number of factors. Parallel analysis is based on [27] and is a simulation technique which compares the size of the eigenvalues with those identified by a set of data of the same size generated randomly. The decision rule is to compare the eigenvalues from the data with the randomly calculated eigenvalues, retaining as factors, all factors where the former is larger than the latter. O'Connor [28] provides SPSS syntax for parallel analysis which was utilised in this case. Table 1 shows that seven factors should be retained.

Common factor analysis was performed in IBM SPSS with a fixed number of factors, as identified by the parallel analysis. The method of extraction was maximum likelihood with oblique rotation. Oblique rather than restricting the factors to be orthogonal was selected because in real life, attitudes are rarely statistically independent. The KMO test of sampling adequacy and Bartlett test of sphericity were both significant suggesting successful extraction of factors as latent or underlying variables. Overall, 59.24% of the total variance in the data is explained. The first three

Table 1 Raw data eigenvalues, and mean and percentile random data eigenvalues

Factor	Raw data eigenvalue	Random eigenvalue
1	4.623566	0.735188
2	2.447426	0.615303
3	1.497743	0.527339
4	1.031411	0.452571
5	0.90761	0.384584
6	0.671444	0.321034
7	0.449144	0.26273
8	0.243274	0.20851
9	0.206211	0.156821
10	0.100836	0.107474
11	0.018734	0.060722
12	−0.0156	0.014331
13	−0.05637	−0.02999
14	−0.09589	−0.07184
15	−0.11599	−0.11341
16	−0.13846	−0.15603
17	−0.1488	−0.19671
18	−0.16733	−0.23655
19	−0.22541	−0.27819
20	−0.26212	−0.32221
21	−0.28082	−0.3748

factors (public transport lovers, car snobs and bike lovers) account for around 40% of the variance with the remaining four factors roughly contributing equally at around 5% each. The principal loadings for each factor are shown in Table 2, which identifies names for each factor in the header row.

4.3.2 Attitudes to Driverless Trains

A first stage in examining whether more general attitudes have an influence in determining attitudes towards driverless trains was by undertaking a two-way analysis of variance (ANOVA). The two-way ANOVA did not reveal any interaction effects between all but one of the latent variables. The factor 'bike lovers' shows just significant interaction effects at the 5% level of significance. For the main, therefore, these results show that there is no reason to suspect that the association between attitudes to driverless trains from the group of users characterised by each factor differs according to their train use. So, for example, if a respondent's general attitudes are heavily influenced by public transport lover attitudes, their attitude to driverless trains is unaffected by whether they are a train user or not. From a marketing or education perspective, this is helpful

Table 2 Factor loadings on the latent variables from the 21 attitude statements

Statement	Factor						
	Public transport lovers	Car snobs	Bike lovers	Car lovers	Walkers	Car safety pragmatists	Travel time minimisers
I like taking public transport	0.87						
I prefer to take public transport rather than drive whenever possible	0.70						
Public transport can sometimes be easier for me than driving	0.65						
It does not matter to me which type of car I drive		−0.98					
To me, the car is nothing more than a convenient way to get around		−0.47					
To me, the car is a status symbol		0.33					
I prefer to ride a bike rather than drive whenever possible			0.90				
Riding a bike can sometimes be easier for me than driving			0.85				
I like riding a bike			0.73				
I like driving				0.83			
I feel free and independent if I drive				0.72			
I like to drive just for fun				0.66			
Getting there is half the fun				0.29			
Travelling by car is safer overall than riding a bicycle				0.28			
I prefer to walk rather than drive whenever possible					0.88		
Walking can sometimes be easier for me than driving					0.74		
I like walking					0.71		
Travelling by car is safer overall than taking public transport						0.89	
Travelling by car is safer overall than walking						0.54	
The only good thing about travelling is arriving at your destination							0.64
Travel time is generally wasted time							0.55

Extraction method: maximum likelihood

Rotation method: Oblimin with Kaiser normalisation

Rotation converged in 9 iterations

since information can be targeted at public transport lovers without worrying about their usage levels.

The two-way ANOVA was followed by carrying out a one-way between groups analysis of variance to investigate the impact of attitude towards driverless trains and each of the latent factors in turn. Starting with the 'public transport lover', there was a statistically significant difference between the reported attitudes to driverless trains and the latent factor with a medium to large effect ($\eta = 0.09$). Post hoc comparisons using the Tukey HSD test showed there were significant differences between the mean values of the public transport factor for the attitude towards driverless trains for those responses shown in the table (and insignificant differences for other combinations of attitudes). This was repeated for the other latent factors with significant differences between the factor means between attitudes being shown in Table 3 (note that the columns that are missing are due to no significance being found). All size effects where the factor showed significant associations are large (shown by eta). Table 3 shows that across this questionnaire response there is a similar pattern for the 'car snob' and 'car safety pragmatists' mean scores and the response to the attitude towards driverless trains suggesting that similar information could be used to target these two groups.

In terms of education and information campaigns to create a more positive attitude towards driverless trains, two-way and one-way ANOVAs can be used with the different aspects of driverless trains although space precludes this being shown here. The two-way ANOVA to screen out potential interaction effects and then one-way between groups ANOVA to investigate associations. This will allow different targeted material to be designed and used against citizens with different underlying transport attitudes.

5 Conclusions and Recommendations

This paper presented results of analyses of data collected in Sydney and related to people's perception of and attitudes to a driverless train. With Sydney introducing a new driverless metro service from 2019, the paper overall adds to the discussion on driverless trains. However, it also reveals that there is much more to be done. A larger scale of research in different locations with different baseline transport modes needs to identify how common or how location dependent are the public views and attitudes to UTO. This paper is a useful starting point for the next drive in investigating public perception of driverless trains.

Table 3 Significant results of post hoc tests for one-way between group ANOVA for latent factors

Level of support for driverless trains	Level of support for driverless trains	Public transport lovers	Car snobs	Car safety pragmatists
Strongly support	Support			
	Neutral			
	Oppose			
	Strongly oppose			
Support	Strongly support			
	Neutral	0.035		
	Oppose			
	Strongly oppose	0.000		
Neutral	Strongly support			
	Support	0.035		
	Oppose			
	Strongly oppose		0.029	0.008
Oppose	Strongly support			
	Support	0.019		
	Neutral			
	Strongly oppose		0.013	0.031
Strongly oppose	Strongly support			
	Support	0.000		
	Neutral		0.029	0.008
	Oppose		0.013	0.031
η		0.09	0.05	0.06

The questionnaire highlights issues of safety and links to driverless trains. This is clearly an area where some information dissemination would be useful. Perhaps more importantly, identifying how different attitudes towards transport more generally (the latent variables identified by factor analysis) are associated with views on safety would allow targeted material to be developed to educate and provide information to users and non-users. It also demonstrates that marketing to the average is likely to be less effective (and more expensive) than understanding the different needs of citizens with specific travel attitudes.

The results of this questionnaire show more research is necessary in the human errors department as the public seems unclear as to what can go wrong and what will be done to overcome these issues when a driverless train system is in operation. With this information and more disaggregate data on general transport attitudes, operators can help the public by explaining how the system works and how human errors will be dealt with.

This questionnaire has identified how little awareness there is of driverless trains. Awareness raising is important if driverless trains are to help drive sustainability outcomes. Repeating the questionnaire in other locations will identify if this is a Sydney-specific or more general outcome.

As with all questionnaires, further research delving more deeply into the 'why' the answers are what they are will be helpful in framing responses to issues that may be critical in determining whether or not a citizen will use a driverless train.

Acknowledgements The authors would like to thank Mr Yan Gao, mechanical engineering M.Sc. student at Newcastle University and Mr Hanif Zaman, an A-level Nuffield Research Placement student from Heaton Manor school in Newcastle, for working on an initial version of the questionnaire and initial analysis of the original data set used in this paper. This research is supported by Transport for NSW through their support for the Chair in Public Transport at the University of Sydney, Australia.

References

1. Handy S, Weston L, Mokhtarian PL (2005) Driving by choice or necessity? Transp Res Part A 39:183–203
2. Mokhtarian P, Salomon I (2001) How derived is the demand for travel? Some conceptual and measurement considerations. Transp Res A 35:695–719
3. Dulbosc A, Vella-Brodrick D (2015) The role of transport in supporting the autonomy of young adults. Transp Res Part F 33:97–105
4. Karvonen H, Aaltonen I, Wahlström M, Salo L, Savioja P, Norros L (2014) Hidden roles of the train driver: a challenge for metro automation. Interact Comput 23:289–298
5. UITP (2013) Metro automation in 2013. Observatory of automated metros world Atlas report. UITP, Brussels
6. Fraszczyk A, Brown P, Duan S (2015) Public perception of driverless trains. Urban Rail Transit 1(2):78–86
7. Payre W, Cestac J, Delhomme P (2014) Intention to use a fully automated car: attitudes and a priori acceptability. Transp Res Part F 27:252–263
8. Commission E (2011) White paper: roadmap to a single European transport area-towards a competitive and resource efficient transport system. COM 2011:144
9. Bansal P, Kockelman KM, Singh A (2016) Assessing public opinions of and interest in new vehicle technologies: an Austin perspective. Transp Res Part C Emerg Technol 67:1–14
10. Kyriakidis M, Happee R, de Winter JCF (2015) Public opinion on automated driving: results of an international questionnaire among 5000 respondents. Transp Res Part F 32:127–140
11. Piao J, McDonals M, Hounsell N, Graindorge M, Graindorge T, Malhene N (2016) Public views towards implementation of automated vehicles in urban areas. Transp Res Procedia 14:2168–2177
12. Harvey J, Thorpe N, Caygill M, Namdeo A (2014) Public attitudes to and perceptions of high speed rail in the UK. Transp Policy 36:70–78
13. Hess S, Shires J, Jopson A (2013) Accommodating underlying pro-environmental attitudes in a rail travel context: application of a latent variable latent class specification. Transp Res Part D Transp Environ 25:42–48
14. Gazette R (2015) Helsinki metro automation called off. http://www.railwaygazette.com/news/urban/single-view/view/helsinki-metro-automation-called-off.html. Accessed 27 Oct 2016
15. Wang Y, Zhang M, Ma J, Zhou X (2016) Survey on driverless train operation for urban rail transit systems. Urban Rail Transit. doi:10.1007/s40864-016-0047-8
16. Grogan A (2012) Automated metro systems set to overtake. E&T Magazine. http://eandt.theiet.org/magazine/2012/05/driverless-trains-its-the-automatic-choice.cfm. Accessed 10 Nov 2015
17. BBC (2014) Driverless tube trains: unions vow 'war' over plan. http://www.bbc.co.uk/news/uk-england-london-26381175. Accessed 9 Dec 2014
18. Brown P (2014) Are driverless trains the future? Railway Technology Magazine Feb/Mar 2014. http://www.railtechnologymagazine.com/Comment/are-driverless-trains-the-future. Accessed 4 Oct 2016
19. Hasham N (2013) Driverless trains plan must overcome public scepticism. http://www.smh.com.au/nsw/driverless-trains-plan must-overcome-public-scepticism-20130607-2nvjq.html. Accessed 9 Dec 2014
20. Transport for NSW (2014) Delivering Sydney's new rapid transit trains. Fact Sheet. http://nwrail.transport.nsw.gov.au/Info?type=Fact-Sheets. Accessed 20 Nov 2015
21. ABC (2015) North West Rail Link: Sydneysiders get first glimpse inside driverless trains. http://www.abc.net.au/news/2015-11-05/driverless-trains3a-sydneysiders-get-first-glimpse-inside/6914786. Accessed 10 Nov 2015
22. North West Rail Link (2015) Delivering Sydney's new rapid transit trains. Fact Sheet. http://www.sydneymetro.info/documents. Accessed 10 Nov 2015
23. Sydney Metro (2015) Project update—Epping. Brochure. http://www.sydneymetro.info/documents. Accessed 10 Nov 2015
24. Gao Y (2016) Public perception of and attitude to driverless train. A case study of Sydney, Australia. M.Sc. thesis, Newcastle University
25. Australian Bureau of Statistics (2016) Census. http://www.abs.gov.au. Accessed 10 Jul 2016
26. Fraszczyk A (2014) An exploration of the characteristics of excess travel within commuting. Ph.D. thesis, Newcastle University
27. Horn JL (1965) A rationale and test for the number of factors in factor analysis. Psychometrika 30:179–185

3

Experimental Study on the Durability of Monorail Track Beam After Acid Rain Corrosion

Ni Yuancheng[1] · Zhu Eryu[1] · Chen Zhuo[2]

Abstract An experimental study is proposed in this paper to investigate about the durability of monorail track beam after acid rain corrosion in a monorail transportation system. Firstly, a 20-m full-scale beam was built to analyze the causes of corrosion of a monorail track beam, and the corrosion pool was used to make a catalyzed corrosion test on the beam. Then the strand steels were contrasted after 3000-h of accelerated corrosion test. Next, experimental study was performed and the characteristics of strand corrosion in the monorail track beam with corrosion is discussed based on experimental observations. The results show that after the acid rain corrosion, the cross-sectional area of the prestressed steel strand was reduced, although the degree of corrosion was not serious compared with the plain bars, it was more dangerous because of high stress, so further research is necessary to clarify the effects of acid rain corrosion on the durability of the track beam.

Keywords Monorail track beam · Durability · Acid rain · Corrosion model

✉ Zhu Eryu
eyzhu@bjtu.edu.cn

Ni Yuancheng
niyuancheng@sina.com

Chen Zhuo
chenzhuo8832@126.com

1 School of Civil Engineering, Beijing Jiaotong University, No. 3, Shangyuan Village, Haidian District, Beijing, China

2 China Railway Fifth Survey and Design Institute Group CO LTD, No. 9, Kangzhuang Road, Daxing District, Beijing, China

Editor: Xuesong Zhou

1 Introduction

The straddle monorail transportation system mainly adopts the PC (prestressed concrete) track beam, and the existing weaknesses are mostly concentrated in the cracks of the component; the grouting deficiency and the corrosion are caused by environmental factors such as steel corrosion. Scholars have done a lot of research on the durability of concrete structures. The primary damage to concrete structure was the deterioration due to corrosion of steel bars, and the main factor affecting the corrosion of the steel bars was the penetration of chloride ions in the environment [1, 2]. Concrete carbonization and acid ions that lead to steel reinforcement was the main reason for corrosion [3]. Through the studies on the determination of the chloride concentration in concrete, it points out that different erosion environments would establish different erosion models [4]. The effects of chloride content on the chemical transmission properties of cement in concrete was studied, and the factors affecting the solidification of chloride ions and the factors causing chloride corrosion were analyzed [5]. Through the experiment, the penetration mechanism of chloride in concrete was analyzed, and the corresponding calculation model was established [6, 7]. Based on the time-dependent effect, the corrosion of reinforcement in concrete was discussed in the literature [8]. Combined with the experimental data at home and abroad, the mechanical properties of corroded reinforcement were studied, and the mathematical model of stress–strain relationship of corroded reinforced bar under different environmental conditions was established [9, 10]. The bond stress in anchorage zone of corroded specimen was obtained by accelerated corrosion test, and the bond slip constitutive relationship of corroded reinforced concrete was established [11]. Based on the theory of elasticity and the theory of fracture mechanics, the calculation formula of corrosion expansion cracking, reinforcement corrosion rate and corrosion expansion cracking time of concrete protective layer were established [12].

It can be seen that the existing research mainly focuses on the durability of ordinary reinforced concrete structures, such as concrete cracking and durability vulnerabilities of reinforced concrete structures in the marine climate and corrosive environments. There were few researches on concealed weaknesses of prestressed structures, especially for the hidden vulnerabilities of PC track beam under special conditions. Due to the environmental conditions, duct deficiency of grout bleeding, and prestressed steel strand, the deficient mechanism of subway prestressed concrete structure differed from that of the ordinary concrete structure. Based on this observation, this paper researches on the hidden weakness of PC track beams after acid rain corrosion in a monorail transportation system and analyzes the durability of PC track beams after the vulnerability is induced, in order to lay the foundation for the life expectancy of PC track beams.

2 Corrosion Experiment

2.1 Experimental Program

A full-scale post-tensioned concrete beam was designed with a rectangular cross section of $b \times h = 625 \times 1500$ mm and a total length of 20,000 mm. One plain bar of HPB235 with 12 mm diameter at the bottom, two deformed bars of HRB335 with 16 mm diameter on the top, and stirrups with 90-mm spacing and 8 mm diameter were used in each beam. A duct with 50 mm diameter was reserved inside the beam during casting in the laboratory to ensure that a single 15.2-mm-diameter strand composed of seven wires can be arranged. The details of the beam are shown in Fig. 1.

The mechanical properties of the steel bars and strands are shown in Table 1. The initial strength of steel strand was 1228 MPa, i.e., 0.70 times of the ultimate strength. The concrete mix of concrete is given in Tables 2.

There were two grouting techniques used in the beam grouting test: ordinary grouting technique and vacuum grouting technique. The bellows used were divided into plastic bellows and metal bellows, with a diameter of 50 mm. The steel strand used was divided into ordinary steel strands and coated steel strands. The durability tests for the 12 groups were carried out on the test beam.

2.2 Corrosion Test Design

The corrosion pool at different ion concentrations was placed on both sides of the beam to simulate different corrosion environments. The corrosion pool shown in Fig. 2a and b with different pH values was, respectively, arranged at each side of the beam body. The plastic foam was placed in the test beam; after the completion of the casting, the plastic foam was removed, and the openings in the bellows allow the corrosion solution to enter the bellows. The acid rain corrosion test adopted periodic immersion method, and every 6 days was a test cycle. In order to keep the pH of the solution constant, the pH value was determined and adjusted to the original acidity everyday. The test temperature was 25 °C, and the humidity of the test environment was the same as that of the operation environment. 3000 h was used as the acid rain corrosion test time interval, so it can effectively reflect the situation of a structure under long-term corrosion. After the corrosion was finished, the steel strand in the center section was intercepted and tested (Figs. 3, 4).

3 Result Analysis

After the corrosion test, the bellows were cut and broken into 3 m. Then the strands that were not severely bent or damaged were taken out, and the dust on the surface was removed by purified water. The diameter of the strand was measured at this time, accurate to 0.01 mm; following this, the corroded strands were cleaned by hydrochloric acid solution and then neutralized by the acid with alkali. Then the measurement procedure was repeated, until the difference of the measured values was less than 0.1 mm.

The experiment results are shown as follows: (1) The steel strands and corrugated pipes of N9 and N12 did not undergo obvious corrosion. It proves that the epoxy-coated steel strand has better durability than ordinary steel strands in environments susceptible to acid rain corrosion; (2) for ordinary steel strands produced by different factories, the corrosion degree is the same; (3) comparing and analyzing the corrosion of strand under stress-free and high

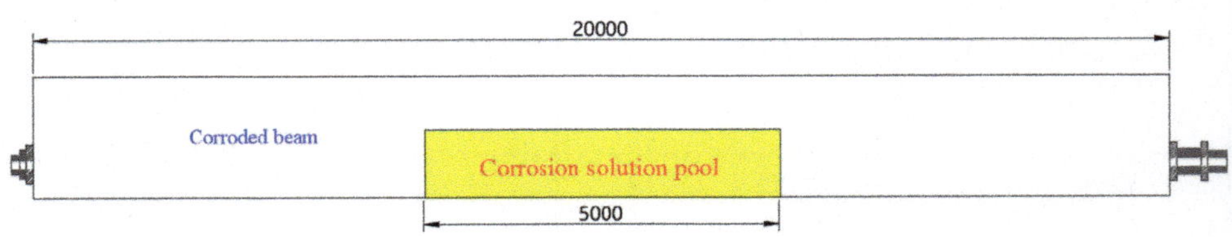

Fig. 1 Details of the corroded beam (unit: mm)

Table 1 Mechanical properties of reinforcement and steel strand

Type	Diameter (mm)	Yield strength (MPa)	Elastic modulus (GPa)
Steel strand	15.2	1830	195
Deformed bars	16	335	210
Plain bars	12	235	210

Table 2 Concrete mixture proportion

Water-to-cement ratio (%)	Cement (kg/m^3)	Water (kg/m^3)	Fine aggregate (kg/m^3)	Coarse aggregate (kg/m^3)
0.4	415	189	685	1109

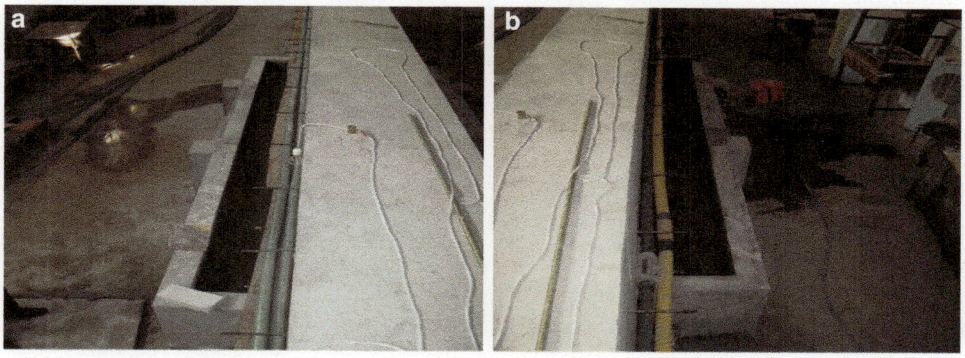

Fig. 2 **a** Location corrosion pool (pH = 1.5), **b** location corrosion pool (pH = 3.5)

Fig. 3 Corrosion of concrete

Fig. 4 Corrosion of steel strand

stress conditions, it is found that there is no obvious change in the corrosion depth of steel strand; (4) for the steel strand without stress, it is found that the lower the pH value of the simulated acid rain, the stronger the corrosion to the steel strand.

4 Damage Model of Steel Strand

4.1 Corrosion Depth of Prestressed Steel Strand

After the prestressed steel strand was corroded, its quality, thickness, mechanical properties, and organizational structure will be changed, and the rate of change in its physical and mechanical properties can be used to indicate the degree of metal corrosion. The corrosion of prestressed steel strand was evaluated by the current index and mechanical performance index of metal corrosion speed.

According to Faraday's law, the quality of the dissolved material on the corrosion electrode is related to the amount of electricity which has passed through the electrode and the number of electrons participating in the reaction. It can be expressed as:

$$\Delta W = \frac{m}{Fn} \times It \tag{1}$$

where ΔW is the quality of corrosion of dissolved substances, I is the intensity of electric current, t is the general time interval, F is Faraday's constant, m is relative atomic mass, n is the number of electrons in the reaction.

This can be introduced as the weight loss of the metal v in the unit surface area per unit time and can be expressed as:

$$v = \frac{\Delta W}{St} = \frac{Im}{FnS} = \frac{iN}{F} \tag{2}$$

where $i = I/S$ is the current density (unit: A/m^2) and S was metal surface area (unit: m^2).

The corrosion thickness of the metal was $d = v/\rho$. The time-varying characteristics of the corrosion current density of prestressed steel strand were considered while the unit conversion was unified, and the corrosion depth of the prestressed steel strand at a certain time D can be expressed as:

$$D = 3.268 \times 10^{-3} \frac{\int i(t)dt\,N}{\rho} \tag{3}$$

where $i(t)$ is the current density relative to the change in time.

In the case of acid rain simulation with pH = 3.2 solution, the corrosion current density can be approximated to 90 $\mu A/cm^3$ if the current density is not considered according to the test results.

The calculation of the corrosion of steel strand is as follows: $t = 3000$ h, $i(t) = 100$ $\mu A/cm^2$, which put into formula (3), and it can be expressed as:

$$D = 3.268 \times 10^{-3} \frac{\int i(t)dt\,N}{\rho} = 3.268 \times 10^{-3} \frac{i(t) \times t \times N}{\rho} = 0.3503\,\text{mm}$$

Compared with the test data obtained in Table 3, the corrosion depth and theoretical value of other common steel strand were close to the theoretical value except for the epoxy-coated steel strand, which shows that the prediction of corrosion depth of steel strand can be carried out by the above formula.

Table 3 The diameter of strand with high stress after 3000-h accelerate corrosion test (cm)

Duct	Corrosion	Steel strand diameter				
		A	B	C	D	Average
N1	Hard	1.540	1.550	1.540	1.550	1.545
N2	Hard	1.540	1.550	1.550	1.550	1.548
N3	Hard	1.550	1.540	1.550	1.530	1.543
N4	Hard	1.570	1.570	1.570	1.530	1.560
N5	Local	1.580	1.580	1.570	1.570	1.575
N6	Local	1.580	1.570	1.580	1.570	1.575
N7	Local	1.590	1.590	1.590	1.590	1.590
N8	Local	1.580	1.580	1.580	1.580	1.580
N9	Obsolete	1.670	1.660	1.680	1.680	1.673
N10	Hard	1.530	1.530	1.540	1.530	1.533
N11	Hard	1.550	1.550	1.560	1.540	1.550
N12	Obsolete	1.670	1.660	1.690	1.660	1.670

4.2 Strength of Prestressed Steel Strand After Corrosion

At present, the study on the mechanical properties of prestressed steel strand was mainly based on the experimental research, and the tensile test was carried out by steel strand of different corrosion degree, which can be used to count the change in mechanical properties. The relation between the loss rate of the tensile strength and the corrosion degree of the steel strand can be expressed as:

$$\gamma(t) = 1 - \frac{0.986 - 1.103\eta(t)}{1 - \eta(t)} \tag{4}$$

where $\gamma(t)$ is the tensile strength loss rate of the steel strand in the year, $\eta(t)$is the loss rate of steel strand section in the year. The comparison between the final test value and the theoretical value is listed in Table 4.

All tensile strength of the steel strand decreased obviously after corrosion, except for steel strand A. The decline of the majority (B, C, D) was around 20 $\sim$ 40 MPa, that is to say, the original tensile strength decreased by about 1.5 $\sim$ 2.0%. It shows that if the corrosion rate of the prestressed steel strand is small (usually within 5% of the cross-sectional loss), the mechanical properties of

Table 4 Loss rate of tensile strength after corrosion

Steel strand	A	B	C	D
Predicted	1.34	1.63	1.78	1.92
Experimental	0.19	1.38	1.56	2.01

the prestressed steel strand will be affected by corrosion, but not very seriously. Corrosion of steel strands leads to the reduction of mechanical properties. On the microlevel, the electrochemical reaction occurred between the steel strand and the corrosion ion, and the microstructure of materials has been changed. On the macrolevel, the main factors may be the cross-sectional loss and the stress concentration. By comparing the theoretical values with test values, it can be seen that the theoretical rate of loss of tensile strength in ordinary steel is basically consistent with the test results, in addition to the epoxy-coated steel strand A. Therefore, formula (4) can be used as a model to evaluate the tensile strength degradation after normal prestressed steel strand corrosion.

5 Conclusions

1. The corrosive environment and poor construction quality may cause the hidden vulnerabilities such as concrete cracking and prestressed steel strand corrosion during service time, and it will affect the quality and safety of the monorail track beam.
2. The corrosion products produced by the corrosion of the tendons in the beam will produce swelling force and cause the cracking of concrete; the cohesive force between the tendons and concrete will be greatly reduced.
3. On the premise of understanding the mechanism of steel strand corrosion, it is necessary to study the mechanism of the deterioration of mechanical properties of corroded tendons and establish a reasonable prediction model of prestressed structures' durability.

Acknowledgements This work reported here was conducted with the financial supports from a grant from the Beijing mass transit railway operation corporation research program (No. 2016000501000005) and a grant from China railway corporation technology research and development plan (No. 2016G002-L). The supports are gratefully acknowledged.

References

1. Mehta PK (1991) Durability of concrete—fifty years of progress. Spec Publ 126:1–32
2. Castel A, Vidal T, François R et al (2003) Influence of steel—concrete interface quality on reinforcement corrosion induced by chlorides. Mag Concr Res 55(2):151–159
3. Saraswathy V, Muralidharan S, Thangavel K et al (2003) Influence of activated fly ash on corrosion-resistance and strength of concrete. Cement Concr Compos 25(7):673–680
4. Yuan Y, Ji Y, Jiang J (2009) Effect of corrosion layer of steel bar in concrete on time-variant corrosion rate. Mater Struct 42(10):1443–1450
5. Bazzoni B, Lazzari L (1995) Potential monitoring in presence of stray currents. Corros Manag 6:9–12
6. Bertolini L, Pedeferri P, Bazzoni B et al (2001) Measurements of osmic drop free potential in the presence of stray current. ACI Mater J 98(4):296–304
7. Schupack M (2004) PT grout: bleed water voids. ACI Concr Int 26(8):69–77
8. Turnbull A, Mccartney LN, Zhou S (2006) Modelling of the evolution of stress corrosion cracks from corrosion pits. Scripta Mater 54(4):575–578
9. Liu Q, Zhang P, Zhao J (2007) Duct grouting investigation and research on mechanical properties of steel strand in PC continuous bridges. Constr Technol 36(2):63–66
10. Vu NA, Castel A, François R (2009) Effect of stress corrosion cracking on stress–strain response of steel wires used in prestressed concrete beams. Corros Sci 51(6):1453–1459
11. Gardoni P, Pillai RG, Reinschmidt K et al (2009) Probabilistic capacity models for corroding posttensioning strands calibrated using laboratory results. J Eng Mech 135(9):906–916
12. Coronelli D, Castel A, Vu NA, François R (2009) Corroded post-tensioned beams with bonded tendons and wire failure. Eng Struct 31(8):1687–1697

4

Rail Transportation Lead Urban Form Change: A Case Study of Beijing

Chun Zhang[1] · Haishan Xia[1] · Yan Song[2]

Abstract In the background of rapid urban metro transit construction in China, rail transportation becomes an important factor in shaping urban form. This case study focuses on the urban form changes caused by urban rail transportation development in Beijing and especially explores the synergetic links between urban planning and transportation from the perspective of a transportation–population–economy research framework. Based on population census surveys and basic economic unit census surveys on street district level, combined with digital urban metro lines on street district level, findings show that the population spatial distribution tends to disperse along the metro lines, and concentrates around metro network connections. The employment distribution turns to be more concentrated in high accessibility areas around metro intersections. The population and employment distributions indicate a job–housing mismatch, where the employment centers are concentrated in the central city, while residential centers are concentrated in suburban areas. The consequence of this mismatch is that regions with extreme job–housing imbalance also see higher degrees of imbalance in urban metro volume. This paper aims to provide insights on how to improve job accessibility and creating a more sustainable urban form, as well as promoting a synergetic development between urban planning and transportation.

Keywords Urban rail transportation · Urban form · Job accessibility · Job–housing spatial relationship · Beijing

1 Introduction

In the second half of the twentieth century, metropolitan urban forms changed dramatically massive investments in transportation infrastructure. In particular, construction of regional and urban rail transit, including urban light rail and metro system, not only shortens the commuting time of urban residents, but significantly alters urban spatial patterns, such as population and employment distributions. From an urban planning perspective, rail transportation not only brings change in commuting speed, but also entails unprecedented challenges and opportunities for a twenty-first-century metropolitan development. For example, the European Union initiative *Polycentric Metropolitan* strategy [1–3], the USA 2050 [4], and several Asian countries promote transportation infrastructure as important factor in supporting sustainable development of metropolitan areas.

Globally, large-scale high-speed rail drives regional metropolitan area changes [5] at the macrolevel, while on a mesoscale urban metro systems also reshape the urban form [6]. In particular, cross-regional transportation infrastructure becomes a key factor in regional cooperation among different administrative jurisdictions. For example, since China's first 5-year plan in the 1950s, national railway construction has represented the nation's advanced industrialization process.

Chun Zhang and Yan Song: UNC-BJTU Consortium for Urban and Regional Transportation.

✉ Haishan Xia
hshxia@bjtu.edu.cn

[1] School of Architecture and Design, Beijing Jiaotong University, Beijing, China

[2] Department of City and Regional Planning, University of North Carolina at Chapel Hill, Chapel Hill, NC, USA

Editors: Haishan Xia and Chun Zhang

The concept of urban form is not only limited to the Chinese vocabulary "spatial shape," but its wider definition focuses on layout and development patterns from both physical and non-physical aspects [7]. From the perspective of a synergetic development of urban planning and transportation, urban form highlights the interaction between economic factors, population statistics and transportation modes, as well as the potential impact of transportation on the spatial distribution in metropolitan areas The two most widely used indicators to analyze urban form and transportation are job–housing spatial relation and job accessibility.

Previous research on urban form has focused on various aspects, such as transportation efficiency, network configuration and job–housing spatial relationship. For example, Frank and Pivo [8] describe how the four-stage traditional transportation model predicts traffic demand and traffic volume based on traffic analysis zones (TAZs). The key issue for the transportation model is to improve efficiency and mobility of transportation systems [7]. Later, as transportation research shifted toward the urban field, more scholars began to emphasize the job–housing spatial relationship in metropolitan areas, using demographic data to simulate current and future travel behavior [9, 10]. The availability of census data and development of GIS technology support the analysis of population and employment density data in correlation with transportation networks.

Furthermore, since the 1990s, transportation-related problems increasingly became a universal challenge in most metropolitan areas. Both transportation planners and urban planners realized that transportation policies alone are not sufficient to tackle highly complex transportation problems. Instead, the New Urbanism movement and Smart Growth policy in the USA aimed to build a comprehensive analytical framework drawing on a wide range of population, economy, land-use and transportation data. Today in Chinese cities, the rapid growth in urban rail transit development occurs in tandem with a fast urbanization process, posing new, complex problems to transportation planners, urban planners and government officials.

The current literature on urban transportation mainly focuses on traffic model characteristics, accessibility and transportation network efficiency. However, comprehensive studies on the spatial, social and economic impacts of transportation infrastructure are still rare. Against this background, this paper analyzes changes in urban form in the metropolitan area of Beijing. By using street district-level GIS data, we especially look into how the rapidly developing rail transit system affects the job–housing spatial relation. Focusing on the interactive relationship between urban rail transit and urban form, this paper will provide policy implications for the synergetic development of transportation, land-use and urban planning in Beijing–Tianjin–Hebei area.

2 Literature Review on Transportation Impact on Urban Form

The current literature on the influence of transportation infrastructure on urban form includes three issues, job–housing spatial relationship, job accessibility and the relation between urban form and travel behavior.

2.1 Job–Housing Spatial Relationship

The issue of job–housing spatial relationship has long been a hot topic and has received wide interest from the urban planning, transportation and sociology field. The spatial mismatch concept originated with Kain [11] who found that while job opportunities in America moved to the suburbs, low-skilled workers continued living in city centers. The original spatial mismatch hypothesis (SMH) has been continuously discussed and challenged in many ways. Some scholars considered it a racial discrimination problem rather than a spatial problem [12] or a language problem [13, 14]. Other scholars have concluded that the mismatch might be caused by differences in travel mode, i.e., car ownership, especially in cities with insufficient public transit systems, such as Detroit [15].

Although transportation technology has greatly improved in recent years, the job–housing mismatch in big cities appears to have worsened [16]. While most research focuses on the job–housing mismatch and related unemployment issues within the city, some case studies provide metropolitan-level analysis. As suburbanization processes provide more employment opportunities in suburban job centers, the urban poor living in the outskirts of cities may face difficulties in accessing jobs due to a lack of public transportation facilities [17]. Thus, the traditional model based on an urban center–suburban binary division might be not sufficient.

The job–housing mismatch in Chinese cities has gained more research attention in recent years. In the planned economy era, the job–housing relationship used to be perfectly balanced on a Work Unit basis, but with the disintegration of Work Units, the job–housing spatial relationship has gradually shifted from a "spatial bond" to "spatial mismatch". In particular, the rapid development of affordable housing in urban fringe locations, while the economic momentum remains in city centers, has led to a widening of the gap between job and housing locations [18]. The rapidly changing job–housing patterns in Chinese cities has drawn attention from both the transportation field and the urban planning field, due to the links between urban form and commuting behavior.

2.2 Job Accessibility

Job accessibility is an important indicator when analyzing urban form performance from a transportation perspective. Traditionally, transportation experts paid more attention to traffic engineering-related issues, such as traffic configuration, the number of intersections and connectivity of local roads, highways, railways and other transportation infrastructure [19]. Later, accessibility to job and other facilities received more research interests from both scholars and policy makers, because of the social implications for disadvantaged groups.

Accessibility can be defined as the efforts to overcome certain fictions to a pointed destination [20]. There are usually two approaches calculating accessibility: the gravity model and the accumulated opportunity model. The gravity model investigates the potential attraction between origin and destination on a TAZ level based on transportation networks [21, 22]. The accumulated opportunity model calculates the number of jobs that can be accessed within a fixed commuting time [23]. Job accessibility is not only a technical index to capture the urban form pattern and transit cost from home to the job location, but also measure the performance on job markets—especially for low-income, women, ethnic minorities and other vulnerable groups [15]. In terms of urban form, job accessibility also includes the consideration of the land-use and spatial distribution information. Currently, better accessibility is listed as the principles of good urban planning.

3 Research Framework, Case and Method

Previous research has mostly focused on the job–housing spatial relationship and job accessibility-related topics, while research on how rail transportation influence urban form is still rare. This paper builds its research framework on the urban form theory, focusing on the interaction between transportation, population and employment (Fig. 1). Correspondingly, this framework highlights three research questions: (1) rail transportation lead population distribution (residential space) change; (2) rail transportation lead employment distribution (economic space) change; and (3) relation between job–housing and travel behavior. By applying the methodology of TAZ-level analysis by GIS in American cities, this paper adopts street district, town and township as basic units, combined with population census data from 2000 and 2010, and basic unit economy survey data from 2001 and 2008. Transportation network data including subway and light rail are also transferred to GIS format.

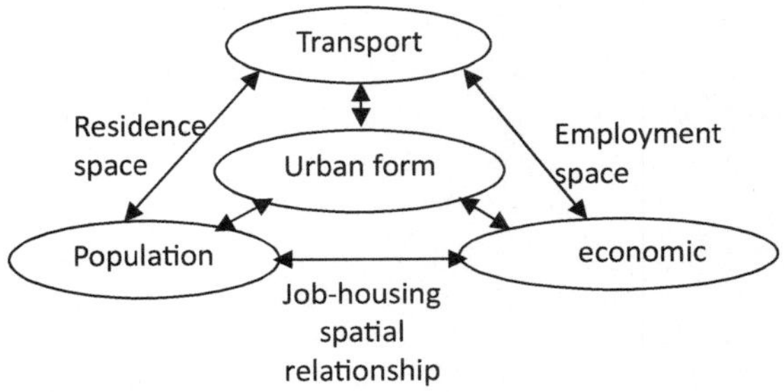

Fig. 1 Research framework of "transportation–population–economy"

This paper selects Beijing as case study. Beijing's rail transit has experienced a rapid growth, with the total number of lines increasing from 3 to 18 (including the airport link) during 2000–2015 (Fig. 2). The total length of rail traffic today amounts to 527 km (2015). At the same time, the job–housing spatial relationship has changed significantly from a relatively balanced job–housing integration before the economic reform, to a job–housing separation in the transition period. The Beijing metropolitan area as analyzed in this case study also includes parts of Tianjin and Hebei (Jing-Jin-Ji).

4 Results: Urban Form Changes on Metropolitan and Urban Scale

4.1 Accessibility Changes in Beijing–Tianjin–Hebei Area

Studies from countries in the West have noted that high-speed rail construction can balance the population distribution in metropolitan areas. In particular, experiences from European cities show that improvements in accessibility can encourage citizens to move from city centers to remote suburban areas [24]. Similarly in the analysis of the Beijing–Tianjin–Hebei region, although the accessibility of urban agglomeration increased significantly in the high-speed rail period, these cities might not benefit by just having rail transportation passing through.

As an important transportation infrastructure, high-speed rail significantly changes the accessibility and population distribution on a metropolitan level compared to regular rail systems. In the year of 2000, a 1-h commuting trip in any given direction would only get the citizen to the boundaries of Beijing's administrative zone and a few nearby cities southeast of Beijing (Fig. 3, left). The average travel time to all the prefectural-level cities was around

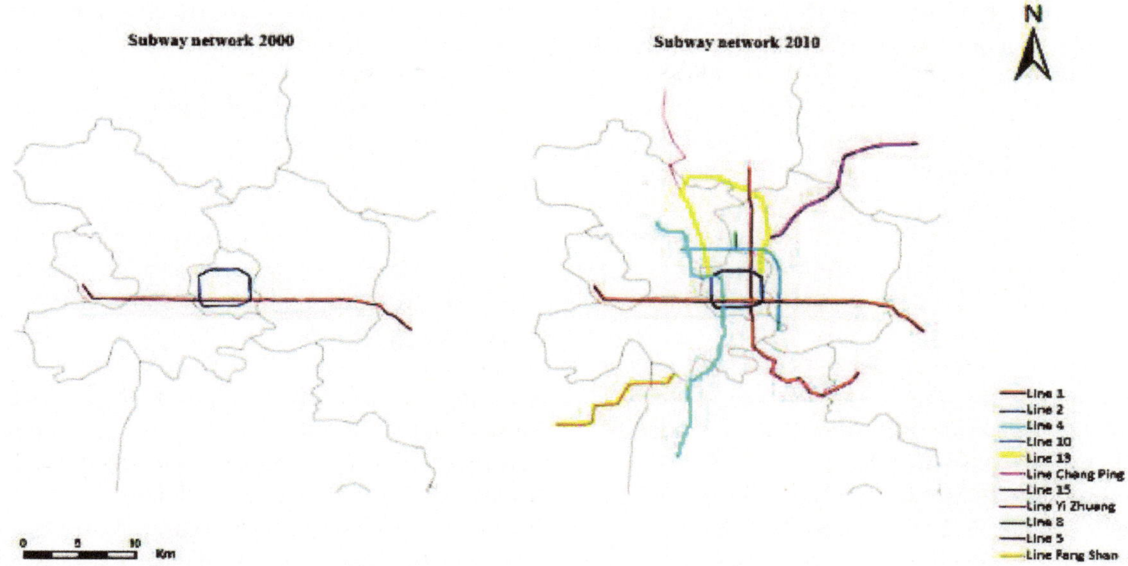

Fig. 2 Urban metro system expansion from 2000 to 2010 in Beijing metropolitan area

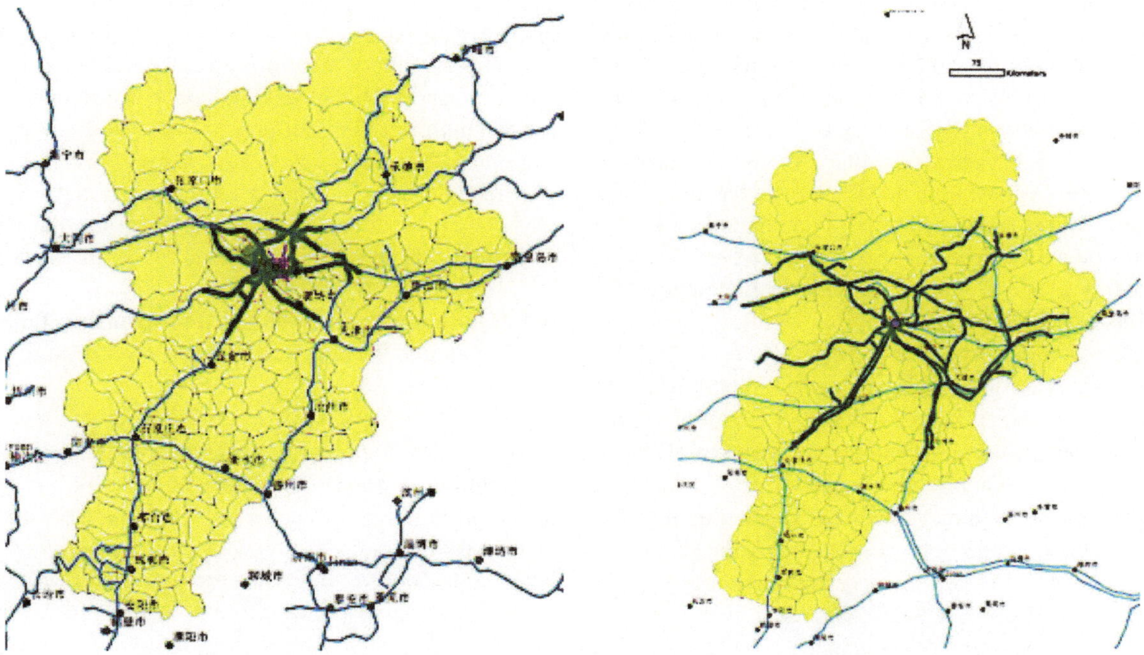

Fig. 3 Beijing Tianjin and Hebei region accessibility change from the county-level units (*left*) to high-speed rail (*right*)

3–3.5 h with regular speed trains (Fig. 3, right). In the year of 2010, after the introduction of high-speed rail lines, the 1-h commuting trip would reach five times as far and reach nine cities. Correspondingly, the average traveling time to all the prefectural-level cities is around 1–1.5 h in high-speed rail era. Hence, high-speed rail has clearly expanded the accessibility for commuters and greatly integrated the Beijing–Tianjin–Hebei region.

Furthermore, county-level analysis shows that cities benefit significantly from being connected to the high-speed rail network. While the traditional railway system connects counties or county-level cities and stops at most of these, the high-speed rail is only accessible from larger cities. As a consequence, smaller towns and cities are increasingly losing their connectivity advantage and turn into locations which the high-speed rail passes through without stopping. The accessibility on city scale depends more on the link with high-speed rail network on regional scale. In short, the changes in accessibility and population redistribution caused by high-speed rail transportation seem to further deepen the gap between the central city and its surrounding areas.

4.2 Residential Density Change and Population Redistribution

Studies on metropolitan scales usually indicate that citizens tend to concentrate in better accessible areas, and construction of highways and other infrastructure in suburbs usually attract more residents [16].

In our case, street district-level analysis of Beijing's metropolitan area shows that with the establishment of the urban metro system from 2000 to 2010, population distribution experienced overall dispersion and local centralization at the same time. On the one hand, there are more new subcenters emerging with the extending urban metro system (Fig. 4, left). For example, the red bar shows emerging residential centers in Tiantongyuan and Huilongguan along subway 13 and also Shuangqiao and Tongzhou New Town along subway 8.

One the other hand, citizens became more concentrated at specific metro connection nodes, due to the improved accessibility. For example, the population density increased around Xizhimen metro station, where subway lines 2 and 4 connect, and around Huangzhuang, where subway lines 4 and 10 connect (Fig. 4, right).

The findings indicate that the impact of rail transit on population redistribution can be a double-edged sword, as it encourages suburbanization because of shorter commuting time, but also causes population concentration at transportation nodes. On urban scale with the establishment of urban rail transit system, population appears both trend of suburbanization along urban metro lines and centralization on transportation nodes.

4.3 Job Density Change and Employment Redistribution

While residential statistics show a twofold process of both increased suburbanization and local population concentration, employment locations also change with due to the urban metro system. Our analysis shows that the areas of employment are become more centralized, which indicates that employment centers emerge in the inner city because of better accessibility by rail transit. In particular, central business district areas tend to emerge at intersections of metro lines, such as Guomao, Fuxingmen and Zhongguancun (Fig. 5, red area). In addition, our analysis found that job density gaps are even larger in 2010 compared to 2000, which indicates a change in location value caused by the development of the urban metro system.

Compared with the population redistribution in suburban areas, the process of job suburbanization is more unclear. Clifton et al. [7] case study in the USA suggests that employment centers move to suburban areas in tandem with transportation infrastructure construction, seeking lower land price. Furthermore, in metropolitan areas such as London, Paris and Tokyo have distinct polycentric urban patterns at different scales [25]. Compared with these western cities, urban rail transit in China plays a more important role in centralizing employment centers.

4.4 Relationship Between the Job–Housing Balance Index and Urban Rail Transit Commuting Behavior

While the development of urban rail transit has changed the distribution of populations and employment space, it has also had dramatic impacts on the job–housing spatial

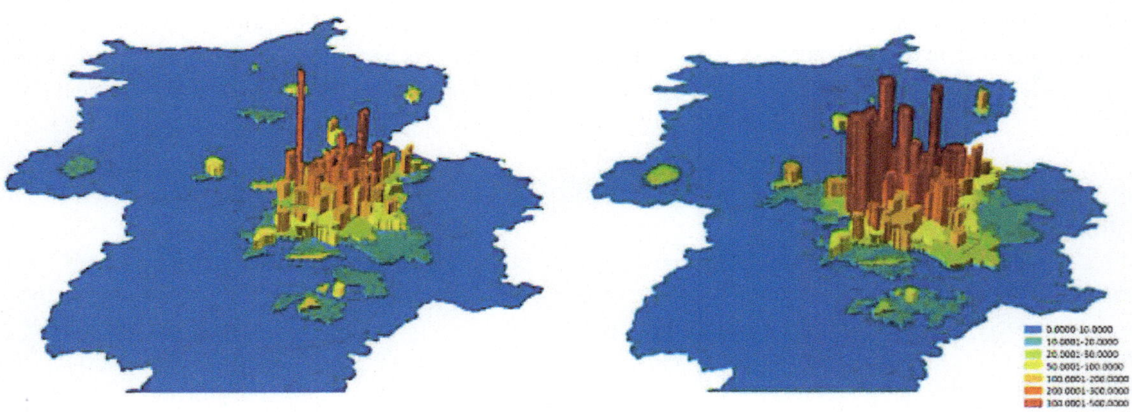

Fig. 4 Development of urban rail transit and the change in population spatial distribution (2000–2010)

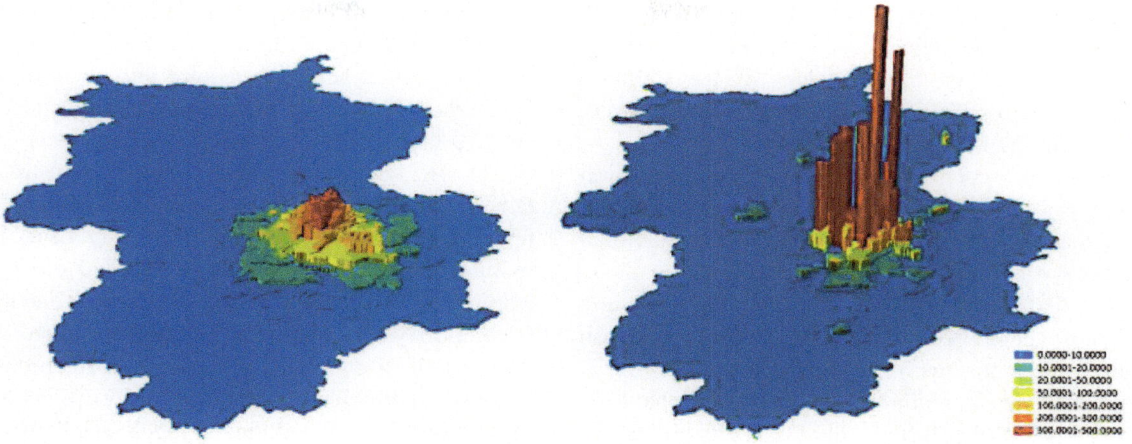

Fig. 5 Urban rail transit development and the change in employment spatial distribution (2000–2010)

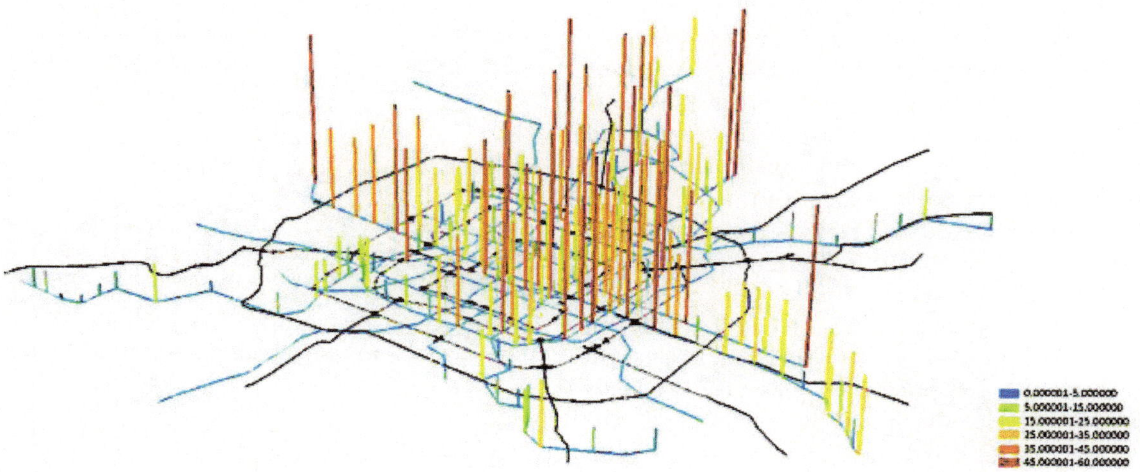

Fig. 6 Passenger volume of urban metro station in Beijing

relationship in Beijing. In terms of job–housing balance (JHB) index, it shows the job–housing mismatch pattern while the JHB index is higher in the city center area with more job opportunities, and JHB index is lower in the suburban area with more residential functions. It usually finds in the USA cities that job–housing spatial relationship has great impact on commuting behavior on metropolitan area [26]; some scholar even taken it as sin to long-distance commuting.

An analysis of the passenger flow of urban metro station shows that the commuting volume is higher in areas with an extreme job–housing mismatch. For example, a job center such as Guomao has a very high JHB index, while residential centers such as Tiantongyuan and Huilongguan score very low on the JHB index (Fig. 6). With an imbalanced job–housing relationship, the urban metro system will face heavy volume pressure locally. Our findings indicate that microlevel rail transit passenger volume of each station correlates with the job–housing spatial relationship. The policy implication for transportation planners and urban planners is that only transportation infrastructure cannot solve commuting problems, but there is a need to also consider the influence of urban form such as job–housing spatial relationships.

5 Conclusion

This paper explores the impact of urban rail transit on urban form by metropolitan and urban scale analysis. Based on the research framework of transportation–population–economics, findings show that the establishment of the urban metro system increased local transportation accessibility, as well as caused spatial redistribution of both residential and employment centers. The introduction of high-speed rail has significantly improved accessibility among cities located on the boundary of the Beijing–Tianjin–Hebei metropolitan area and now allows a 1-day commute within the metropolitan area.

The analysis of residential density shows that population distribution is scattered along the urban metro lines in suburban areas, and tended to concentrate around urban metro stations where metro lines connect. In terms of job density, our findings show that employment distribution becomes more centralized and the gap between urban centers and suburban areas becomes even larger.

The accumulated literature on the relation between transportation and urban form in the USA, experience the shift of research topics from transportation network configuration, job accessibility to job–housing spatial relationship in a larger scale. Currently, the research has been focus on the influence of existing urban form on travel behavior, while high-density, mixed land-use planning principals are promoted to calm down traffic. However, as its counterpart in Chinese cities, rapid urban rail transportation induces urban population and employment space redistribution, while it also induces job–housing spatial mismatch on a larger scale. While vast transportation investment promotes accessibility and provides convenience for urban residents, it also brings new transportation problems and heavier traffic pressure.

This paper shows different model in terms of urban form change under rail transportation between China and the US cities. In most American cities, it continues the job–housing mismatch pattern brought up by John Kain in the 1960s, where the urban rail transit serving as important facilities in the inner city for the urban poor. On the contrary in most Chinese cities, rail transportation mostly benefits the middle- and low-income group who lives on the urban fringe, but has to commute between employment location in the urban center and residential location in the far suburban area. With the rapid growth rail transportation system, it usually shows dynamic, interactive and complex changes on urban form, and also appears new problems, new opportunities and challenges under the unique period of mass production of urban rail in China.

Based on the analysis in this paper, future studies on the impact of urban rail transportation on urban form can add important insights on how to widen the scope of transportation planning to also include the links between population, economy and land use.

The limitation of this paper lies in that transportation is not the only reason driving urban form changes, which may occur with or without metro construction. For example, in most American cities, urban form is more shaped by highways than by urban rail transit. Future research is needed to further explore the interaction between urban form and other elements to clarify the relation between urban form and rail transportation. Besides the physical aspects of urban form, social impacts also require due attention, i.e., analyzing how the travel behavior of disadvantaged groups (such as the elderly, poor, women) is impacted by urban rail transit. The social impacts of rail transit construction contain issues relating to social equity and sustainable development. With increasing access to big data, POI data, mobile phone data, Baidu heat map and other new data sources, the relationship between rail transit and urban form is potentially a vibrant field for future research.

The will provide policy implication for promoting the healthier and more sustainable urban form, while increase job accessibility and improve job–housing balance on a larger scale, making transportation, land-use and urban planning integration in the future in rapid developing cities.

Acknowledgements Funding was provided by National Natural Science Foundation of China (Grant Nos. 41401651, 51678029), the China Fundamental Research Funds for the Central Universities (Grant No. 2016JBWZ002).

References

1. Goodstadt V (2008) European perspectives on megaregions and spatial planning. In: Metropolitan area and space planning seminar, Lincoln Research Center, Peking University, Beijing
2. Pain K, Hall P (2006) People and places: interrelating the space of flows and the space of places. In: Hall P, Pain K (eds) The polycentric metropolis: learning from mega-city regions in Europe. Earthscan, London, pp 113–121
3. Hall P (2006) Firms and places: inside the mega-city regions. In: Hall P, Pain K (eds) The polycentric metropolis: Learning from mega-city regions in Europe. Earthscan, London, pp 57–94
4. Lang R, Lefurgy J, Nelson AC et al (2006) The six suburban eras of the United States research note. Opolis Int J Suburban Metrop Stud 2006(1)
5. Wang J, Lin Z (2011)The impact of high-speed railway on urban spatial evolution: an analysis based on China's characteristics [J]. Int Urban Planning 2011(1):16–23
6. Todorovich P, Schned D, Lane R (2011) High speed rail: international lessons for U.S. policy makers. Doc De Treball 76(2):537–553
7. Clifton K, Ewing R, Knaap G-J, Song Y (2008) Quantitative analysis of urban form: a multidisciplinary review. J Urban Int Res Placemak Urban Sustain 1(1):17–45
8. Frank L, Pivo G (1994) Relationships between land use and travel behavior in the Puget sound region. Single Occupant Vehicles 86(10):1361–1362
9. Cervero R (1994) Transit-based housing in California: evidence on ridership impacts. Transp Policy 1(3):174–183
10. Messenger T, Ewing R (1996) Transit-oriented development in the sun belt. Transp Res Rec J Transp Res Board 1552(1):145–153
11. Kain JF (1968) Housing segregation, negro employment, and metropolitan decentralization. Q J Econ 82(2):175–197
12. Ellwood DT (1986) The spatial mismatch hypothesis: are there teenage jobs missing in the ghetto? In: Freeman RB, Holzer HJ (eds) The black youth employment crisis. University of Chicago Press, Chicago, pp 147–190

13. Ley D (1993) Social geography of the city. Harper & Row Press, New York
14. Mattingly DJ (1999) Job search, social networks, and local labor-market dynamics: the case of paid household work in San Diego, California. Urban Geogr 20(1):46–74
15. Grengs J (2010) Job accessibility and the modal mismatch in Detroit. J Transp Geogr 18(1):42–54
16. Weitz J, Schindler T (1997) Are Oregon's communities balanced? A test of the jobs–housing balance policy and the impact of balance on mean commute times. Unpublished manuscript, Department of Urban Studies and Planning, Portland State University
17. Hu L (2015) How does job access affect the employment outcomes of different income groups? In: Transportation Research Board annual meeting
18. Zhang C, Man J (2015) Examining job accessibility of the urban poor by urban metro and bus: a case study of Beijing. Urban Rail Transit 1(4):183–193
19. Loutzenheiser D (1997) Pedestrian access to transit: model of walk trips and their design and urban form determinants around Bay area rapid transit stations. Transp Res Rec J Transp Res Board 1604:40–49
20. Handy SL, Niemeier DA (1997) Measuring accessibility: an exploration of issues and alternatives. Environ Plan A 29(7):1175–1194
21. Hansen WG (1959) How accessibility shapes land use. J Am Inst Plan 25(2):73–76
22. Huff DL (1963) A probabilistic analysis of shopping center trade areas. Land Econ 39(1):81–90
23. Wachs M, Kumagai TG (1973) Physical accessibility as a social indicator. Socio Econ Plan Sci 7(5):437–456
24. Vickerman R (1999) Is transport infrastructure effective? J Transp Geogr 7(3):230–231
25. Zhang C, He C (2010) Metropolitan area and spatial planning: international experience. Int Urban Plann 4:85–91
26. Boarnet M, Crane R (2001) The influence of land use on travel behavior: specification and estimation strategies. Transp Res Part A Policy Pract 35(9):823–845

5

A Safety Management and Signaling System Integration Method for Communication-Based Train Control System

Fei Yan[1] · **Chunhai Gao**[2] · **Tao Tang**[3] · **Yao Zhou**[4]

Abstract The safety and the correctness of signaling system not only relate to the safety and efficiency of the rail transit operation, but also link with the life safety of passengers. In order to guarantee the safety of a signaling system for metro, the safety certificate for the trial operation with carrying passengers must be obtained. In this paper, a suitable safety management and signaling system integration model are explored according to the CENELEC standards and applied in China. With taking account of the strict safety requirements for the Communication-Based Train Control (CBTC) system, a safety assurance and assessment method based on safety verification and validation process was put forward. This method was applied in every phase of the CBTC system development life cycle to monitor and control each activity in the life cycle and to review each document in system development process. At the same time, this method is also used to ensure the traceability of relevant documents and to test all the functions of the whole system sufficiently and completely. So that the safety operation of train control system can be ensured. Up to now, the independently developed CBTC system with the safety management had been applied in many urban rail transit lines of Beijing, such as Yizhuang Line, Changping Line, Line No. 14, and Line No. 7. The CBTC signaling systems of these projects have been authorized by the safety certification from a third party, e.g., Lloyd Register which is a British company and famous for the safety verification and validation process.

Keywords Rail transit system · CBTC · Safety management · Signaling system · Safety verification and validation · System safety assurance

1 Introduction

While technology is changing faster than the engineering techniques to cope with the new technology are being created, software is pervasive in our society: Its scope is widening more and more in many critical domains such as avionics, space, railway, automotive, nuclear, medical, and air traffic control. As a consequence, it is of utmost importance to assure the society at large that deployment of a given software-intensive system does not contribute to pose an unacceptable risk of harm. For metro system, communication-based control system now has been widely used in new building metro lines. And most of operational functions are realized by software.

Lessons learned over centuries about designing to prevent accidents may be lost or become ineffective when older technologies are replaced with new ones. Digital technology has created a quiet revolution in most fields of engineering, but system engineering and system safety engineering techniques have not kept pace. Digital systems introduce new failure modes that are changing the nature of accidents.

✉ Fei Yan
fyan@bjtu.edu.cn

[1] National Engineering Research Center for Rail Traffic Operation and Control, Beijing, China

[2] Beijing Traffic Control Technology Co., Ltd., Beijing, China

[3] State Key Laboratory of Rail Traffic Control and Safety, Beijing Jiaotong University, Beijing, China

[4] School of Electronic and Information Engineering, Beijing Jiaotong University, Beijing, China

Editors: Jiaqi Ma, Yihui Wang, Xiaolu Rao, Shane Wu and Alireza Khani

Traditional methods in safety engineering therefore struggle to keep up with the development of new technologies and usages of the same. The main drawback of such methods is that they are founded on an assumption of a certain degree of predictability and decomposability. These approaches have lately been labeled "Safety I." New concepts, such as resilience engineering, have gained attention. A common foundation for these concepts is that they all focus on adaptive capacity to stay in control when facing unforeseen disturbances or events, labeled "Safety II."

With the rapid development of urban rail transit systems, it is urgent to develop high-performance train control system with high safety, high reliability, and high efficiency to ensure the safety of train operations. The Communication-Based Train Control (CBTC) system is the trend for the most advanced train control system all over the world.

Compared to the traditional train control system based on track circuits, the CBTC system has the following characteristics: (1) the CBTC system could supply much larger capacity to support the continuous bidirectional train–ground communication; (2) the safety computers platform has been used in both the ground equipment and the on-board equipment to handle train status and control commands, to realize the continuous interval control, route control, speed protection, and automatic driving; (3) the train position can be achieved with high precision; (4) the number of the cables, train control equipment, and first-stage costs, and operational and maintenance cost could be reduced; (5) the CBTC information can be superimposed on the existing signaling system to facilitate the transformation of the existing line and to enable the interconnection of urban rail transit lines.

The CBTC system consists of ground zone controllers (ZC), the on-board equipment, the bidirectional train–ground information transmission system, and the train positioning system (as shown in Fig. 1). In particular, ZC, the core of the CBTC system, calculates the movement authority (MA) for the rear train based on the position of the front train, the status of the line obstacles, the interlock condition, and the speed limit of the train. The on-board equipment compares the running speed of the train with the MA received in real time. If the speed of the train exceeds the speed limit in the MA, the on-board equipment will automatically implement the common brake or emergency brake to ensure the train stops in front of the safety point. The train–ground information transmission system could adopt wireless communication, cross-ground induction loop, waveguide, or other media to achieve the bidirectional communication. To ensure safety, the train must accurately determine its position and direction, where the on-board computer, the speed meter/speed sensor/accelerometer (for measuring distance, velocity and acceleration), and the trackside balise can be operated collaboratively to obtain the accurate position [1].

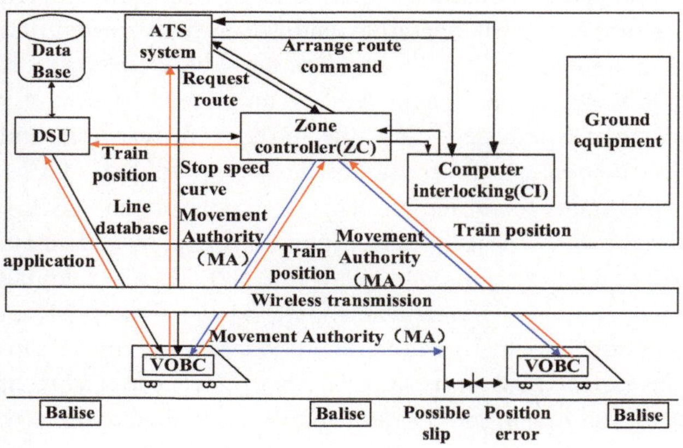

Fig. 1 CBTC system structure

Beijing Jiaotong University set up an independent innovation project to carry out the research about the CBTC core technology and key equipment by the way of Industry–University Research (IUR) proposed by the Chinese government. The demonstration of the independent innovation CBTC project was carried on the Beijing Yizhuang Line, which was formally approved and started its operation on December 30, 2010. [2].

By referring to the advanced safety standards from other countries and the actual situations of domestic rail transit engineering in China, the safety management and certification activities have been designed special for rail transit project in China and carried out and tested in practice.

Mature safety management concepts of the rail transit were formed in electronic and electrical standard IEC61508 [3], aviation industry standard ARP4761 [4], and the related European railway signal standards CENELEC [5–7]. According to these standards, British railway safety and standards board (RSSB) built the safety risk model (SRM) based on the accumulated data of railway operating for decades to evaluate the safety state of railway operation and to provide guidance [8]. The European commission proposed the SAMRail [9] and SAMNET [10] projects, which aimed at railway risk management promotion in 2002 and 2003. These projects have unified the railway safety operation strategy and established general safety targets and indicators and finally formed the railway operation safety performance indicators and safety target allocation methods [11], based on the common safety target, the common safety indicator, and the common safety method.

2 Challenges in Innovation of CBTC Systems

CBTC system is a kind of train control system which adopts advanced communication and computer technologies to continuously control and monitor the operation of trains. CBTC systems were independently studied and developed by Simens, Alstom and Thales according to the development process of the safety-critical control system.

In China, we started to study CBTC in 1995 and have experienced more than 20 years' development since the beginning of 1995. The CBTC system consists of the control, communication, network, integration, transportation organization, material technology, and other multidisciplinary integrated system technologies. All related research works were carried out in accordance with the requirements of the whole life cycle.

The structure diagram of the CBTC system developed by us is shown in Fig. 2. The data transmission between trackside equipment and on-board equipment is achieved by the data communication network. In addition, the CBTC on-board equipment controls the operation of trains and the CBTC trackside equipment sends data to the ATS system, which supervises and controls the train through the interlocking systems.

The real-time two-way communication between on-board equipment and wayside equipment is the cores of the CBTC system, which can provide a large amount of information transmission with fast transmission speed to achieve moving automatic block easily. So the expansion of information utilization and function in CBTC system can greatly improve the ability of the interval and the capacity of the tracks. The flexible organization of two-way operation and one-way continuous operation reduces the number of wayside equipment, which makes the CBTC system easy to install and maintain. Moreover, it is also easy for the CBTC system to adapt to different speeds, different traffics, and different types of traction train operation control, etc. Wireless communication methods are mainly divided into wireless AP transmission mode, waveguide transmission mode, and leaky cable transmission mode as well as inductive loop mode.

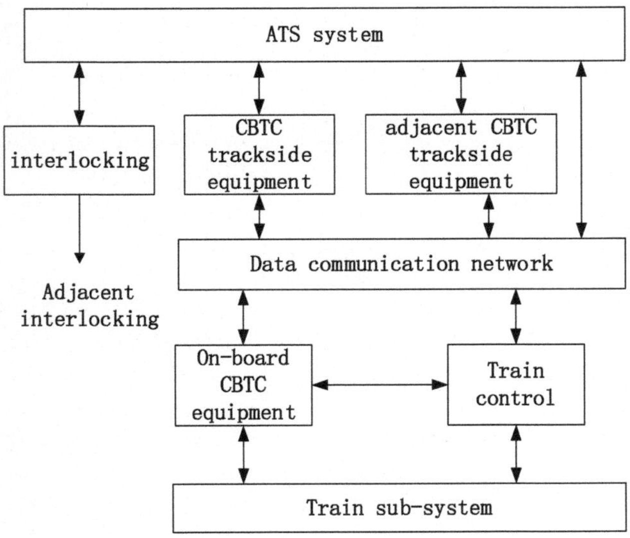

Fig. 2 CBTC system structure diagram

2.1 Complexity of the CBTC System

To solve the technical problems of the CBTC system in principle is not equal to achieve the feasibility; there is a long distance between reliability and safety in engineering. Rail transit CBTC signaling technology's research and development, engineering and industrialization is a typical complex system. As an intelligent complex safety control system, CBTC systems are required to realize a safe and reliable operation in different weather conditions along the whole life cycle, as well as to achieve the collaborative control (among human, trains, and railroad) with complex multivariate parameters characteristics.

Urban rail transit operations are affected by the weather and the passenger flow, and in such situation we need to ensure a continuously safe and efficient operation for 30 years. Even when there is rain/snow or trains are overloaded, e.g., 150%, the CBTC system must provide safe and reliable services. The CBTC system is a complex safety critical system with perfect function and clear hierarchy according to the CENELEC international safety standards. The SIL4 level train control equipment must adopt the risk-control development method and the fault-tolerance redundancy computer platform in the whole life cycle. Hence, the traditional R & D methods and safety analysis theory are facing difficulties to ensure the safe operation of the CBTC system.

The failure propagation model based on the complex scenarios of train operation and the system design and development method that covers the whole life cycle is significant for the system safety requirements. The safety management system and the integrated research and development platform should be set up to meet CENELEC standards of the highest safety integrity level SIL4. Furthermore, the train and ground two signal dedicated, portable safety computer platform, and CBTC complete set of technical equipment should be developed. Moreover, the mechanism and evolution law of the hazard causations (accident causes) and the establishment of the hazard log and safety integrity level requirements can eventually control the system risk within acceptable limits.

2.2 Distributed and Collaborative Control of the CBTC System

The equipment of the CBTC system is distributed in the train, trackside, station, and control center. Those devices are connected by many different types of fiber optic cable and have numerous interfaces. For example, to equip 1-km track route, there are dozens of kilometers fiber optic cable laying, thousands of connection points and nearly ten thousands components. The failure of any component will affect the stability of the CBTC system since it is required to have high reliability.

The CBTC system relies on wireless communication technology to achieve train–ground communication, which could use wireless LAN technology based on ISM band. There are some problems in this technology, such as vulnerable to interference and the frequent handoff because the wireless transmission environment consists of the underground tunnel and the ground section in urban rail transit systems. So the single transmission mode is difficult to meet the high reliability requirements for the CBTC train–ground communication.

Since the doors open automatically, the safety protection of passengers is an important requirement for the CBTC systems. In addition, the CBTC system is also closely related to the operations staff; for example, dispatchers need to use the ATS equipment, and drivers need to use ATP/ATO equipment. Therefore, CBTC system is very important to improve the service quality of the operation.

2.3 Multivariate Parameters Adjustment

Moving block signaling system can achieve a larger line capacity when compared with the fixed block signaling system because a train can be allowed to move to the tail of the front train with a certain safety distance. The CBTC system with moving block could secure safe, fast, reliable, punctual, comfort, and energy-efficient operation of trains.

The objectives for the operation of trains are contradicted with each other, such as the safety and the fastness. For passengers, they want trains to run as fast as possible; however, the speed of trains is limited by the signaling systems, train characteristics, etc., to ensure the safety. So the CBTC systems need to find a trade-off between these objectives to achieve the maximum capacity allocation.

Compared with the traditional signaling system, the control precision of the CBTC system is greatly improved. For example, the train positioning error can be limited within centimeter level. The parking control precision of a train with hundreds of tons can be less than 30 cm. The solution of each problem requires careful theoretical deduction and a large number of field tests. Besides, the location of the train needs to be sent to ZC by wireless communication to generate MA. The delay of wireless communication impacts the correctness and precise position of the train because the train always moves forward. When ZC calculates the MA for next train, it needs to consider the delay.

3 Processes of Safety Management and System Integration for CBTC

In order to deal with the challenge of CBTC system development, we try to find best practice in the world. As we know, the biggest safety challenge of CBTC System is that system safety cannot be simply achieved by testing and need to take safety management measures to deal with the human errors and software failures. And the systematic features of CBTC system need a proper integration method along its product life cycle. The safety management system during the railway development, construction, operation, maintenance is the effective way to conduct the risk control and set up the implementation framework of railway safety management.

European EN series standards are important guideline in the railway industry, but simply apply the method described in it cannot solve the management and technical problem encountered in our project process. In the practice, we summarized our own methodology.

3.1 European Safety Standards and Principles

SC9XA Committee of the European electrification Standards Committee (CENELEC) established a set of standards which includes s IEC62278, IEC 62279, IEC 62280 and IEC 62425 (as shown in Fig. 3) for rail transit train control system based on computer control [12, 13]. Because these standards have strong pertinence and better practicality, they have been promoted in many countries and been adopted by the IEC Committee as an international standard.

IEC62278 (EN 50126) defines the reliability, availability, maintainability, and safety (RAMS) for rail transit train control system and the relationships among them. In

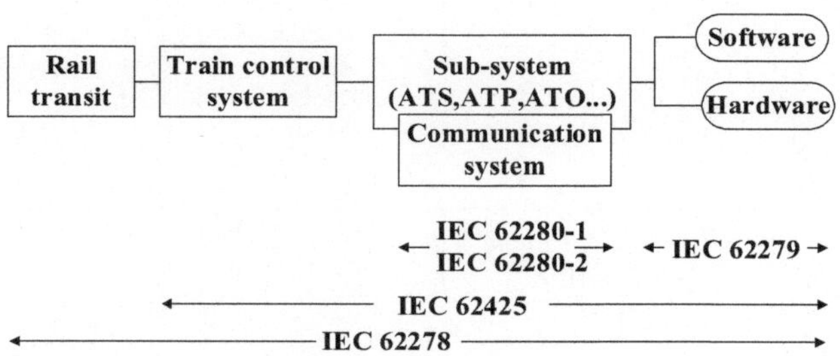

Fig. 3 International safety standards for rail transit

addition, it also standardizes the credibility system's framework of the life cycle and the works that need to be done at every stage of the life cycle to ensure the RAMS of the system. IEC62425 (EN 50129) makes the IEC62278 improved according to safety case of train control and system safety integrity level, etc. Furthermore, IEC62279 (EN 50128) is the safety-related design standard for train control system software and it complements the content of software engineering under the framework of IEC62278. IEC62279 also provides the technology and management for different safety-level software design. IEC62280 (EN 50159) is a rail transit communication system safety standard, which involves two parts: open communication system and closed communication system.

In rail transit, IEC62278 recommends three international common safety principles: the Globalement Au Moins Aussi Bon (GAMAB) principle of France, the Minimum Endogenous Mortality (MEM) Principle of Germany and the British As Low as Reasonably Practicable (ALARP) principle. GAMAB requires all new systems must provide the best safety performance, and at least it should be equivalent to the safety performance of the existing similar systems.

MEM principle means that the application of the new system cannot increase the probability of casualties. In Germany, casualties caused by rail transit systems are below the country's lowest natural mortality rate (the natural mortality rate for the 5–15 years old is the lowest, which is 1/20000 per year in Germany). Thus, the rail transit system in general set the safety objectives RM $= 10^{-5}$ catastrophic risk/person years.

The ALARP principle means that a reasonable and feasible method is adopted to reduce the risk as far as possible in the case of cost. This principle divides the risk into three levels, as shown in Fig. 4. The focus of the ALARP principle is to make a final overall judgment about the risk associated with the system and try to make every

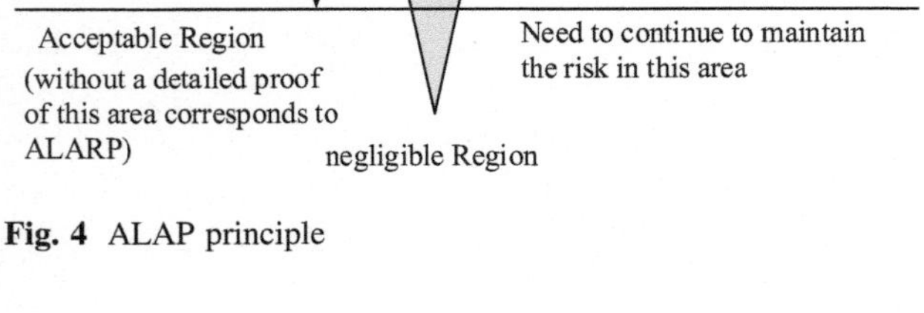

Fig. 4 ALAP principle

risk acceptable or tolerable. In addition, the overall risk of the system should be reduced as much as possible.

From a formal perspective, the GAMAB principle and the MEM principle clearly define the safety requirements in the form of the risk acceptance thresholds, which is represented by the "existing similar system" and "minimum national population natural mortality" and have a strong universality and objectivity. The limitation of the MEM principle is that it does not consider the differences in risk perception among different types of systems. For example, the acceptance of the risk for rail transit and civil aviation, highways and other means of transport systems could be different.

The ALARP principle encourages the maximal reduction of risk to improve safety by taking the cost into account. To a certain extent, it promotes the research and application of risk reduction technology. Compared with the GAMAB and MEM principles, the ALARP principle is more stringent since it needs to analyze a variety of risk reduce measures and then choose the most reasonable solution during the design process. British Health and Safety Executive (HSE) issued a number of regulations on the interpretation and constraints of the ALARP principle and provided many court cases as references. Rail transit regulatory authorities also issued the ALARP principle guidance, the engineering safety management (Yellow Book) and other industry standards to provide detailed application procedures and steps for the ALARP principle in system design, operation, maintenance, and safety demonstration.

During the development process of the CBTC system for China practice, the requirement analysis, risk analysis, system design, product realization, system testing, and other works were carried out according to the CENELEC standard. The "standards, methods, processes, people" of each task is analyzed based on its own characteristics. The final product obtained the SIL4 safety certification from a foreign third-party certification company, and it can be applied in the actual project.

3.2 The General Method of Safety Management for Rail Signaling System

The life cycle of a system refers to the various stages of the whole process from the initial concept to retirement and tasks at each stage (as shown in Fig. 5). The entire phase includes a plan, management, control, and supervision of a safety-related system. System life cycle (Lifecycle System) is a description of the process of system development and usage. System life cycle provides a unified framework for planning, management, control and supervision of the quality, and the performance, construction period, cost and other aspects of the system. System safety is the result of

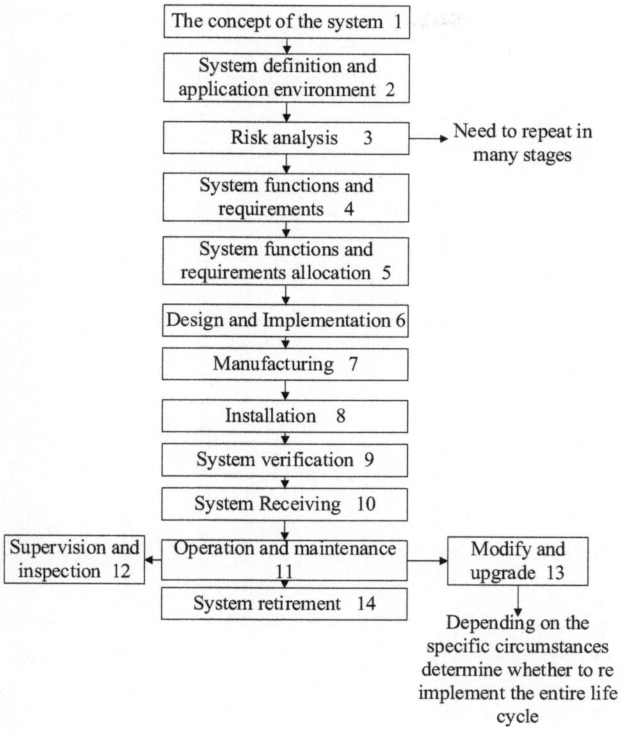

Fig. 5 Life cycle of the system

many factors working together, and these factors are distributed in all the system life cycle stages. In particular, the factors involved in different stages are different. Therefore, we need to carry out the system safety work in system life cycle to control and monitor the different factors at different stages. The system life cycle that includes the system safety work is called the safety life cycle, also called Safety Lifecycle.

From the activities of the whole life cycle, it could be found that the management concept-based risk is adopted to control the risk of the system within the system or project life cycle. In the process of the actual rail transit project implementation, the whole life cycle is usually divided into 5 milestones, as shown in Fig. 6. The objective of the first milestone is to generate the system requirements, and this includes the first 4 phases. The second milestone needs to distribute the safety requirements to subsystems and give detail design outline. Before the third milestone, implementation and installation of the system need to be finished. The main activity of milestone 4 is system certification and safety inspection. The final milestone is to state commercial operation.

3.3 The Processes of Safety Lifecycle for CBTC

In October 2004, Beijing Scientific and Technological Committee carried out the project "Communication based on Urban Rail CBTC system research" for the first time. The team at Beijing Jiaotong University was assigned to develop the automatic train protection (ATP) and automatic train operation (ATO) system which was once dependent on technology import and restricted the further development of signaling system in China.

By the end of 2007, the team overcome the difficulties of the CBTC system's key technologies and completed the integrated test of the whole CBTC system. The function and performance test were carried out on the 1.3-km test line of Beijing metro afterward. Finally, the CBTC systems were checked and accepted by the experts. Moreover, the various functions and performance indicators have reached the international advanced level. The resulting CBTC system also broke the technical monopoly in China [14].

The system design theory was established and the self-developed complete sets of systems and equipment were developed, which include: (1) the system design theory of moving block; (2) the design method of safety critical system in the whole life cycle; (3) self-developed safety computer platform; (4) technology and equipment based on the integrated design, e.g., the two-level scheduling model and three kinds of control level [11].

In view of the complexity of CBTC system, we use the scene analysis method to analyze. According to the EN50129, combined with the actual characteristics of the Beijing metro project, a safety life cycle model for the self-developed CBTC products was established and the details are shown in Fig. 7 [15].

3.3.1 Design of Safety Management Process

A general overview of the system/subsystem/device design describes the main functions of the system, as well as the internal and external interface, so that the relevant staff can clearly understand the principles and techniques used in the system.

Safety management process is composed of a number of stages and activities, which are connected to the safe life cycle. The design and validation of the system life cycle can be seen as a "top-down" stage and is accompanied by a "down-top" phase, as shown in Fig. 7, which looks like a V-type.

3.3.2 Safety Organization and the Safety Plan Establishing

The safety management process should be performed under the control of a suitable safety organization via a personal that is competent for the relevant work; Fig. 8 shows the organization structure of CBTC project. According to the requirements of the relevant system's safety integrity level, there should be a proper independence between different personals.

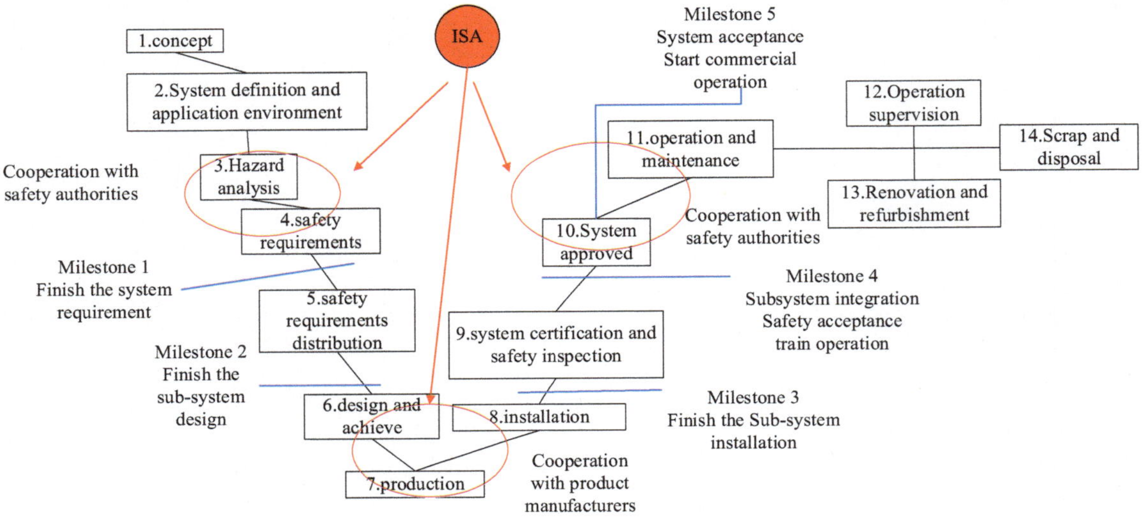

Fig. 6 Life cycle milestone diagram

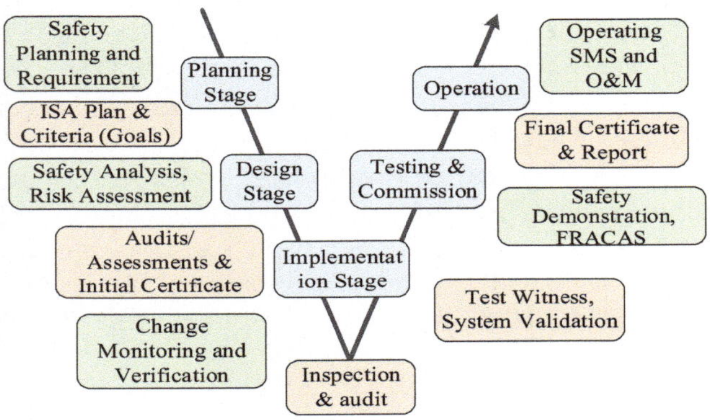

Fig. 7 Independent research and development of CBTC product life cycle

A safety plan should be made in the early stage of the system life cycle. This plan should determine the organizational structure of the safety management which involves the entire life cycle, the need for a review of safety plans within the appropriate intervals, and all aspects of the system/subsystem/device (including hardware and software).

In the early stage of the system design, we should create a hazard source log based on the results of hazard analysis (including PHA, SHA, SSHA, IHA, and OSHA) and maintain the hazard log throughout the entire safety life cycle. Safety requirements can be considered from two aspects, i.e., the functional safety requirements and the safety integrity requirements. Safety functional requirements are the actual safety-related functions which the system, subsystem, or equipment is required to carry out. Safety integrity requirements define the level of safety integrity required for each safety-related function.

3.3.3 Safety Audit, Verification and Validation

At the appropriate stage of the system life cycle, safety audit should be carried out to monitor the safety management process in accordance with the requirements of the safety plan and the relevant standards.

Verifying the content of each phase for the life cycle should meet the safety requirements specified in the previous stage and finally confirm that the system/subsystem/equipment meets the initial safety requirements. According to the opinions of the safety authorities, the assessment staff should be authorized by the safety authorities, be completely independent from the project team, and report directly to the safety authorities. Furthermore, the evidence for a system/subsystem/device meeting the safety acceptance criteria shall be listed in a structured document called "Safety Case." Before Safety Case sent to independent company, internal verification and validation activities need to be done.

3.3.4 Safety Management for System/Subsystem/Equipment Transfer, Operation and Maintenance

System/subsystem/equipment should meet the conditions of the safety acceptance before the delivery to the railway authorities. The delivery also includes the submission of "Safety Case" and "safety assessment report." After the system is put into operation, it should meet the process, support system and safety monitoring requirements that are defined in the safety plan and technical safety report.

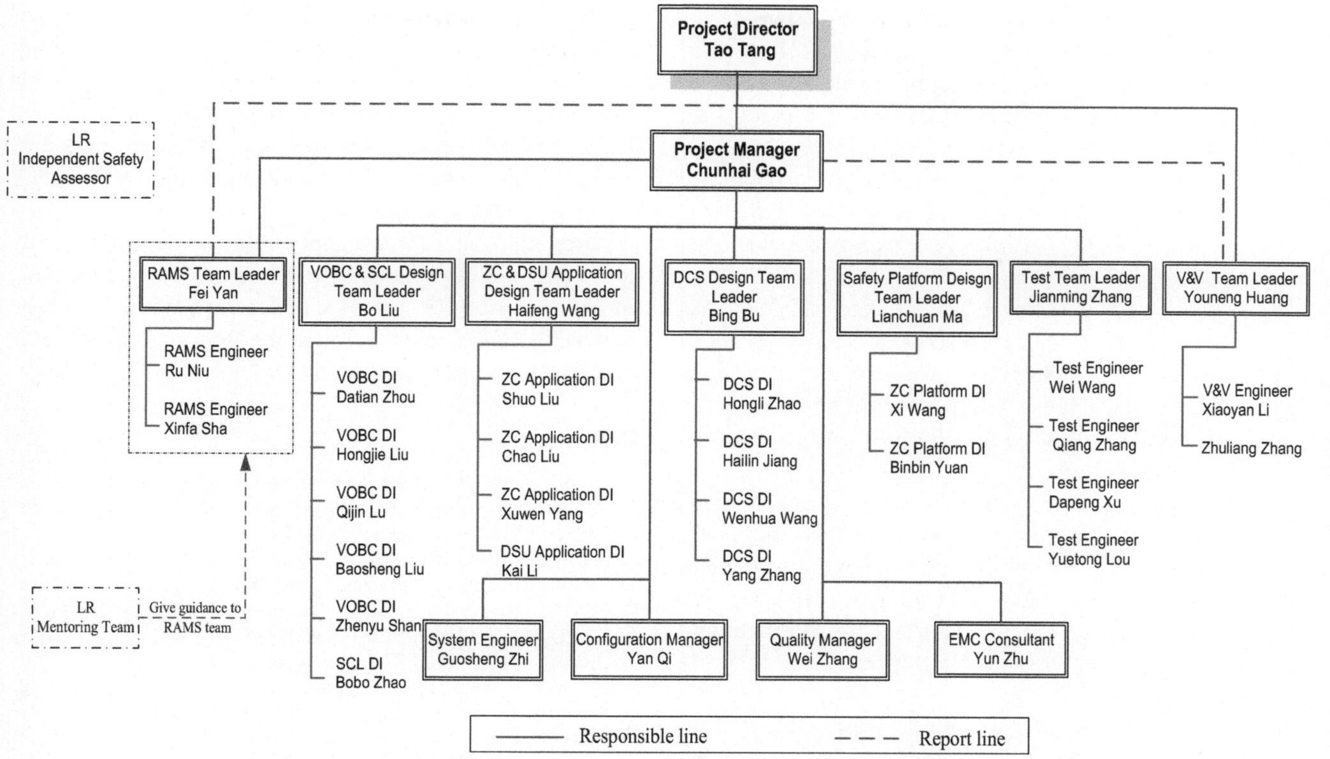

Fig. 8 Organization structure for CBTC project

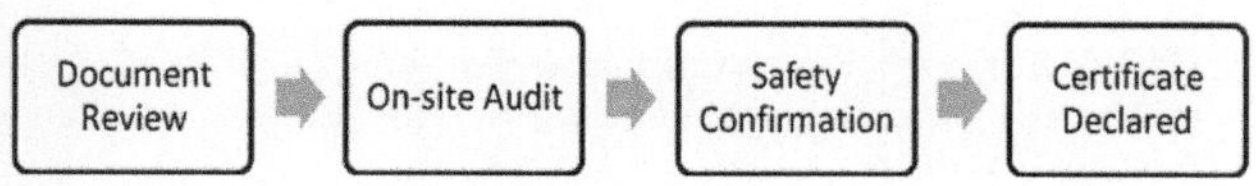

Fig. 9 Process for independent certification

3.3.5 Processes of Safety Certification

Safety certification needs a proper process, as Fig. 9 describes. Before conducting an independent evaluation, a checklist is required about user requirements and related standards. This checklist can provide a systematic method to record the results of independent evaluation and evidence and save as the evaluation document for traceability.

(1) Document Review

In the beginning of the project, the independent safety assessment (ISA) staff will review the project plan, safety plan, quality plan, configuration management plan, test plan, and system requirements specification, etc. The staff then put forward written improvement opinions.

After the integrators and subcontractors finish the development of related subsystems, ISA staff will review the test specifications of the integration testing and system testing for each subsystem, the test plans and test reports and then put forward written improvement opinions.

After the revision of the document and passing the ISA review, the improvement opinions will be closed.

(2) On-site audit

According to EN50128, EN50129, and other standards, the ISA staff applies the on-site audit to the systems integrators and the core technology supplier during the stage of the system requirements and system design, as well as the implementation stage and system testing stage from the aspects of the system assurance, software quality assurance and management, etc. The staff then put forward a written audit report. System integrators and core technology supplier need to respond to the audit opinions presented by the ISA staff and provide the relevant evidence. After that, the ISA staff can close their views.

For the problems existing in the audit process, the owners, the independent safety assessment staff, as well as systems integrators and core technology provider will hold a weekly safety meeting to communicate and resolve the problems.

(3) Safety confirmation

In order to ensure the safety of the signaling system after testing, debugging and operation, the single train commissioning, multi-train commissioning, trial operation, operations need to be carried out under special monitoring mechanism. System integrators and core technology suppliers provide related design documents, test report and safety documents. ISA then conduct the safety assessment by means of document review, on-site audit, and witness testing according to the requirements of EN50128 and EN50129. After that, ISA provides the safety assessment report and publish the safety certificate. When achieving the certificate, the testing, debugging, system operation, etc., could be carried out.

In the safety assessment of metro projects, ISA will check each of the safety critical system products provided by suppliers to make sure that they have obtained the product safety certification. The safety certification process for the CBTC system is as follows:

a. Products design
The design and dev med by the design team, and they make a detailed record for the works that completed in each stage of the system design process and development process. The record could be the system and subsystem requirements specification, system architecture specification, etc.

b. System test
The implementation of the system, subsystems, components, and modules needs to be verified by testing according to the corresponding test specifications at every stage of the CBTC system development [16]. In the testing process, the engineers use the test cases that described in the specifications and test each function and performance individually.

c. Verification and validation (V&V)
This work is done independently from the project team and is performed by the V&V staff. In each phase of the project life cycle, the V&V staff needs to ensure that the work done at each stage is correct [17] and generates a validation report for each phase.

The safety certification for CBTC covered the certifications in the test line, single train commissioning, multi-train commissioning, trial operation, and the carrying passenger trial operation, etc. At each stage of the whole life cycle of the system, the safety is measured and evaluated by the international advanced and practical standards and scientific methods. In addition, an independent third-party assessment team is hired to carry out safety audits and assessments in each process to ensure the CBTC system with independent intellectual property rights can be used safely and reliably in the application of urban rail transit in China.

4 Conclusions

After absorbing the essence of international safety standards and adapting China urban transit construction practice, the safety management and system integration method of CBTC system came into being. The method was proved in the process of pilot test on test line, single train commissioning, multi-train commissioning, trial operation on mail line, and the carrying passenger trial operation, and finally the CBTC system and related products developed by Beijing Jiaotong University has obtained the safety integrity level 4 (SIL4) certificate.

After the successful application of independently developed CBTC system, the advanced fully automatic operation system is currently under development and the new technologies will be used in the Beijing Yanfang Line which is expected to be opened by the end of 2017.

References

1. Yan F, Tang T (2005) Development of safety technology of rail transit signaling system and the research status of foreign countries. J China Saf Sci J 15(6):94–99
2. Gao CH (2014) Research on the core technology of independent innovation CBTC system. J Modern Urban Rail Transit 24(1):7–10
3. IEC61508-2010 Functional safety of electrical/electronic/programmable electronic safety-related systems
4. Society of Automotive Engineers (SAE) (1996) Guidelines and methods for conducting the safety assessment process on civil airborne systems and equipment (ARP4761). SAE International: Warrendale, PA. [Aerospace Recommended Practice]
5. IEC 62278-2002 Railway applications—specification and demonstration of reliability, availability, maintainability and safety (RAMS)
6. IEC 62279-2002 Railway applications—communications, signalling and processing systems—software for railway control and protection systems
7. IEC62425-2007 Railway applications—communication, signalling and processing systems—safety related electronic systems for signalling
8. RSSB (2011) Safety risk model risk profile bulletin (version 7) [R]. http://www.rssb.co.uk/risk-analysis-and-safety-reporting/risk-analysis/safety-risk-model-(srm)
9. European Commission-2004 (2004) Safety management in railway, WP.2.2.2: guidance for the safety management system
10. EI Koursi EM, Tordai L (2003) SAMNET: Safety management and interoperability thematic network in railway systems. WCRR2003, Edinburgh, pp 198–202
11. Common safety targets, common safety indicators and common safety methods. A joint paper of the Safety Systems Harmonisation Working Group and the Technical Interfaces Working Group of the UIC Safety Platform, Sept'2002
12. CENELEC EN50129-1999 (2003) Railway applications:safety related electronic systems for signalling [S]
13. CENELEC EN50126-1999 (1999) Railway applications:the specification and demonstration of reliability ,availability,maintainability and safety (RAMS) [S]
14. Tang T, Niu YM, Gao CH (2010) Research and innovation of CBTC system for rail transit. J Munic Technol S2:349–353
15. Yan F, Gao CH, Tang T (2011) Safety management and assessment mode research for rail transit signalling engineering project. Urban Rapid Rail Transit 24(4):12–16
16. Gao CH, Yan F, Tang T (2005) Research on safety assessment method of rail transit signaling system. J China Saf Sci J 15(10):74–79
17. Morisio M (2000) Commercial-off-the-shelf (COTS): a survey a DACS state-of-the-art report[R]

6

Challenges in the Integration of Light Rail and Land Use: A Case Study of Yizhuang Line in Beijing

Haishan Xia[1] · Chen Shen[2] · Chun Zhang[1,3] · Xinhao Wang[3,4] · Daizong Zhang[1]

Abstract Transit-oriented development (TOD) is advocated by urban planners as a tool to promote transportation and land use integration, and it has especially been applied in new town developments in Chinese cities. Since 2000, urban metro system construction in Beijing has accelerated and many new towns in suburban areas have emerged due to rail transit connections with the central city. This paper is a case study of Yizhuang new town, which is located in the southeast suburban area of Beijing. It explores whether light rail is successful in promoting transportation and land use integration. First, by comparing the land use within a 1 km radius of three rail stations, findings suggest that the TOD model in terms of land use may not be implemented. Second, by tracing land use changes over time, we found that commercial functions are critical to support TOD development. Third, the mismatch of urban functions between low-skilled labor employment opportunities and high-skilled residential neighborhoods emphasizes the failure of planned reverse commuting patterns. In conclusion, our findings suggest that without a proper land use policy, TOD based on light rail only may not be a successful model in the case of Yizhuang new town. The policy implication is that the planning code of land use should be more flexible and adaptable, in order to promote the integration of transportation and land use in a synergetic way.

Keywords Transit-oriented development · Light rail · Integration · Beijing · Yizhuang

1 Introduction

The concept of transit-oriented development (TOD) can be traced to American cities around the turn of the twentieth century when trams, buses, and streetcars became the dominating mode of urban transportation. Most urbanized areas were situated along transit lines. As automobiles became the main mode of transportation, especially after World War II, middle-class families consequently moved to suburban areas, leading to a substantial decline of public transportation facilities. Driving to work gradually becomes a necessary part of typical American life. In response to the problems posed by increased car use, such as traffic congestion, increased energy consumption, and air pollution, several American cities invested in public transit such as light rails.

Under these circumstances, TOD emerged as the main model of transportation development [1]. TOD advocates that neighborhoods around transit stations should have high-density and mixed land uses [2, 3]. Peter Calthorpe popularized the term TOD in his 1993 book, "The New American Metropolis," to encourage public transportation and mixed land use in nearby neighborhoods (Calthorpe 2001). Other scholars later developed more TOD definitions and adopted TOD principles in their practice (Table 1). For example, Bernick and Cerbvero pointed out

✉ Chun Zhang
zhangc@bjtu.edu.cn

1 School of Architecture and Design, Beijing Jiaotong University, Beijing, China

2 School of Civil Engineering, Beijing Jiaotong University, Beijing, China

3 UC-BJTU Future City Co-Lab, Beijing, China

4 School of Design, Architecture, Art and Planning, University of Cincinnati, Cincinnati, OH, USA

Editors: Haishan Xia, Chun Zhang

Table 1 Literature on TOD in the USA

TOD in the USA	
Peter Calthorpe [1]	TOD is a kind of mixed land use of the community and plays a role in encouraging public transport objectively
Bernick and Cerbvero [2]	TOD is a compact, functional mix of community, with a public transport site for community centers, through a reasonable design, to encourage people to use less cars, more public transport. Bus stations are the centers of communities within a 400-m radius
Freilich [3]	TOD is similar to the new development (neo-traditional development, NTD); it aims to improve use of high-capacity transportation, as well as new city development mode, mixed use of land, and diversification of architectural design to replace the monotonous form of land use
Maryland Department of Transportation [4]	TOD has a relatively high density of development, including living, employment, commercial, and public facilities, such as the function of mixing in a bus or rail transport site within walking distance
California Department of Transportation [5]	TOD is a moderate- or high-density land use around the main public transportation site, usually will include living, employment, commercial facilities within the scope of mixed arrangement, and is suitable for walking, and the coordination of public transport facilities

that bus stations should be within 1/4 miles from residential locations in order to create a compact and multi-functional neighborhood [4]. Similar to planned unit development (PUD) and traditional neighborhood planning (TND), TOD promotes walkability [5]. The New Urbanism movement in the 1990s removed debate of whether or not to promote TOD in urban design and neighborhood planning. Empirical studies support the advantages of TOD neighborhoods compared with other suburban neighborhoods: Residents drive less by 33% [6] and have more social interaction with neighbors [7] than those in low-density, cul-de-sac neighborhoods. Many state transportation bureaus also adopted TOD. For example, Maryland Department of Transportation promotes high-density development, accessible facilities and amenities, and bus or rail transportation systems at walking distance [8]. California Department of Transportation highlights land use development and commercial facilities around transportation facilities and promotes a residential–employment–commercial integrated land use model [9]. TOD not only improves the use of high-capacity transportation, but also encourages mixed land use and building style diversification. In conclusion, TOD provides more travel options of a neighborhood the choices of life style. It helps to reduce car use and thus also decreases greenhouse gas emissions and improves air quality. According to the New Urbanism advocators, TOD also promotes healthier and sustainable life style because it encourages citizens to shift from driving to walking [10].

TOD was introduced in China in the late 1990s against the background of rapid urbanization and suburbanization [11]. While large-scale affordable housing communities emerged on the outskirt of large cities [12], urban rail transportation infrastructure became critical given that it provides the essential connection between suburban housing and urban center jobs [13]. In particular, among middle- and low-income groups living in affordable housing, rail transportation became the main affordable travel mode supporting large volumes of mass transit.

TOD has been widely used in China and remains popular in urban planning practice, especially with the acceleration of rail transit development. In contrast with most of American cities, Chinese cities generally have higher density, and as such they are very suitable cases for the application of the TOD model [8]. Also, the car-oriented development model does not have a deep-rooted historical influence in most Chinese cities, which makes the TOD principles more easily accepted by Chinese planners and decision makers [14].

On a national scale, many cities emphasize TOD in their comprehensive planning, such as Beijing, Dalian, Shenzhen, Wuhan, and Tianjin [15]. In 2016, the Ministry of Housing and Urban–Rural Development (HURD), the National Development and Reform Commission, and the Ministry of Finance jointly selected 127 towns in China as the first group of national featured towns [16]. The TOD model was highlighted by both urban planners and government officials in the featured town planning and urban designs.

Some scholars point out difficulties in fully realizing the TOD model in the short run. For example, in cases of extremely rapid urbanization, public transit facility construction usually lags far behind mass-produced suburban housing projects [17]. An extremely high passenger volume during peak hour is described as a failure of the fulfillment of TOD [17].

To contribute to the literature of assessing TOD in the context of contemporary urbanization in China, this paper analyzes the connection of urban rail transit and urban planning, using the light rail line in Yizhuang as a case. Yizhuang is selected because it is the only New Town connected to the city center with a rail transit infrastructure, which makes Yizhuang a suitable case to assess the success of the TOD model. The purpose of the study is to clarify use and expectations of transit-oriented development in China.

2 Methods and Data

The study starts with a description of the Yizhuang development and the purpose of developing the Yizhuang light rail line. Land use and travel data in three stations in Yizhuang are analyzed. The analysis is conducted at three scales. At the microscale, it identifies the land use types around metro stations to evaluate the matching degree between transportation and land use. At the mesoscale, it identifies the layout of urban functions along the urban rail transit lines to evaluate the coordination of different urban functions. At the macroscale, it examines the commuting pattern to evaluate the effectiveness of the urban rail transit network on the job-housing spatial distribution (Fig. 1).

Data used in the study include the Beijing comprehensive plan, obtained from Beijing Municipal Government, the detailed planning from Beijing Urban Planning Information Center. The land use image from the years of 2008 and 2015, outlining the before and after situations of the construction of the three stations, is collected from the Beijing Urban Planning Bureau and Google map images.

A small sample survey was conducted near the three metro stations. In total, 150 questionnaires were handed out and 108 valid questionnaires were collected. The structured questionnaire asked about the personal information, commuting behavior, employment, and residential experiences among the respondents. Land use data within 1 km of the three Yizhuang stations are extracted and classified using the China Urban Planning land use classification system. The 1-km distance is based on average walking distance to metro stations in Beijing.

3 Results and Discussion

3.1 New Town Development Plan and Light Rail Line in Yizhuang

Triggered by the preparation for the Olympics in 2008, the urban rail transit system in Beijing has experienced dramatic expansion. Since the turn of the century, it has increased from 2 lines of 51.4 km to a well-established network including 19 lines, 288 stations, and 574 km in 2015. According to the latest Beijing comprehensive plan in 2004, the initiative of massive rail transportation construction was decided upon to support the new urban form plan, namely "two-axis, two-belt, and multiple centers" [18]. In particular, three suburban new towns Tongzhou, Shunyi, and Yizhuang were designed as three new job centers of the polycentric structure.

This paper is a case study of the light rail system in Yizhuang New Town, because it is the only national level economic development and also undertakes the task of evacuating industries. Unlike other new towns such as Tongzhou and Yizhuang, Yizhuang is designed to be a suburban job center. It makes the light rail in Yizhuang a unique case, which is designed as a reverse commuting urban metro line. Historically, Yizhuang was a small rural village until 1994 when it was selected as a National Economic Development Zone. According to the 2004 Beijing master plan, the Yizhuang New Town was planned

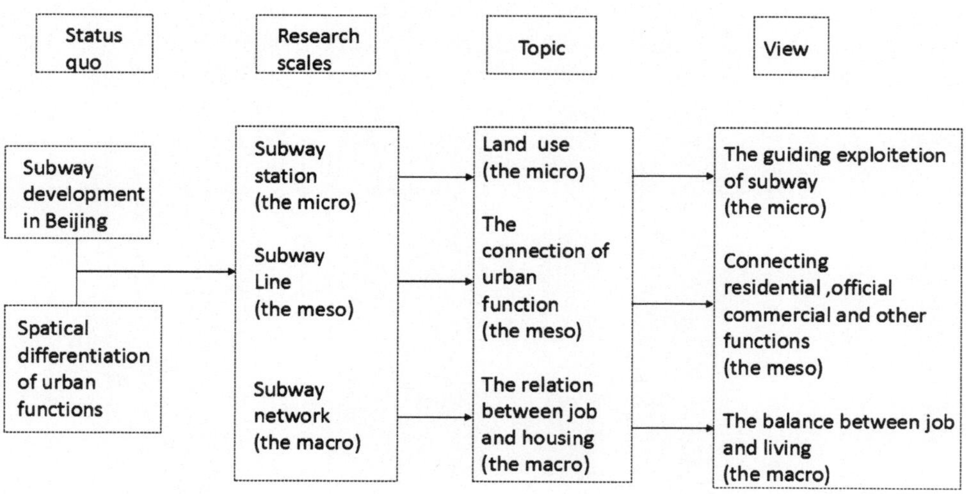

Fig. 1 Research framework of integrated development between rail transit and land use

as a high-tech industrial center and an advanced manufacturing center, with a total population of 300,000 in 2010 and estimated population of 700,000 in 2020; it is the result of evacuation of both jobs and population from the city center. The light rail line is designed to be the essential transportation facility to draw employees from urban center to economic development zone in the new town, achieving the objectives of population evacuation.

The Yizhuang line commenced operation in 2010 and functions as an extension of metro line 5. Yizhuang light rail starts at the Songjiazhuang metro station where it connects to the subway line 10, stretching 23.3 km toward southeast to the edge of Tongzhou district. It has 13 stations and passes through the central area of Yizhuang New town (Fig. 2).

Along the Yizhuang light rail, three stations are inside Yizhuang New Town. Wanyuanjie station is one of the earliest stations in the old administrative center of Yizhuang. Rongjingdongjie station is near the new administrative center. Jinghailu station is located in the new developed area near the end terminal (Fig. 3).

Both the 2004 master plan and detailed planning highlight TOD principles, which advocates arranging land use density and land use types according to the distance to light rail stations. Commercial buildings are located in the inner circle just outside the stations, office and residential buildings constitute the outer layer of the circle along streets, while industrial buildings are located on the periphery of the 1-km radius area. The intent of the planning is to integrate transportation infrastructure with surrounding land uses and facilities and to provide better accessibility and convenience for rail transit users.

3.2 Comparison of Land Use Between Plan and Reality

Microscale comparison of the land use within a 1-km radius around the metro station highlights the differences between plan and reality. Although the original plan was prepared according to TOD principles, not all of the actual land use follows the plan. Figure 4 illustrates the planned land uses in the 2004 master plan and the observed land uses in 2015 for the three stations. A visual comparison shows that the development in Wanyuanjie (Fig. 4a) and Rongjingdongjie (Fig. 4b) stations mostly followed the plan. However, Jinghailu station (Fig. 4c) turned out to be a low-density, manufactural function-oriented industrial zone.

The inconsistency between planned and actual land uses reveals the difference in TOD performance in new town development in Chinese cities. Without suitable planning implementation tools, the area around light rail station cannot naturally grow in a TOD way. It also shows that in the early stage of new town development, new development may not automatically take place near transportation facilities, which may have negative consequences for

Fig. 2 Rapid development of urban rail transportation in Beijing and Yizhuang light rail line

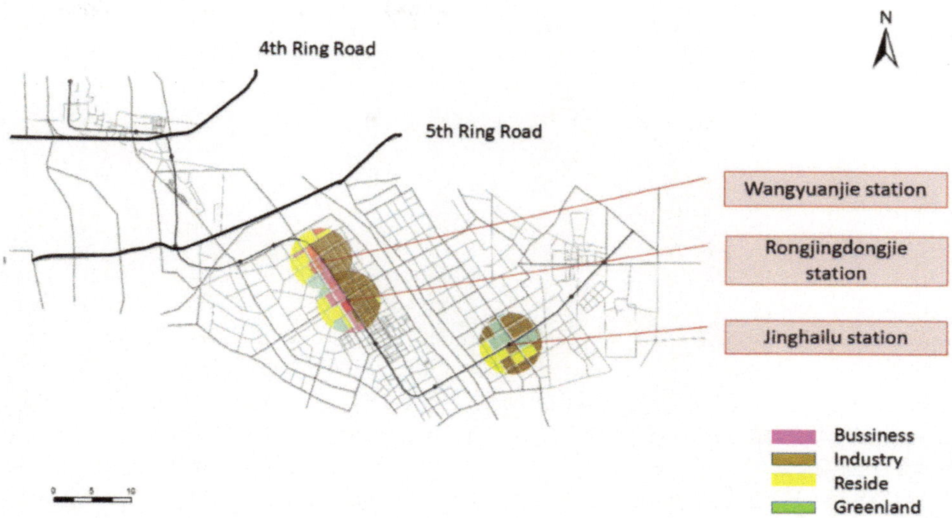

Fig. 3 Location and study area of the 1 km land use around station

future TOD planning. The failure in fulfilling the TOD model around Jinghailu station, suggests that industrial-oriented development may not effectively comply with the TOD model in Chinese cities.

3.3 Land Use Change Over Time Around Metro Station

Transportation planners and urban planners tend to assume that with the opening of an urban metro station, the land use of the surrounding area will develop naturally according to planned TOD models, and that any mismatch between planning and reality is only temporary. In particular, commercial facilities and amenities are usually neglected in the early stage of planning. By comparing the land use pattern in 2008 and 2015 by Google map, this section traces the land use change over time around urban rail stations.

By examining land use change from 2008 to 2015 within a 1-km radius around three stations, our findings suggest that a certain proportion of commercial land is critical for the TOD model (Table 2). For example, there are 14.52% commercial land within the 1-km radius around Wanyuanjie, which provides enough space for various facilities and amenities such as shopping centers, convenient stores, restaurants, bars, and cafes. In contrast, commercial land proportion is only 3.62% near Rongjingdongjie. In the last case of Jinghailu, there is no commercial land around the station which makes boring landscape. The comparison among three stations shows the importance of commercial land as the attractive core in TOD model.

Land use change from 2008 to 2015 reflects different influence of urban rail stations. In the area around the first case, Wanyuanjie commercial land increased between 2008 and 2015, making it into a multi-functional center including business, commercial, and recreation (Fig. 5a). In the case of Wenjindongjie, residential and administrative land increased, making it a semi-employment, semi-residential area. At the Jinghailu station, industrial land increased making it a manufacturing center (Fig. 5b). These findings show that even with similar urban rail transit stations along the Yizhuang line, the impact of transportation facilities on land use change may be significantly different (Fig. 5c). With suitable urban planning and proper urban design, more facilities and amenities may develop around rail stations with time and thus increase the livability and vitality of the neighborhood. However, these facilities and amenities might not naturally appear without featured public space reserve ahead and a flexible zoning code.

3.4 Urban Functions and Commuting Pattern as Suburban New Town

As one of the suburban new towns, Yizhuang is planned to be an employment center which enables the evacuation of both industry and population from the central city. The light rail line is the essential transportation facility to draw employees from the urban center to the economic development zone in the new town, achieving the objectives of population evacuation.

However, in the perspective of urban functions, there is large low-density industrial land concentrated along the light rail line, which is not conducive in making Yizhuang New Town an attractive employment center (Fig. 6), as it mainly provides low-skilled job opportunities, while failing to provide job opportunities for the high-skilled residents in the surrounding neighborhoods. Thus, as an economic and technological development zone, it attracts employees either from the urban center or local neighborhood.

Fig. 4 Actual (*left*) and planned (*right*) land use around three light rail stations

Table 2 Land use surrounding three stations in 2015

Land use station	Commercial (%)	Administrative (%)	Residential (%)	Industrial land (%)
Wanyuanjie	14.52	7.52	27.67	41.15
Rongjingdongjie	3.62	17.25	18.71	49.45
Jinghailu	0	0	23.82	44.28

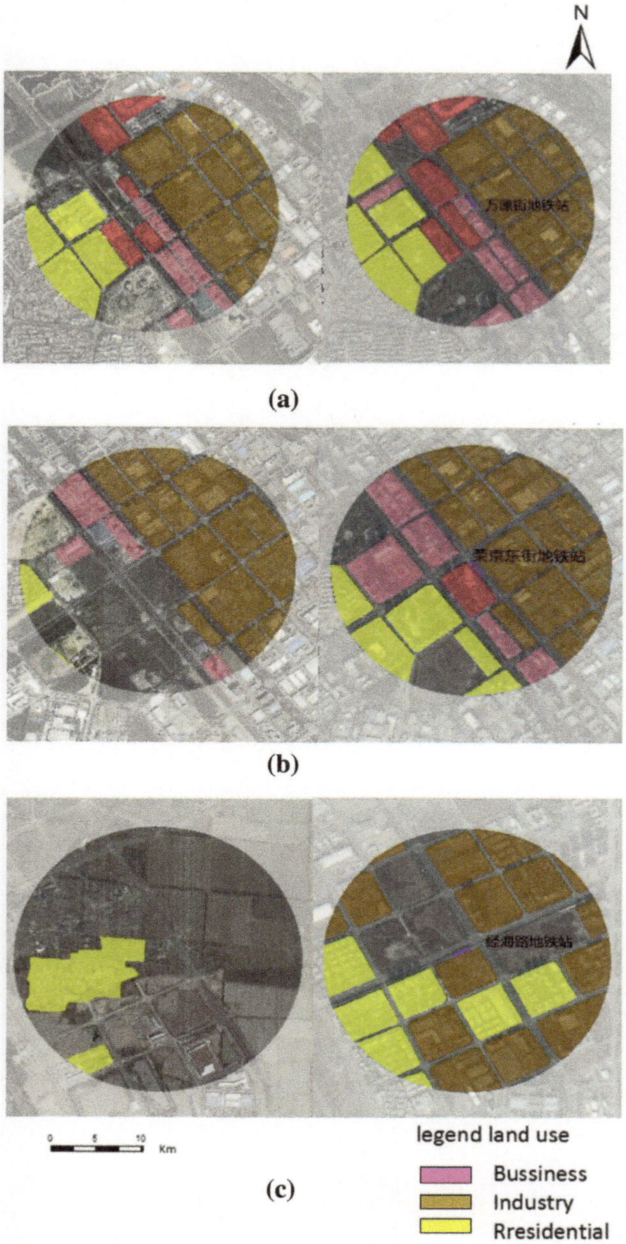

Fig. 5 Land use in 2008 (*left*) and 2015 (*right*) around three light rail stations

The findings from the small-scale questionnaire collected near the light rail stations support the assumption of a mismatch between low-skilled job opportunities and high-skilled residential neighborhood. Among the total 108 samples, 50.9% of local residents still work in the city center, while in terms of local employment rate, 46% of local employees come from other, more distant suburban neighborhoods (Fig. 7). The gap between the ideal and the actual commuting patterns implies that the objective of making Yizhuang New Town a job center has not been successful. Ideally, most local residents should commute in a reverse way to the suburban industrial areas, while a limited volume of local residents commute to the city center. However, the actual situation in Yizhuang suggests that most of the residents still commute to the urban city center for employment, while the industrial-oriented economic development zone attracts only a small proportion of commuters from the city center and surrounding suburban neighborhoods (Fig. 7).

In addition, the small-scale questionnaire survey reveals that the Yizhuang light rail line does not fulfill its purpose to transport citizens from the center to the suburbs. In contrast, residents living in the suburb instead choose the subway to transit to the city center for employment. The mismatch between the ideal and real commuting patterns reveals the failure of light rail as a method to evacuate job opportunities to Yizhuang New Town. Instead, the findings of this case study suggest that the light rail line aggravates traffic congestion of the city center, in conflict with the original master plan of Beijing.

4 Conclusion

While transit-oriented development originated in the USA, it is also promoted in China, especially for new town developments. With the trend of rapid urban rail transportation construction, it is generally assumed that the existence of transportation facilities will automatically lead to a TOD model. Selecting Yizhuang New Town in suburban Beijing, this paper evaluates the performance of light rail in Yizhuang in supporting the integration between transportation and land use. This paper will also explore whether attractive job center emerges and evacuates population and jobs as the master plan advocated in the case of Yizhuang.

Yizhuang in Beijing is a suitable case for studying the impact of light rail systems on land use in a suburban new town. The Yizhuang light rail line is planned to support the city center and near-suburban residents commuting to far suburban new town. Exploring the GIS-based spatial analysis, this paper compares the planned and actual patterns of land use within a 1-km radius area of three stations along the Yizhuang line. It finds that the gap between plan and reality in the land use pattern is significant, especially around the Jinghailu station at the end of the rail line, where an industrial-oriented model is adopted. The land use change over time also reveals the importance of commercial land in supporting a successful TOD model. Without attractive facilities and amenities provided by commercial land, the daily needs of resident and employees are not met. Additionally, the planned reverse commuting pattern fails to realize the aims of evacuation

Fig. 6 Land use map around light rail line of Yizhuang

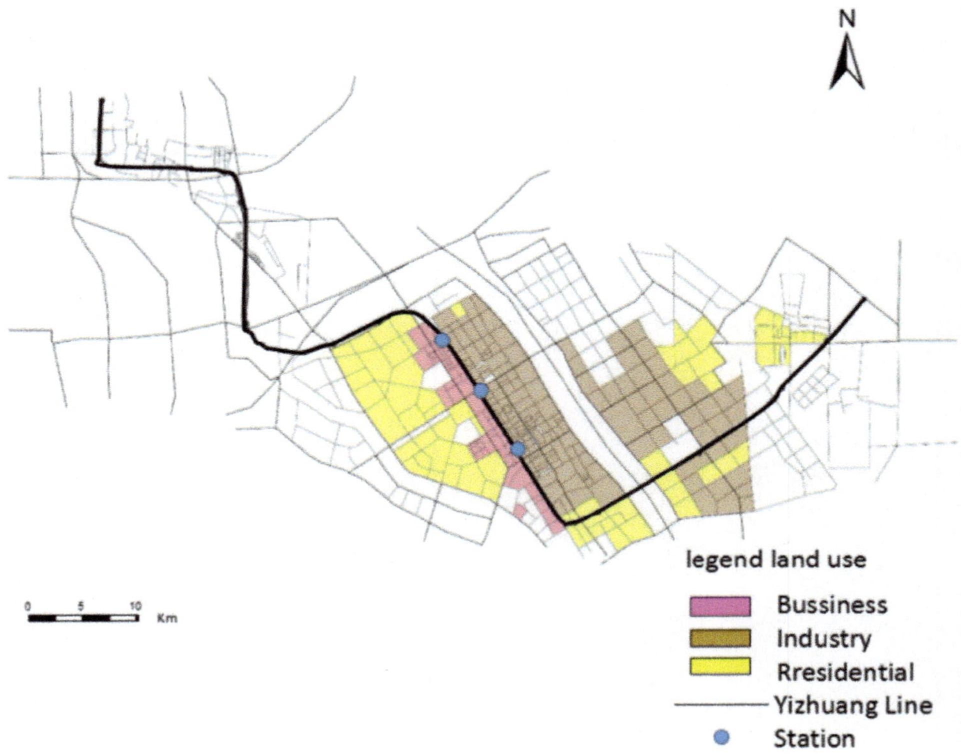

Fig. 7 Planned (*above*) and actual (*below*) commuting pattern

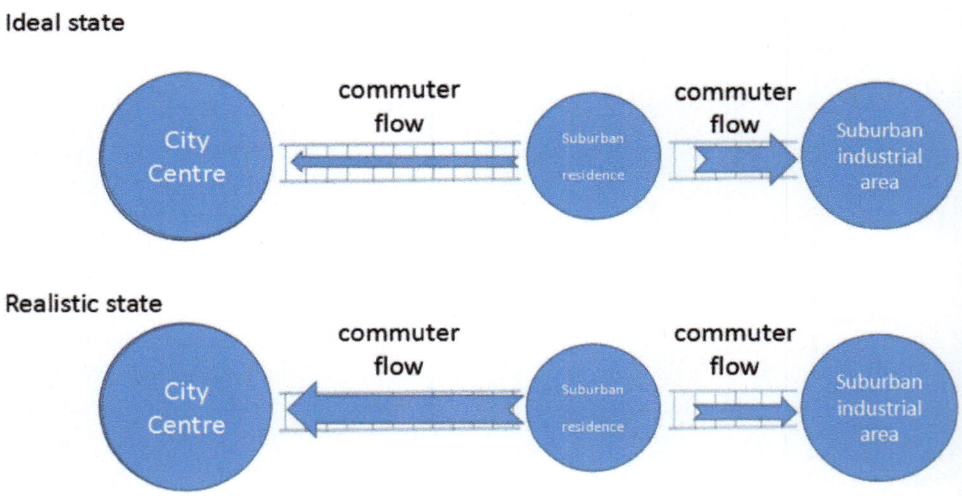

population and jobs from city center, due to the mismatch between the existing low-skilled job opportunities and the high-skilled residents in the area as illustrated in Fig. 7.

The policy implication for this study is that the urban planning code for land use should be flexible and adaptable to the transportation infrastructure in TOD model. While there are many successful cases in China, the case of Yizhuang may not be considered as one. Our findings suggest that applying the concept of TOD into light rail development is not entirely straightforward, and instead an integrated approach between rail transportation and land use may be more successful. Proper land use is essential in successful TOD development, and to single-handedly focus on urban rail line development does not automatically lead to a successful TOD model in surrounding areas. Surrounding area naturally grows in a TOD way. Transportation and land use coordination are critical in implementing integrated planning. Additional studies on activity based and demand driven may be needed to further explore the needs for transportation and land use integration.

Acknowledgements This study was supported by the National Nature Science Foundation of China (41401651, 51678029).

References

1. Dueker K, Bianco M (1999) Light-rail-transit impacts in Portland: the first ten years. Transp Res Rec J Transp Res Board 1685:171–180
2. Bookout W (1992) Neotraditional town planning: a new vision for the suburbs? Urban Land 51(1):20–26
3. Bookout W (1992) Neotraditional town planning: cars, pedestrians, and transit. Urban Land 51(2):10–15
4. Bernick M, Cervero R (1997) Transit villages in the 21st century. McGraw-Hill, New York, pp 377–398
5. Calthorpe PG, Fulton WB (eds) (2001) The regional city: planning for the end of sprawl. Books. In: The regional city: planning for the end of sprawl. Island Press, New York, pp 107–139
6. Ewing R, Rong F (2008) The impact of urban form on US residential energy use. Hous Policy Debate 19(1):1–30
7. Dittmar H, Ohland G (eds) (2012) The new transit town: best practices in transit-oriented development. Island Press, New York, pp 99–111
8. Cervero R, Arrington GB, Smith-Heimer J et al (2004) Transit Oriented Development in America: experiences, challenges, and prospects, TCRP report 102. Transportation Research Board, Washington
9. Brown Barbara B, Cropper Vivian L (2001) New urban and standard suburban subdivisions: evaluating psychological and social goals. J Am Plan Assoc 67(4):402–419
10. William R (2008) From local to global: one hundred years of neighborhood planning. J Am Plan Assoc 75(2):209–230
11. Lin JJ, Gau CC (2006) A TOD planning model to review the regulation of allowable development densities around subway stations. Land Use Policy 23(3):353–360
12. Ministry of Housing and Urban–Rural Development (2016) Notice of the Ministry of Housing and Urban–Rural Development, the National Development and Reform Commission, and the Ministry of Finance on the Development of Featured Towns, No. 147
13. Parker T et al (2002) Statewide transit-oriented development study: factors for success in California. Final report. California Department of Transportation, Urban Land Institute, Washington
14. Zhao P, Lü B (2009) Transportation implications of metropolitan spatial planning in mega-city Beijing. Int Dev Plan Rev 31(3):235–261
15. Zhang C, Man J (2015) Examining job accessibility of the urban poor by urban metro and bus: a case study of Beijing. Urban Rail Transit 1(4):183–193
16. Wang J, Jin F, Mo H et al (2007) Research on review of TOD development mode analysis. Transp Transp Acad J 12(z2):19–22 **(in Chinese)**
17. Ma L, Wu F (eds) (2005) Restructuring the Chinese city: diverse processes and reconstituted spaces. In: Restructuring the Chinese city: changing society, economy and space. Routledge, London, pp 1–20
18. Beijing Master Planning (2004) Final Report Beijing municipal government and Beijing urban planning Bureau, Beijing. http://zhengwu.beijing.gov.cn/ghxx/ztgh/. Retrieved on Mar 1 2017.

7

Enhancing Beijing's Resilience by Improving Tongzhou's Access to High-Speed Rail Transportation

Christopher Auffrey[1] · Xin Fu[1] · Xinhao Wang[1] · Alyssa Wissel McClearnon[1]

Abstract The relocation of 400,000 Beijing municipal employees to Tongzhou promises to reduce congestion and pollution in the city center. Further, the relocation could facilitate economic cooperation and integration of the Beijing–Tianjin–Hebei megaregion. Together, these initiatives can help make Beijing and Northern China more resilient, though critical issues of the connectivity of Tongzhou to the rest of Beijing and the megaregion need to be addressed. This paper focuses on two questions related to the success of these efforts: (1) How many additional commuters can be expected to use the existing subway stations in Tongzhou as a result of the relocation; and (2) How can Tongzhou be better connected to the megaregion's high-speed rail network and a major airport. To answer the first question, a GIS-based model was used to analyze subway ridership data to estimate changes in rush hour commuters at Tongzhou's subway stations under three scenarios. The results estimate substantial increases in commuters unless large proportions of the workers move their residences to Tongzhou. To answer the second question, a Moving Platform Infrastructure Network (MPIN) was conceptualized as an innovative alternative to provide needed access to high-speed rail and a major airport. The resulting MPIN provides fast and environmentally sensitive connections, while potentially reducing congestion at the Beijing South railway station. In addition, the MPIN concept could be used in other parts of the megaregion.

Keywords Subway · High-speed rail · Megaregion · Resilience · Regional planning

1 Introduction

As the capital of China, Beijing (2015 est. pop. 22 million) serves as the political and cultural center of the world's most populous country [1]. Over the past two decades, and especially since the 2008 Olympics, Beijing has established itself as a contemporary world capital, with an efficient, modern subway system, an agglomeration of distinguished educational institutions and technology firms, and world-class arts, culture, and architecture. Yet, Beijing faces serious ongoing challenges from air pollution, traffic congestion, and rapidly rising housing costs as well as an ongoing flow of migrants from within China who are seeking to share in the capital's superior services and job opportunities [2].

Beijing also is the anchor for northern China's Beijing-Tianjin-Hebei economic megaregion (Fig. 1). Commonly referred to as "Jing-Jin-Ji" (JJJ) in China, the region covers an area of 216,000 km^2 with a population approaching 150 million [3]. Other than Beijing, major cities in the region include the autonomous municipality of Tianjin (est. pop. 15.5 million), and the largest cities in Hebei province, Baoding (est. pop. 11.2 million) and Shijiazhuang (est. pop. 10.7 million) [1]. The goal of the megaregion is to promote more spatially balanced development throughout Northern China. It is envisioned that the region will link the research facilities and creative culture of Beijing with the economic dynamism of the port city of Tianjin and the development

✉ Christopher Auffrey
chris.auffrey@uc.edu

[1] School of Planning, University of Cincinnati, PO Box 210016, Cincinnati, OH 45221-0016, USA

Editors: Haishan Xia and Chun Zhang.

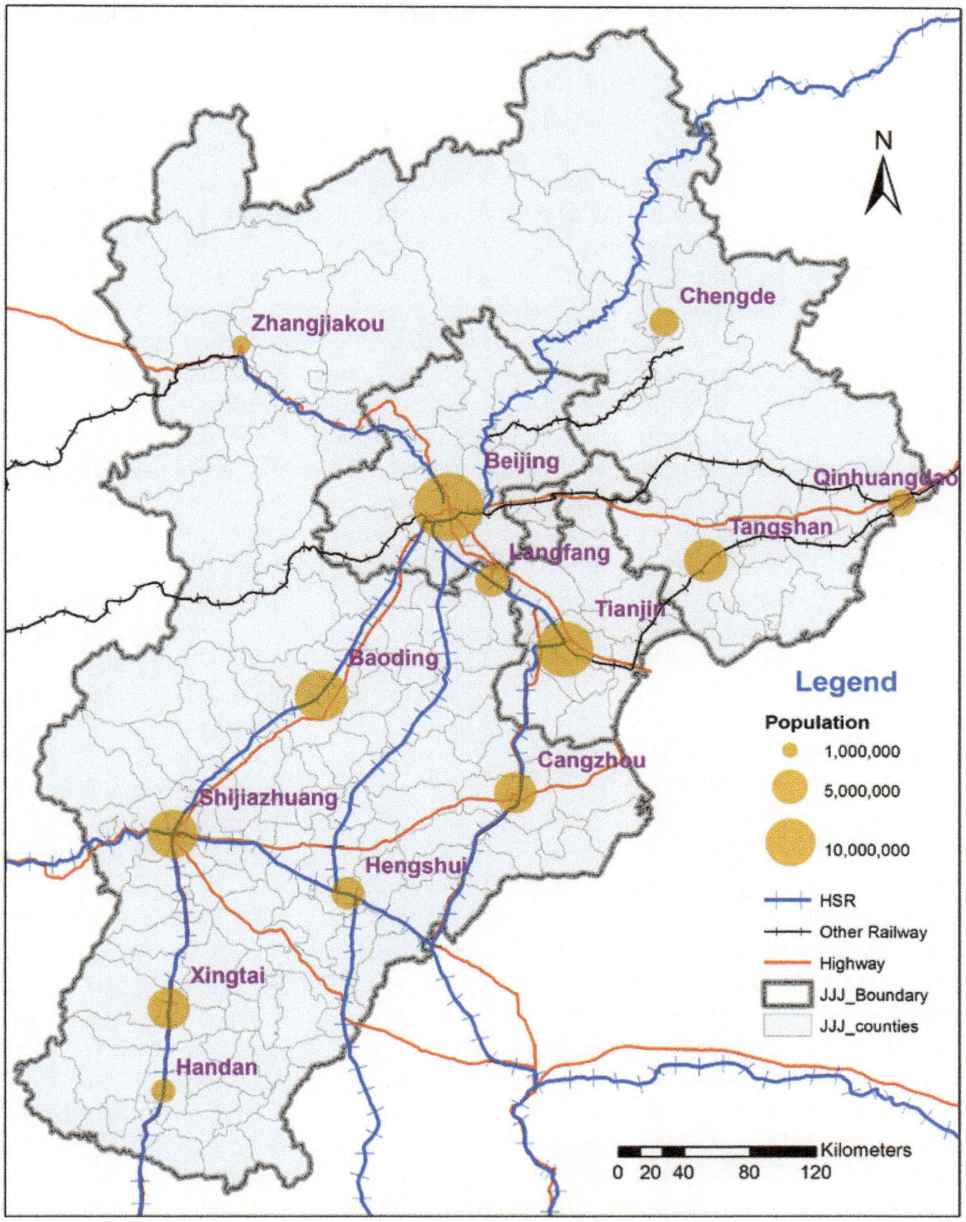

Fig. 1 Major cities in the "Jing-Jin-Ji" megaregion

potential of Hebei province, bringing together jurisdictions that have not worked together in the past [4]. Challenges for the region include building infrastructure linkages to support economic cooperation and integration, which are considered weaker than that those found in China's Yangtze River Delta and Pearl River Delta regions [5].

In the 1990s, the declining quality of Beijing's natural environment, especially air quality, moved the central government to take action. As a result, some manufacturing plants were relocated from the city center to suburban areas. As the environmental challenges continued, a second wave of relocations were initiated in the early 2000s that included the transfer of the Shougang (Capital Steel) Group from Beijing to the city of Tangshan, a location considered to be more appropriately situated to accommodate emission impacts, affordable workforce housing and social infrastructure needs and offered enhanced potential for economic growth and job generation [3]. Yet, Beijing still faces important environmental and quality of life challenges. Traffic congestion and pollution remain as serious problems, and the flow of migrants from second and third tier cities, and rural areas continues. Beijing's population is still rising and the city's developed area continues to absorb farmland [6]. As a result of the ongoing challenges, Beijing's most recent strategy for addressing congestion and pollution is to relocate the municipal government center from the city center to the Tongzhou district on the southeast periphery. This initiative will move 400,000 city employees from offices within the second ring road to what is currently a relatively low density area outside the sixth ring road [7].

The experience of earlier relocation efforts suggests that careful planning is needed for the municipal center relocation to fully achieve its intended purposes. This paper assesses this major shift of jobs from the perspective of sustainability and resilience. The second section provides essential background information for understanding the resilience context of this research and emphasizes efficient transit infrastructure, including direct access to high-speed rail (HSR) and a major airport, as a critical component for the resilient development of Beijing, Tongzhou and the JJJ region. Section 3 assesses the potential impact of the municipal center relocation on Tongzhou's subway stations and housing. Section 4 conceptualizes a Moving Platform Infrastructure Network (MPIN) as it might be used to more directly connect Tongzhou and other areas to HSR. Conclusions are offered in Sect. 5.

2 Background

2.1 Resilience

The concept of resilience comes from the field of ecology which considers resilient systems to have the ability to withstand disturbances and still maintain their essential functions; the ability to self-organize in response to disturbances; and the capacity for learning and adaptation in response to disturbances [8]. In recent years, urban planners have increasingly turned to the concept of resilience as a means of fortifying communities against the growing prospect of natural and human disturbances that threaten to disrupt essential urban services [9].

In Beijing, resilience planning is especially timely, as the city and its surrounding region prepare for continued rapid population growth in the face of uncertain social, climatic, seismic, and economic conditions. The 2004 Beijing plan estimated the city's population would reach 18 million by 2020, but that level was reached before 2010 [10]. The current official estimate is 22 million, with a plan to limit the 2020 population to 23 million residents [11]. The growing population puts added pressure on already stressed local services such as water, sanitation, education and health care, and worsens traffic congestion and air and water pollution [12]. In 2014, the average journey to work for Beijing commuters was 19.2 km, taking 52 min, the longest average commute of any Chinese city [13]. Also, global climate change is expected to produce in Beijing higher temperatures and humidity with little wind, as well as erratic rainfall, resulting in more frequent periods with hazardous air pollution with periodic drought and flash flood conditions [14]. Further, since 2003, there have been ten 4.0 or greater earthquakes near Beijing [15] with an estimated 10% probability of a major earthquake in the Beijing-Tianjin area in the next 50 years [16]. In addition, threats to global financial stability have increased during 2016, requiring innovative approaches to support China's (and Beijing's) continued economic expansion [17].

The spatial aspects of resilience become especially important for addressing the increasing disparities in access to goods, services and opportunities across Beijing and Northern China. Spatial resilience deals with the durability of interconnected social-ecological systems at a citywide and regional scale and recognizes that vulnerability to disturbance, reflected in access disparities, varies over space [18]. Ultimately, spatial properties like enhanced connectivity are essential determinants of a region's resilience capacity [18]. The relocation to Tongzhou is intended to improve regional resource distribution, equality of development, and access to affordable housing. In addition, the resulting reduced congestion in the Beijing city center can contribute to achieving a regional pattern of low-emission mobility. All of this is premised on the ideas that encouragement of non-motorized travel, transit-oriented development, and equality of access are essential for a resilient transit system [19, 20].

2.2 Challenges for Tongzhou

While moving Beijing's municipal offices to Tongzhou is intended to decongest Beijing's city center, there will be significant challenges for Tongzhou to accommodate the changes to its social, natural, and built environments. Tongzhou's transit infrastructure will need to be expanded to accommodate the influx of municipal employees and potential demand for new housing. Between 2008 and 2012, Tongzhou added 252,000 residents but only 28,000 jobs, putting significant strain on the existing bus and subway services that provide access to jobs elsewhere [21]. It has been reported that residents of peripheral districts like Tongzhou are more likely to drive a car to work because the demand for the limited public transit far exceeds capacity, making the use of transit time-consuming, uncomfortable, and something to be avoided whenever possible [13].

2.3 Role of High-Speed Rail

In addition to expanding bus and subway transit services, if the residents and non-resident workers of Tongzhou are to be fully integrated into the JJJ region, efficient access to HSR is essential. Well-designed and fully implemented HSR service has been shown to increase real-estate values, improve quality of life, reduce air pollution and traffic congestion, and provide a "safety valve" for crowded cities [22]. In China, HSR has improved market access, expanding labor markets, and enhanced spatial agglomerations [22]. In effect, by providing additional location options to firms and workers, HSR has reduced congestion and pollution within megacities and has stimulated the economic growth of nearby second-tier cities [23].

In Northern China, HSR has played an especially important role. Currently, there are five high-speed rail lines connecting Beijing to the northeast (Jing-Shen line), southeast (Jing-Hu line), southwest (Jing-Guang line), northwest (Jing-Zhang line), and south (Jing-Gang line). As part of the Jong-Hu line, the Beijing-Tianjin HSR is considered to have significantly increased integration and commercial interaction between northern China's two largest municipalities, improved overall inter-city transport within the JJJ megaregion, and contributed to better regional cooperation [22].

2.4 Role of Airports

In addition to HSR, efficient access to other domestic and international cities via commercial airlines is essential for JJJ's success. Currently, there are two airports in Beijing, the Beijing Capital International Airport (Capital Airport) in the northeast, and Nanyuan Airport, a smaller airport in the southern part of the city with limited expansion capacity and few commercial flights. In 2015, the Capital Airport was the second busiest airport in the world with nearly 90 million total passengers [24]. Of concern is that in 2015 nearly one-third of Chinese flights were late with air traffic control measures accounting for 30.7% of air delays [25]; this issue is compounded by the rapid increase in passengers (68%) and flights (46%) since Capital Airport's last expansion in 2007. Anticipating that congestion could become a serious problem for the Capital Airport, a plan was approved in 2013 for the construction of another Beijing international airport in Daxing [26].

The Beijing Daxing international airport (Daxing airport) is currently under construction 46 km south of the Beijing city center, in the southern part of the Daxing district, near the boundary with Hebei province. It is expected to be complete in 2019. In addition to relieving congestion at the Capital Airport, the Daxing airport is intended to serve the expanding urbanization south of Beijing, as well as provide expanded international service for Tianjin and Hebei province [26]. Plans call for a high-speed rail stop at the airport for the Jing-Gang line which connects Beijing with Bazhou in Hebei province and then continues on to Hong Kong [26]. Also, planned is a Daxing airport subway link to the Beijing South station, which has a high-speed rail service to Tianjin and Shanghai [27].

3 Impacts on Tongzhou

3.1 Importance of Transportation Infrastructure

Located approximately 22 km. east of central Beijing, the Tongzhou district covers an area of 906 km^2, or 6% of Beijing's total area. In the 2010 census, the population of Tongzhou was 1.184 million residents, almost twice its 674,000 population in 2000. The Beijing municipality has set a population cap of 2 million residents for the district [28].

Currently, Tongzhou's central business district has direct transportation connections with central Beijing via several modes of transportation. These include two subway lines (Batong Line and Line 6), a conventional railway (with service to Beijing Station), and several major highways (G1 and S15). There are no direct connections to the Beijing Capital International Airport or Beijing's two HSR stations, Beijing South and Beijing West, as access to both requires transfers to other subway lines. As such, the relocation of Beijing municipal government offices may add significant congestion to both Tongzhou's limited transit options and the transportation infrastructure accessing Beijing's HSR stations.

3.2 Scenario Model

The added passenger volume on the two subway lines connecting the city center to Tongzhou may lead to much higher levels of congestion during the rush hour commutes. Estimates of the magnitude of these changes are needed to support planning efforts. To estimate the subway eastbound ridership on the two subway lines leading to Tongzhou, a GIS-based model was developed using CommunityViz, an ArcGIS extension [29]. The model analyzes smart card data for commuters using the Beijing subway stations on or inside the second ring road to estimate the number of subway commuters under a range of scenarios.

Using this scenario modeling approach, the number of commuters was estimated using four assumptions. First, because the available ridership data only provides the total number of commuters exiting a station, an assumption was needed about the portion of commuters who are municipal employees. An assumption, *PropCityEmp*, was devised to represent the proportion of municipal employees among the commuters who currently disembark from the central city subway stations. Second, some of the municipal employees can be expected to relocate their homes to Tongzhou after the move of their workplace. Another assumption, *PropCommute* was devised to represent the proportion of municipal employees who will commute from their current residence to Tongzhou after the municipal office relocation. Third, since there are two subway lines connecting Tongzhou, an assumption, *PropBaton*, was devised to represent the proportion of municipal employee commuters who will use the Sihui East station on the Batong line to get to their jobs in Tongzhou, with the remainder expected to use the Jintai station on Line 6. Potentially, all the proportions associated with these three assumptions could vary from 0% to 100%. In addition, a fourth assumption was necessary to estimate additional housing demand in Tongzhou. A "Job-Home Ratio" assumption was devised to represent the number of municipal employees per household. For example, a ratio of 1.5 is assumed for this analysis, meaning that for every three municipal employees relocating to Tongzhou, two housing additional housing units will be needed.

Using the information from the four assumptions, three possible scenarios were devised with respect to the proportion of the relocated municipal employees who will live in Tongzhou. The first scenario establishes that all municipal employees will continue to live in their current homes and will commute to Tongzhou (PropCommute = 100%). The second scenario establishes that 50% of the municipal employees will move to new homes in Tongzhou (PropCommute = 50%). The third scenario establishes that all of the municipal employees will move to new homes in Tongzhou (PropCommute = 0%). In each of the three scenarios, it is established that the proportions of city center to Tongzhou commuters will split 60% on the Batong line and 40% on Line 6, as is the approximate current commuter split between these lines (PropBatong = 60%).

A final separate assumption was necessary to establish the proportion of subway commuters currently exiting the city center subway stations who are municipal employees and whose jobs will be relocated to the new municipal offices in Tongzhou. The assumed proportion could not be too low, or the decongestion benefits of the relocation would be insignificant. It was assumed that 50% of the subway commuters currently exiting the city center subway stations are municipal employees who will be moved to the new municipal offices in Tongzhou (PropCityEmp = 50%).

The ComunityViz software, a planning support system extension of ArcGIS, was used because it allows easy adjustment of the values of the four assumptions. As new information becomes available to support assumption different values, the scenarios can be quickly re-evaluated simply by moving the assumption slider bars to the new values.

3.3 Scenario Model Results

Table 1 shows the estimated subway ridership and housing demand in Tongzhou under the different scenarios. The model estimates there are 220,786 morning rush hour commuters traveling to the Beijing city center. Assuming half of them are municipal office employees and they all continue to live in their current residences, about 110,000 additional commuters are expected to take the two subway lines to Tongzhou (Scenario 1). Under this scenario, commuters using the Sihui East station on the Batong Line will increase fivefold, from 15,862 to 82,098. At 82,092 commuters, the Batong Line will have more morning rush hour commuters than are currently using Line 1 in the Beijing city center (e.g. eastbound from Xidan) (Fig. 2). A similar impact is estimated for Line 6 when comparing the number of commuters using the Jintai station with commuters using the Dongsi station. Depending on the capacity of the Tongzhou stations, such large increases in commuter volume may result in significant congestion delays, and associated direct and indirect costs.

Under Scenario 2, it is assumed that half of the municipal employees will move their homes to Tongzhou, resulting in an estimated 48,980 morning commuters on the Batong Line. This ridership is just below the average ridership for the city center on Line 1 and Line 6 (about 52,000). The estimated additional demand for housing in Tongzhou is 36,798 U, about 7.5% of the estimated 490,000 current residential units [30].

Under Scenario 3, there is no expected increase in subway ridership, assuming all the relocated municipal employees will be able to walk, bike or take local buses to work, while the demand for housing in Tongzhou will increase by 73,595 U, a nearly 15% increase.

The results of the scenario modeling show that the relocation of the Beijing municipal government offices is likely to substantially increase the subway ridership (for both the Batong Line and Line 6) and/or demand for housing in Tongzhou, depending on how many of the relocated workers move their residences so they do not need to take the subway to get to work. The next part of this paper conceptualizes the design and use of an MPIN for enhancing Tongzhou's (and the relocated municipal workers') access to HSR. This conceptualization also takes into consideration another major development in Beijing that will impact the use and efficacy of the JJJ regions transit infrastructure, the new Daxing airport.

4 High-Speed Railway Accessibility for Tongzhou Residents and Workers

4.1 Need for Alternative HSR Access

The subway ridership analysis illustrates that relocating a workplace without making provisions for relocating residences may create excessive demand on existing transit. This could be mitigated if a substantial number of the relocated employees move their residence to Tongzhou.

Table 1 Morning rush hour ridership on subway from Beijing to Tongzhou under different scenarios

Scenario	Assumptions	Ridership to central Beijing	Ridership from Beijing to Tongzhou		Additional housing unit demand in Tongzhou
			Batong	Line 6	
	Existing conditions (based on smart card data)	220,786	15,862	9710	0
1	PropCityEmp = 50% PropCommute = 100% PropBaton = 60%	110,393	82,098	44,157	0
2	PropCityEmp = 50% PropCommute = 50% PropBaton = 60% Job-Home Ratio = 1.5	110,393	48,980	31,789	36,798
3	PropCityEmp = 50% PropCommute = 0% PropBaton = 60% Job-Home Ratio = 1.5	110,393	15,862	9710	73,595

Scenario # Batong ridership = existing ridership to central Beijing × PropCityEmp × PropCommute × PropBaton + existing Batong ridership (e.g. Scenario 1 Batong ridership = (220,786 × .5 × 1 × .6) + 15,862 = 82,098)

Scenario # Line 6 ridership = existing ridership to central Beijing × PropCityEmp × PropCommute × (1 − PropBaton) + existing Line 6 ridership (e.g. Scenario 1 Line 6 ridership = (220,786 × .5 × 1 × .4) + 9710 = 44,157)

Scenario # additional housing demand in Tongzhou = existing ridership to central Beijing × PropCityEmp × (1 − PropCommute)/Job-Home Ratio (e.g. Scenario 2 additional housing demand in Tongzhou = 220,786 × .5 × (1 − .5)/1.5) = 36,798)

Fig. 2 Ridership comparison (Sihui East and Xidan on *Line 1*; Jintai and Dongsi on *Line 6*)

However, this may create other problems due to the lack of direct accessibility of Tongzhou to HSR stations and airports. The new Tongzhou residents and worker's additional demand for HSR travel may add to the already significant subway and surface motor vehicle congestion near the Beijing South and Beijing West HSR stations, adding time delays, reduced productivity, and increased air and water pollution.

If a new HSR station more accessible to Tongzhou were added, this congestion may be mitigated. One possible solution would be an additional HSR station near Tongzhou. However, another problem may arise from adding the new HSR station. It takes a considerable amount of time and energy for high-speed trains to accelerate to operating speed. This time and energy is lost when trains must slow down to drop off and pick up passengers and then re-accelerate. This is why HSR stations are more widely spaced compared to slower conventional passenger rail service. Consequently, passengers on average are required to travel greater distances to reach HSR rail stations, compared with traveling to conventional rail stations which can efficiently be located closer together. Therefore, HSR stations draw passengers from an expanded area resulting in higher transit and motor vehicle volumes near HSR stations. This can exacerbate station-area congestion and air pollution, especially in already congested urban areas such as Beijing [31]. Yet, both existing HSR station congestion reduction, and travel time and energy reduction benefits could be achieved if the high-speed trains could avoid stopping to pick up and drop off passengers. The MPIN offers just such a means for transferring passengers on and off the HSR [32].

4.2 Potential of MPIN to Address Tongzhou's HSR Access Needs

The MPIN concept is intended to allow passengers to embark and disembark from trains while the trains are moving, thereby making a high-speed rail network more efficient. The MPIN uses a tram system to collect passengers from local stations. The tram accelerates to match the speed and link to the HSR train, at which point passengers can transfer between the tram and train. The tram and train would be connected for approximately the same length of time as a train would normally spend stopped at a station. The tram separates from the high-speed train and returns to the local station to provide passengers with better access to their destinations than if they used a conventional boarding process from a stopped train [33]. Yet despite MPIN's potential, to date, there are no reports of the use of an MPIN design in a real-world application [34]. The contribution of this paper is to fill that knowledge gap by conceptualizing an MPIN design as it could be used to address the JJJ region access needs of Tongzhou workers and residents.

4.3 Conceptualizing a MPIN for Tongzhou

A conceptualization approach was used to as a first step in assessing the impact of an MPIN linking Tongzhou and the new Daxing airport with the Beijing-Tianjin HSR line. Conceptualization is the process of development and clarification of complex concepts [35] and serves as a necessary preliminary step before conducting a more detailed analysis. This approach is considered to be appropriate for the initial assessment of emerging innovations, such as the MPIN, where real-world applications have yet to be implemented. The approach was chosen after reviewing the available literature and recognizing that sufficient data for a conventional impact assessment did not exist and could not be collected [36].

As part of the conceptualization, an illustrative MPIN route was selected and mapped. This has been done based on the assumptions that:

- A significant number of Beijing municipal government employees working in Tongzhou, as well as others in Tongzhou, will want to board the HSR southbound toward Tianjin;
- A significant number of passengers arriving at the Daxing airport will want to board the HSR southbound toward Tianjin;
- A significant number of passengers on the HSR northbound toward Beijing will want to disembark to go to the Daxing airport; and
- A significant number of passengers on the HSR northbound toward Beijing will want to disembark to go to Tongzhou.

In addition, it is assumed that the following factors would be considered in siting the MPIN corridor:

- Distance along the Beijing-Tianjin HSR line required for MPIN acceleration, docking, and deceleration;
- Potential impacts on environmentally sensitive natural and historic areas;
- Potential impacts on current and future local economic development (jobs, housing) and property values;
- Potential impacts on current and future social connections;
- Potential for needed environmental amenities related to greenways, scenic walking and bike paths, storm water retention, etc.; and
- Total travel distance and time for all passengers, given other constraints.

The Tongzhou MPIN was conceptualized to include two linked but separate rail components (Fig. 3). The first rail component would establish a feeder tram line along the Beijing-Tianjin HSR line with a north tram station at Daxing and south tram station at Langfang, a distance of approximately 40 km. Southbound passengers would board the feeder tram at the Daxing station. Coordinated with the southbound HSR train coming out the Beijing South station, the feeder tram would accelerate to the same speed and link to the HSR train for 3 min to allow passengers to transfer. Then, the tram would delink, decelerate, and stop at Langfang station. It is estimated that the feeder tram requires approximately 20 km. for acceleration and 8 km. for deceleration. At a traveling speed of 250 km/h, the 3-min boarding period requires a distance of 12.5 km. Northbound passengers on the Beijing-Tianjin HSR line with Tongzhou, Daxing, the Daxing airport or Langfang as their destination could disembark onto a northbound feeder tram to Daxing. The feeder tram rail component would add an additional set of rails on each side of the Beijing to Tianjing HSR line for the 40 km. distance from Daxing to Langfang.

The second rail component of the Tongzhou MPIN would be a light-rail loop connecting Tongzhou, Daxing, the Daxing airport and Langfang. At an average speed of 120 km. per hour, the travel time on the light-rail between Tongzhou and Daxing or between Tongzhou and Langfang would be approximately 20 min. The travel time between Daxing and the Daxing airport would be approximately 15 min.

The alignment of the light-rail line would follow a newly established rail corridor designed to be sensitive to local features of the natural environment, such as along the Yongding River near the Daxing airport and the Grand Canal near Tongzhou, where the light-rail corridor would be part of a riverfront conservation green corridor development. This design would follow a biophilic approach, based on the concept of Biophilia for designing buildings and facilities that enhance nature, as popularized by Harvard myrmecologist and sociobiologist E.O. Wilson [37]. Biophilic cities can have a positive influence on psychological, physical, and social well-being [38]. Residents in biophilic cities are directly and actively engaged in

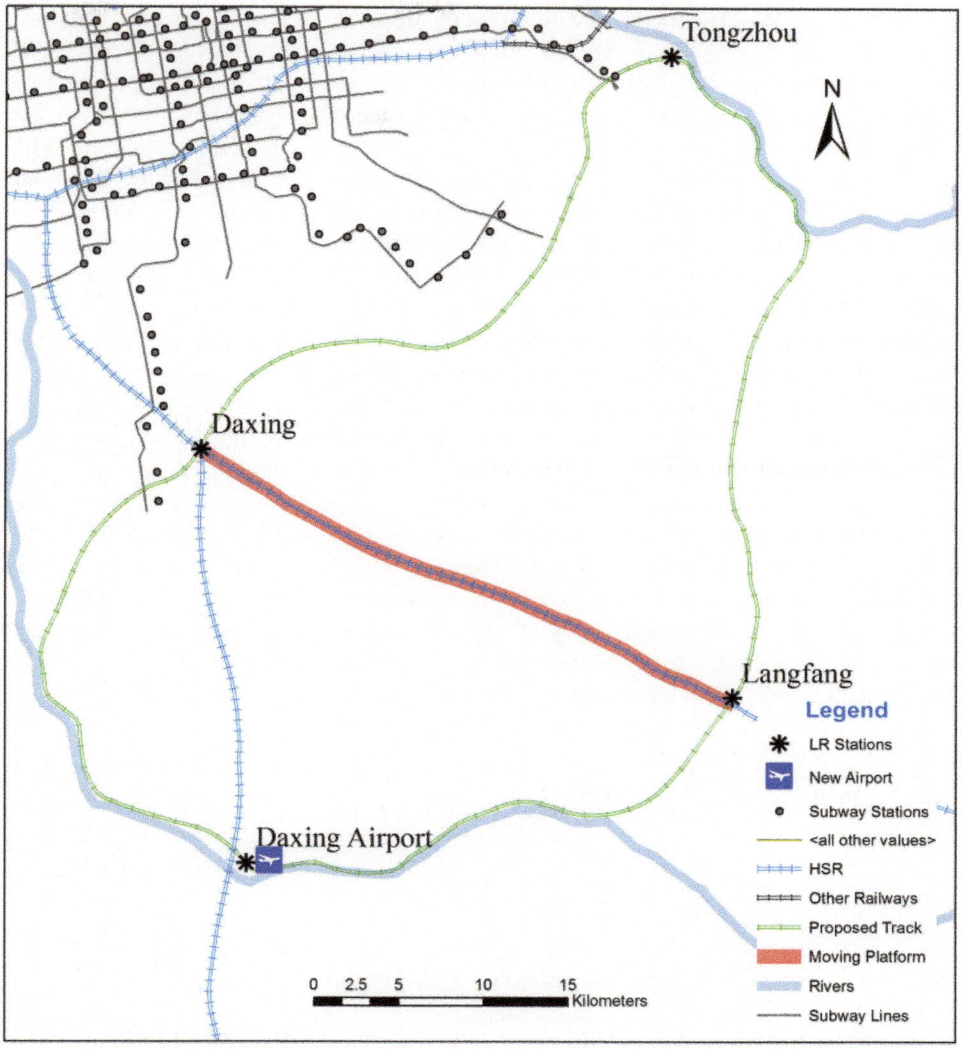

Fig. 3 Illustrative MPIN alignment for Tongzhou HSR Access

learning about, enjoying, and caring for the nature around them and have developed important emotional connections with this nature [37]. The light-rail loop ecological corridors would be designed to increase opportunities for passengers to explore and appreciate the natural environment. The light-rail corridor also would provide green walking and biking trails for nearby residents, understanding that the aesthetics may be diminished by the frequent passing of the light-rail trains. The light-rail alignment along the Yongding River also would be designed to negate existing sources of stormwater runoff pollution and bolster the area's freshwater resources.

5 Conclusions

The sustainability and resilience of a region are dependent on the patterns and dynamics of its natural, built, and social environments. In the complex context of the global marketplace, these patterns and dynamics increasingly depend on the creation and maintenance of conditions that promote economic cooperation and integration through the efficient movement of people, goods and services. With the establishment of the Beijing-Tianjin-Hebei (JJJ) economic megaregion, China has recognized the need for and potential benefits of extending the development of Beijing and Tianjin to a much larger part of Northern China. This approach recognizes that Beijing and Tianjin are reaching the physical limits of their sustainability and that a resilience strategy requires the responsive adaptability that the megaregion potentially offers. Yet to accomplish this, the efficient movement of people, goods, and services across the region is critical.

At the same time, Beijing has been required to take steps to aggressively counter problems of congestion in its city center. The relocation of 400,000 municipal employees to Tongzhou promises to reduce congestion and pollution in the center of Beijing. However, a scenario model based on the analysis of subway ridership data suggests that the relocation could overwhelm Tongzhou's existing subway infrastructure unless it is substantially expanded or a large proportion of the relocated workers move their residences to Tongzhou.

This study also recognizes how the relocation of workers to Tongzhou could contribute to the needed cooperation and integration for the success of the JJJ megaregion, though this largely depends on Tongzhou's improving access to HSR and a major airport. To this end, this research conceptualized an environmentally friendly MPIN that would provide just such access. The resulting MPIN would provide substantial potential for providing fast and efficient connections to both the Jing-Jin HSR line and the new Daxing international airport, while reducing congestion at and near the Beijing South railway station.

In addition, there is potential for the future use of the MPIN concept in other parts of the JJJ region that could contribute to the region's success as well as enhance its sustainability and resilience goals (Fig. 4). For example, a second MPIN could be used to connect Fangshan and Daxing with the Beijing-Guangzhou HSR line and the Daxing airport. This also would provide additional connections for Tongzhou to other Beijing districts in the west and southwest portions of the municipality, such as Mentougou and Fangshan, and to Zhuozhou in Hebei province. In addition, a third MPIN in the Shunyi district to the northeast of Beijing could be used to connect the Beijing Capital International Airport with the Beijing-Shenyang HSR line to northeast China.

Finally, it is important to acknowledge that this study has been premised on the importance of creating sustainable and resilient cities. The characteristics of sustainable cities include compact form, planned growth areas, and walkable and bikeable neighborhoods which contain a mix of residential, commercial, and recreational land uses integrated with highly connected public transit networks. Resilient sustainable cities have these same characteristics, and in addition have the capability to absorb natural and human disturbances while maintaining essential services. It has been important to consider how Beijing can best emerge both more sustainable and more resilient as the city attempts to resolve the two significant issues addressed in this paper. The rapid development that can be expected in

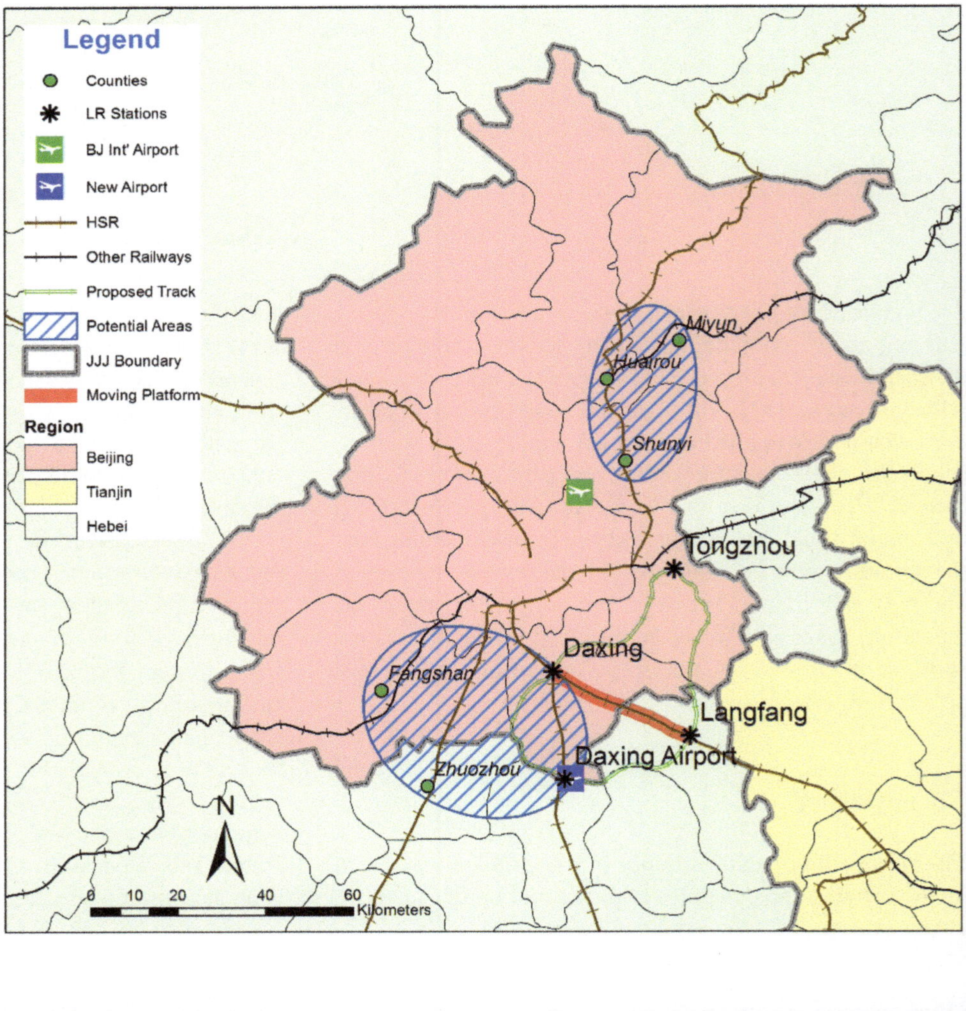

Fig. 4 Other potential MPINs in the Jing-Jin-Ji megaregion

Tongzhou presents an opportunity for planners to require compact, walkable, mixed-use development with strong public transit connections that can sustain a high quality of life for 3 million people. Likewise, the emergence of the JJJ region offers opportunities for both building resilience capacity and sustainability for both Beijing and Northern China, though neither will be automatic. Clearly, fast and efficient access and connectivity are essential. Ultimately, this study suggests that an MPIN has potential as a sustainable innovation for allowing the JJJ megaregion to cope with rapid urbanization, and the need to relocate workers and expanding high-speed rail access.

Compliance with Ethical Standards

Conflict of interest The authors declare that they have no conflict of interest.

References

1. National Bureau of Statistics of China (2015) China statistical yearbook 2015. National Bureau of Statistics of China. http://www.stats.gov.cn/tjsj/ndsj/2015/indexeh.htm. Accessed 12 Dec 2016
2. International Labour Office (2011) Employment policies report—China: promoting decent employment for rural migrant workers. http://www.ilo.org/wcmsp5/groups/public/—asia/—ro-bangkok/—ilo-beijing/documents/publication/wcms_184814.pdf. Accessed 12 Oct 2016
3. Baily T (2015) Jing-Jin-Ji: China's new mega-region. World Finance. http://www.worldfinance.com/infrastructure-investment/jing-jin-ji-chinas-new-mega-region. Accessed 22 Sep 2016
4. Johnson I (2015) As Beijing becomes a super city, the rapid growth brings pains. New York Times. https://www.nytimes.com/2015/07/20/world/asia/in-china-a-supercity-rises-around-beijing.html. Accessed 22 Sep 2016
5. Lockett H (2016) Together as one: is JingJinJi all that it is touted to be? CKGSB Knowledge. http://knowledge.ckgsb.edu.cn/2016/01/20/china/together-as-one-is-jingjinji-all-that-it-is-touted-to-be/. Accessed 22 Sep 2016
6. Luo X (2015) Internal migration in China. In: Stone J et al (eds) The Wiley Blackwell encyclopedia of race, ethnicity, and nationalism. Wiley, Hoboken, pp 1–4
7. Fong D (2016) A housing frenzy bedevils Beijing. Wall Street Journal. https://www.wsj.com/articles/a-housing-frenzy-bedevils-beijing-1480069808. Accessed 25 Nov 2016
8. Walker B, Salt D (2006) Resilience thinking: sustaining ecosystems and people in a changing world. Island Press, Washington
9. Mehmood A (2016) Of resilient places: planning for urban resilience. Eur Plan Stud 24(2):407–419. doi:10.1080/09654313.2015.1082980
10. Yu W, Mao M, Wang B, Liu X (2014) Implementation evaluation of Beijing urban master plan based on subway transit smart card data. In: Proceedings of the 22nd international conference on geoinformatics. Kaohsiung, Nov 2014. doi:10.1109/GEOINFORMATICS.2014.6950823
11. Chen K, Patton D (2015) Beijing to try to cap population at 23 million by 2020. Reuters News Agency. http://www.reuters.com/article/us-china-parliament-capital-idUSKBN0M215620150306. Accessed 22 Sep 2016
12. Wen K, Zhu E, Zhang G, Ye T, Wu Q, Zhu H (2015) Basic situation of carrying capacity of Beijing, Tianjin and Hebei Province, and development countermeasures. In: Wen K, Zhu E (eds) Report on development of Beijing, Tianjin and Hebei Province: measurement of carrying capacity and countermeasures. Springer, New York, pp 3–29
13. Chen S (2015) Beijing workers have longest daily commute in China at 52 minutes each way. South China Morning Post. http://www.scmp.com/news/china/article/1692839/beijingers-lead-chinas-pack-longest-daily-commute. Accessed 10 Oct 2016
14. Lai EC (2011) Climate change impacts on China's environment: biophysical impacts. Wilson Center, China Environment Forum. https://www.wilsoncenter.org/publication/climate-change-impacts-chinas-environment-biophysical-impacts. Accessed 5 Dec 2016
15. Earthquake Track (2016) Recent earthquakes near Beijing, China. Earthquake track. http://earthquaketrack.com/cn-22-beijing/recent?mag_filter=4. Accessed 24 Nov 2016
16. Chan J (2011) In case you were also wondering: how earthquake prone is Beijing? The Beijinger. https://www.thebeijinger.com/blog/2011/03/24/case-you-were-also-wondering-how-earthquake-prone-beijing. Accessed 5 Oct 2016
17. International Monetary Fund (2016) Global financial stability report. International Monetary Fund. http://www.imf.org/external/pubs/ft/gfsr/index.htm. Accessed 23 Oct 2016
18. Cumming GS (2011) Spatial resilience in social-ecological systems. Springer, Dordrecht
19. Cervero R (1998) The transit metropolis. Island Press, Washington
20. Clercq FL, Bertolini L (2003) Achieving sustainable accessibility: an evaluation of policy measures in the Amsterdam area. Built Environ 29(1):36–47. doi:10.2148/benv.29.1.36.53949
21. Xu H (2015) Moving 2 million people for Beijing's urban reset. Caixin Media. http://english.caixin.com/2015-10-16/100863713.html. Accessed 22 Nov 2016
22. Zheng S, Kahn M (2013) China's bullet trains facilitate market integration and mitigate the cost of megacity growth. Proc Natl Acad Sci 110(14):E1248–E1253. doi:10.1073/pnas.1209247110
23. The World Bank (2014) People's Republic of China: regional economic impact analysis of high speed rail in china, main report. The World Bank. http://www.worldbank.org/content/dam/Worldbank/document/EAP/China/high_speedrail-%20in-china-en.pdf. Accessed 22 Nov 2016
24. Airports Council International (2016) ACI releases preliminary world airport traffic rankings. Airports Council International. http://www.aci.aero/News/Releases/Most-Recent/2016/04/04/ACI-releases-preliminary-world-airport-traffic-rankings. Accessed 27 Oct 2016
25. Zhao L (2016) Measures planned to reduce commercial flight delays. China Daily. http://www.chinadaily.com.cn/china/2016-07/11/content_26036456.htm. Accessed 29 Sep 2016
26. McGregor T (2011) Beijing builds world's biggest airport due to necessity. China Radio International, english service. http://english.cri.cn/6909/2011/09/13/1781s658034.htm. Accessed 29 Oct 2016
27. Zhang Y (2016) Beijing Daxing International Airport to link railway, subway systems. China Daily. http://europe.chinadaily.com.cn/china/2016-07/05/content_25978583.htm. Accessed 29 Sep 2016
28. Zheng J (2015) Tongzhou population to be capped at 2 million. City of Beijing. http://www.cityofbeijing.gov.cn/2015-12/11/content_22689879.htm. Accessed 2 Dec 2016
29. Placeways (2016) CommunityViz: analyze, visualize, and engage. City Explained, Inc. http://placeways.com/communityviz/index.html. Accessed 23 Nov 2016
30. Knoema (2016) Beijing—average size of family households. World Data Atlas, China, Beijing. https://knoema.com/atlas/China/Beijing/Average-Household-Size. Accessed 2 Dec 2016

31. Armstrong J, Hood I, Preston J (2014) Planning and prioritizing conventional and high-speed train services. AET papers repository, association for European transport. http://abstracts.aetransport.org/paper/index/id/4227/confid/19. Accessed 29 Oct 2016
32. PriestmanGoode (2016) Moving platforms. http://www.priestmangoode.com/project/movingplatforms/. Accessed 10 Oct 2016
33. Environment Clean Generations (2016) Moving platforms' could make rail travel more efficient. Environment Clean Generations. http://environment-cleangenerations.blogspot.com/2011/11/moving-platforms-could-make-rail-travel.html. Accessed 18 Sep 2016
34. White C (2011) Nonstop high-speed trains enabled by docking trams. Mashable. http://mashable.com/2011/06/23/nonstop-high-speed-trains/#DGS0CpGFDuqg. Accessed 10 Sep 2016
35. Trochim W, Donnelly J (2006) The research methods knowledge base, 3rd edn. Cengage Learning, Independence
36. Newman L (2012) Basics of social research: qualitative and quantitative approaches, 3rd edn. Pearson Publishing, Upper Saddle River
37. Beatley T, Newman P (2013) Biophilic cities are sustainable, resilient cities. Sustainability 5:3328–3345. doi:10.3390/su5083328
38. Windhager S, Atzwanger K, Bookstein F, Schaefer K (2011) Fish in a mall aquarium-an ecological investigation of biophilia. Landsc Urban Plan 99:23–30. doi:10.1016/j.landurbplan.2010.08.008

8

Dynamic Response Analysis of the Straddle-Type Monorail Bridge–Vehicle Coupling System

Hongliang Wang[1] · Eryu Zhu[1] · Zhuo Chen[2]

Abstract An analytical procedure of dynamic interaction analysis of the straddle monorail bridge–vehicle coupling system is proposed in this paper based on the finite element method and energy method. The calculation procedure is programmed with VB language for the solution of the governing motion equations of the straddle-type monorail bridge–vehicle coupling system. The effects of speed, three kinds of loads and different radius of curvature on dynamic responses of the monorail bridge–vehicle coupling system are analyzed. The simulation indicates that vertical vibration amplitude of the track beam decreases while the lateral amplitude increases with the increase in the radius of the curvature; the maximum value in lateral and vertical direction is 0.075 and 0.43 mm, respectively; and the maximum amplitude (lateral and vertical) and acceleration (lateral and vertical) are 0.69, 0.046 mm, 0.15 and 0.62 m/s^2, respectively, at the speed of 80 km/h. The vibration amplitude (lateral and vertical) and vertical acceleration increase with the increasing load, and the maximum values are 0.041, 0.43 mm and 0.44 m/s^2, respectively. The lateral acceleration is not easily affected by the load conditions.

✉ Eryu Zhu
eyzhu@bjtu.edu.cn

Hongliang Wang
whl.981@163.com

Zhuo Chen
chenzhuo8832@126.com

1 School of Civil Engineering, Beijing Jiaotong University, No. 3, Shangyuan Village, Haidian District, Beijing, China

2 China Railway Fifth Survey and Design Institute Group CO, LTD., No. 4, Kangzhuang Road, Huangcun Town, Daxing District, Beijing, China

Editor: Baoming Han

Keywords Monorail · Bridge–vehicle coupling system · Track beam · Dynamic interaction analysis · Finite element method · Energy method

1 Introduction

Since urban traffic conditions become heavier, monorail transportation system has been introduced in many cities in China, which has the advantages of low costs, short construction period, little influence on the surrounding environment and strong adaptability to the complex terrain and line type. The first monorail transportation line in China has been put into operation in Chongqing City, which is imported from Japanese technology. The mechanism of this monorail transportation system should be explored in depth; however, the publications of studies on the monorail transportation system are scarce. Lee [1], Kim [2, 3] and Naeimi et al. [4] investigated the dynamic responses of the straddle monorail train and the degree of riding comfort based on the multi-rigid-body principle, respectively. Trahair [5] proposed a calculation method which takes the beneficial lateral load into account in the analysis of the lateral buckling of the monorail steel beam. Kim and Kawatani [6] analyzed the dynamic responses of the improved monorail steel track beam with a transverse support system under seismic activities. Ren et al. [7, 8], Zhao [9], Du and Wang [10], Ma [13], Liu [11, 12] and Shan [14] also studied the responses of the straddle monorail vehicle from different aspects. All the studies above studied the mechanism of the straddle-type monorail train, but there were no publications that recorded the mechanism of the monorail track beam under different factors.

The influence of the various factors on dynamic responses of the straddle-type monorail track beam is investigated in this paper. The mechanism obtained in the analytical procedure provides the necessary theoretical accumulation for the development of the straddle-type monorail traffic technology.

2 Straddle-Type Monorail Track Beam System

The straddle-type monorail track beam is not only the load-bearing structure but also the guild way for the straddle-type monorail train. At present, in the monorail system constructed in Chongqing, track beams are mostly built with prestressed concrete or reinforced concrete, which are simply supported. The longest span of the track beams in Chongqing monorail transportation system is 24 m which has been developed with a length longer than the longest one used in Japan. All the track beams of the whole line are 0.85 m wide and 1.5 m high which are limited by the structure of the vehicle and the weight of the material. The straddle-type monorail track beam in Chongqing is shown in Fig. 1.

To simulate dynamic responses of the straddle-type monorail track beam system, the mathematical model is deduced first and a finite element model of track beam is set up (see Fig. 2). Equations of the track beam are derived in a curved local coordinate system (*XYZ*).

According to the principle of the finite element method, the vertical displacement (Δy), lateral displacement (Δx), angle with respect to axis *X* ($\Delta\theta$) and angle with respect to axis *Y* ($\Delta\phi$) of the beam element can be expressed by Eqs. (1)–(4), respectively. Axial displacement (Δz) and angle with respect to axis *Z* ($\Delta\psi$) of the track beam are expressed by Eqs. (5) and (6), respectively.

$$\Delta y = a_0 + a_1\xi + a_2\xi^2 + a_3\xi^3 + a_4\xi^4 + a_5\xi^5 \tag{1}$$

$$\Delta x = b_0 + b_1\xi + b_2\xi^2 + b_3\xi^3 + b_4\xi^4 + b_5\xi^5 \tag{2}$$

$$\Delta\theta = c_0 + c_1\xi + c_2\xi^2 + c_3\xi^3 + c_4\xi^4 + c_5\xi^5 \tag{3}$$

$$\Delta\phi = d_0 + d_1\xi + d_2\xi^2 + d_3\xi^3 + d_4\xi^4 + d_5\xi^5 \tag{4}$$

Fig. 1 Track beam used in monorail transportation in Chongqing

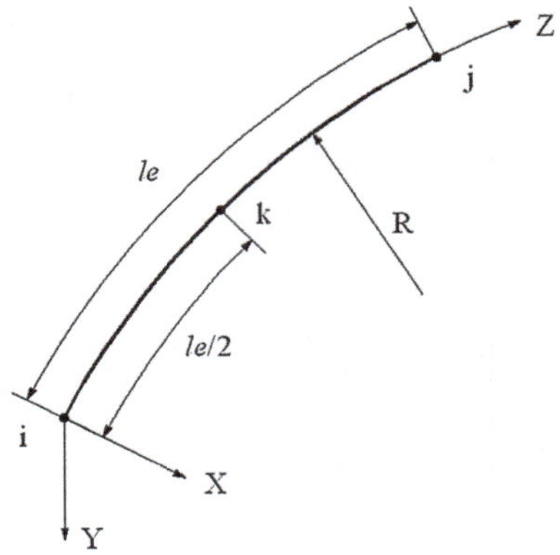

Fig. 2 FEM model of track beam

$$\Delta z = e_0 + e_1\xi + e_2\xi^2 + e_3\xi^3 \tag{5}$$

$$\Delta\psi = f_0 + f_1\xi + f_2\xi^2 + f_3\xi^3 \tag{6}$$

In Eqs. (1)–(6), $a_0 \sim a_5$, $b_0 \sim b_5$, $c_0 \sim c_5$, $d_0 \sim d_5$, $e_0 \sim e_3$ and $f_0 \sim f_3$ are coefficients, and ξ can be represents as:

$$\xi = \frac{2z}{l_e} \tag{7}$$

where l_e is the length of the track beam element.

The mass matrix can be expressed by Eq. (8) based on the energy method.

$$M = \int_{\Omega} \rho N^T N \mathrm{d}\Omega \tag{8}$$

In Eq. (8), ρ denotes the linear density of the beam; *N* denotes the shape function of the element which is derived from Eqs. (1) to (6).

According to the principle of virtual displacement, the expression of the element stiffness matrix can be expressed by Eq. (9) where [*B*] denotes the matrix of strain, [*D*] indicates the matrix of beam modulus.

$$[K] = \int [B]^T[D][B]\mathrm{d}z = \frac{l_e}{2}\int_{-1}^{1} [B]^T[D][B]\mathrm{d}\xi \tag{9}$$

The damping matrix of the track beam is given based on Rayleigh damping principle. It is expressed by Eq. (10) where a_0 and a_1 are the constants expressed by Eqs. (11) and (12), respectively.

$$[C] = a_0[M] + a_1[K] \tag{10}$$

$$a_0 = \frac{2\omega_m\omega_n(\omega_n\xi_m - \omega_m\xi_n)}{\omega_n^2 - \omega_m^2} \tag{11}$$

$$a_1 = \frac{2\omega_m\omega_n\left(-\frac{\xi_m}{\omega_n} + \frac{\xi_n}{\omega_m}\right)}{\omega_n^2 - \omega_m^2} \tag{12}$$

In Eqs. (11) and (12), ω_m and ω_n are the first-order and the second-order frequencies, respectively; ξ_m and ξ_n are the first-order and the second-order damping ratios of the track beam, respectively.

The governing vibration equations of the monorail track beam can be expressed by Eq. (13), where M indicates the mass matrix expressed by Eq. (8); C indicates the damping matrix expressed by Eq. (10); K indicates the stiffness matrix of the track beam expressed by Eq. (9); w is the deformation vector of the track beam; f_b is the external force on the track beam caused by the moving straddle-type monorail vehicle; $(\cdot)$ indicates the derivative with respect to time.

$$M\ddot{w} + C\dot{w} + Kw = f_b \tag{13}$$

3 Monorail Vehicle System

The straddle-type monorail train is different from all the railway and highway vehicles, as it has a special bogie which consists of four steering wheels, four driving wheels and two stabilizing wheels, as shown in Fig. 3. Based on these characteristics, the running mechanism of straddle-type monorail train is also different from any other vehicles on the railway and roads. In order to analyze the dynamic responses of the monorail vehicle, each straddle-type monorail vehicle is simplified as a system which consists of three rigid bodies, i.e., vehicle body, the front bogie and the rear bogie, as shown in Fig. 3, according to the principle of multi-rigid-body dynamics. In this multi-rigid-body system, the rigid bodies are connected by the spring–damping device. The calculation model of the straddle-type monorail train is shown in Fig. 4.

In the system, each rigid body has the six degrees of freedom: axial displacement (Z), vertical displacement (Y), radial displacement (X) and angular displacement about the axis X, axis Y and axis Z (φ, ψ and θ) (see Fig. 4). As each vehicle consists of three rigid bodies, one vehicle calculation model has 18 degrees of freedom. All the parameters of the straddle-type monorail vehicle model are summarized in Table 1.

Based on Lagrange's formulation, the equations of motion of the vehicle can be obtained. The Lagrange's formulation can be expressed by Eq. (14).

$$\frac{\mathrm{d}}{\mathrm{d}t}\left(\frac{\partial T_v}{\partial \dot{q}_k}\right) - \frac{\partial T_v}{\partial q_k} + \frac{\partial U_v^e}{\partial q_k} + \frac{\partial U_v^q}{\partial \dot{q}_k} = 0 \tag{14}$$

In Eq. (14) T_v indicates the kinetic energy which is expressed by Eq. (15); U_v^e indicates the elastic potential energy which is expressed by Eq. (16); U_v^q indicates the damping potential energy which is expressed by Eq. (17); q_k denotes the generalized coordinates.

$$\begin{aligned} T_v = \frac{1}{2}\Bigg\{ & \sum_{i=1}^{nv}\left[M_{ci}(\dot{X}_{ci}^2 + \dot{Y}_{ci}^2 + \dot{Z}_{ci}^2) + J_{c\theta i}\dot{\theta}_{ci}^2 + J_{c\psi i}\dot{\psi}_{ci}^2 + J_{c\phi i}\dot{\phi}_{ci}^2\right] \\ & + \sum_{j=1}^{2}\left[M_{tij}\left(\dot{X}_{tij}^2 + \dot{Y}_{tij}^2 + \dot{Z}_{tij}^2\right) + J_{t\theta ij}\dot{\theta}_{tij}^2 + J_{t\psi ji}\dot{\psi}_{tij}^2 + J_{t\phi ij}\dot{\phi}_{tij}^2\right]\Bigg\} \end{aligned} \tag{15}$$

In Eq. (15), nv is the number of vehicle which is running on the track beam; notation "·" denotes the derivative with respect to time; all the other parameters are obtained from Table 1.

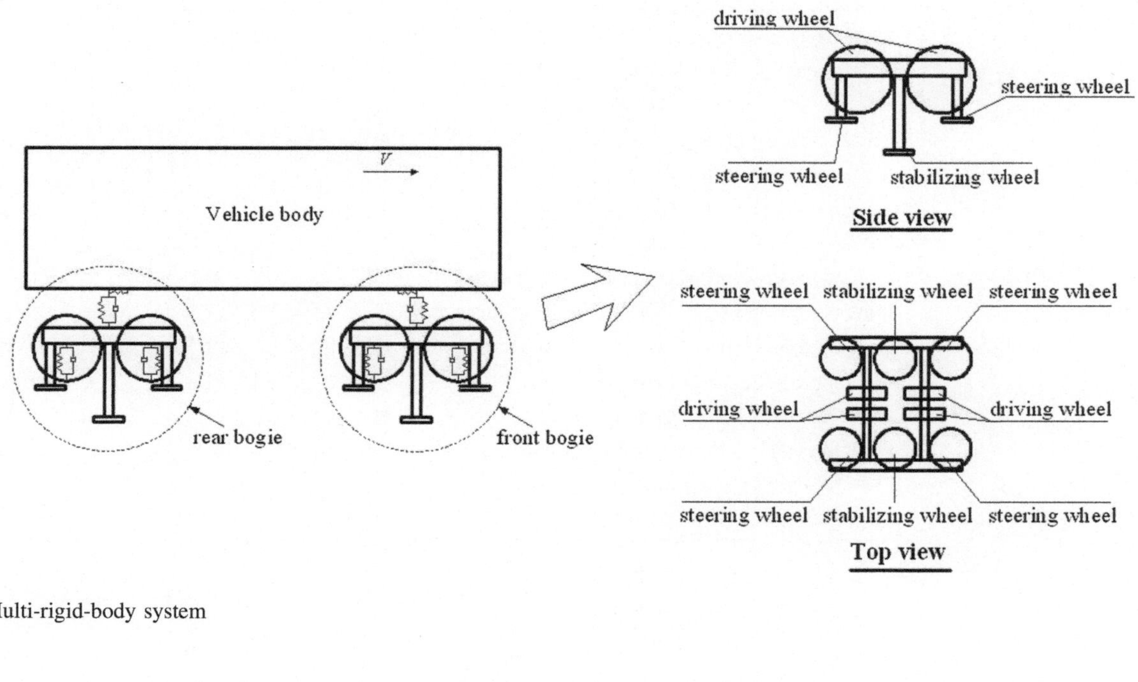

Fig. 3 Multi-rigid-body system

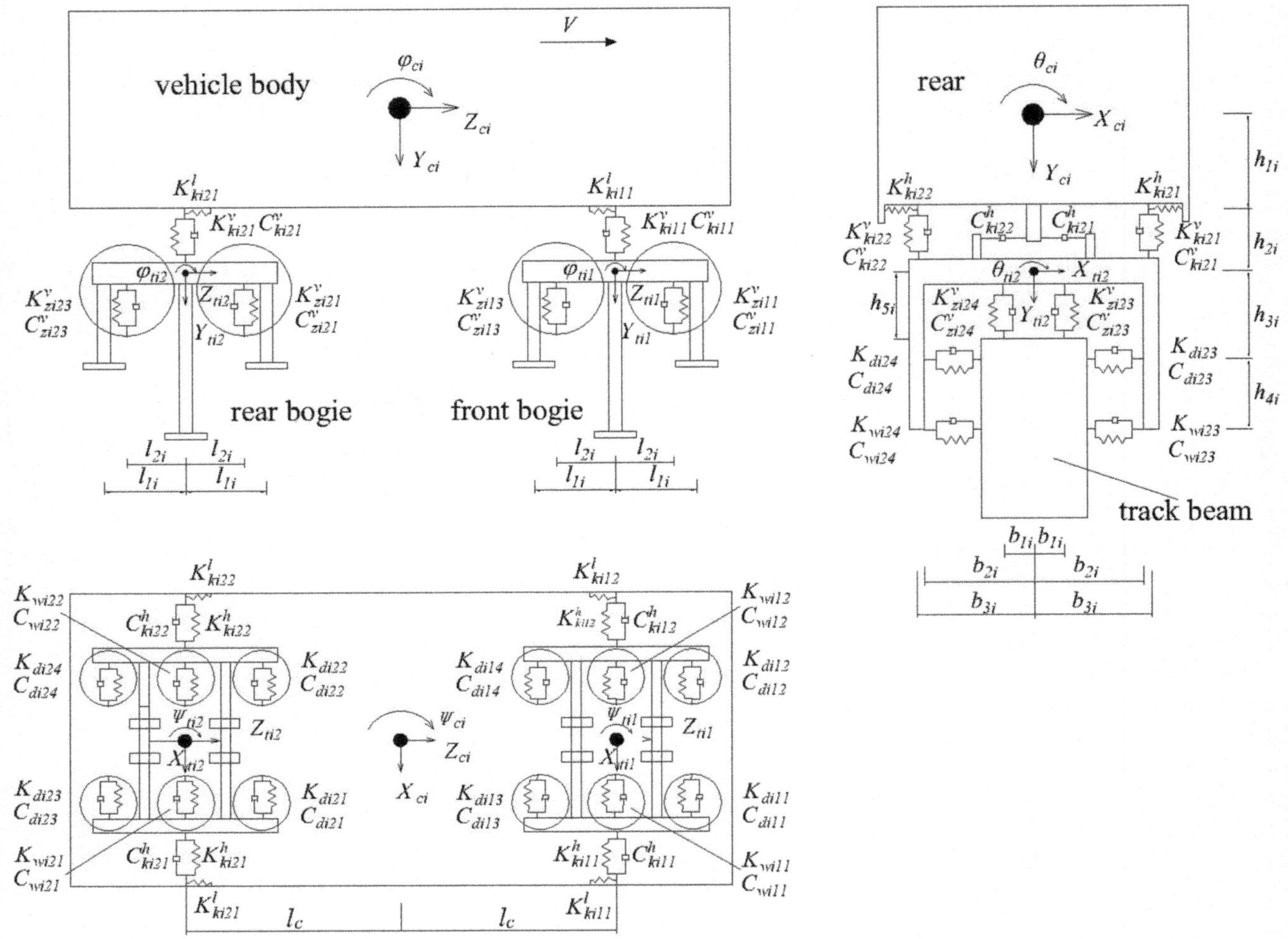

Fig. 4 Diagram of monorail vehicle simplified dynamics model

Table 1 Notations of monorail vehicle dynamics model

Descriptions	Notations
Mass (body and bogie)	M_c, M_t
Spring constant of vertical air suspension	K^v_{ki11}, K^v_{ki12}, K^v_{ki21}, K^v_{ki22}
Spring constant of longitudinal air suspension	K^l_{ki11}, K^l_{ki12}, K^l_{ki21}, K^l_{ki22}
Spring constant of lateral air suspension	K^h_{ki11}, K^h_{ki12}, K^h_{ki21}, K^h_{ki22}
Spring constant of driving wheel	K^v_{zi11}, K^v_{zi12}, K^v_{zi13}, K^v_{zi14}, K^v_{zi21}, K^v_{zi22}, K^v_{zi23}, K_{di24}
Spring constant of steering wheel	K_{di11}, K_{di12}, K_{di13}, K_{di14}, K_{di21}, K_{di22}, K_{di23}, K_{di24}
Spring constant of stabilizing wheel	K_{wi11}, K_{wi12}, K_{wi21}, K_{wi22}
Damping coefficient of vertical air suspension	C^v_{ki11}, C^v_{ki12}, C^v_{ki21}, C^v_{ki22}
Damping coefficient of longitudinal air suspension	C^l_{ki11}, C^l_{ki12}, C^l_{ki21}, C^l_{ki22}
Damping coefficient of lateral air suspension	C^h_{ki11}, C^h_{ki12}, C^h_{ki21}, C^h_{ki22}
Damping coefficient of driving wheel	C^v_{zi11}, C^v_{zi12}, C^v_{zi13}, C^v_{zi14}, C^v_{zi21}, C^v_{zi22}, C^v_{zi23}, C^v_{zi24}
Damping coefficient of steering wheel	C_{di11}, C_{di12}, C_{di13}, C_{di14}, C_{di21}, C_{di22}, C_{di23}, C_{di24}
Damping coefficient of stabilizing wheel	C_{wi11}, C_{wi12}, C_{wi21}, C_{wi22}
Mass moments of inertia (vehicle)	$J_{c\theta i}$, $J_{c\psi i}$, $J_{c\phi i}$
Mass moments of inertia (bogie)	$J_{t\theta ij}$, $J_{t\psi ij}$, $J_{t\phi ij}$

Fig. 5 Monorail track beam (unit: mm)

The elastic potential energy and damping potential energy of the straddle-type monorail vehicle can be expressed by Eqs. (16) and (17), respectively.

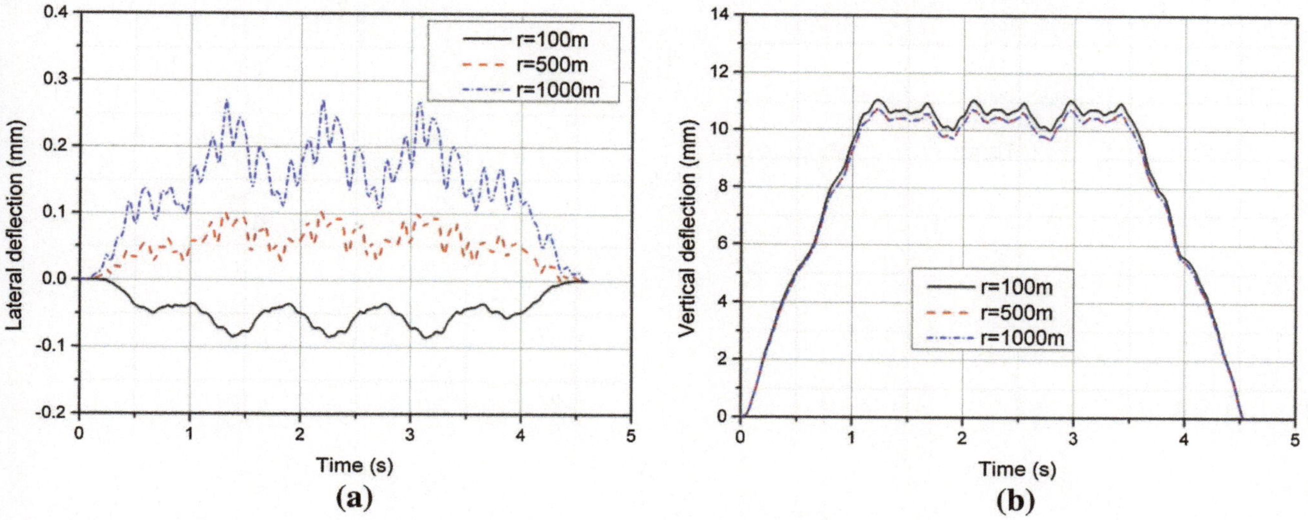

Fig. 6 Record of vibration displacement at middle span (unit: mm). **a** Lateral, **b** vertical

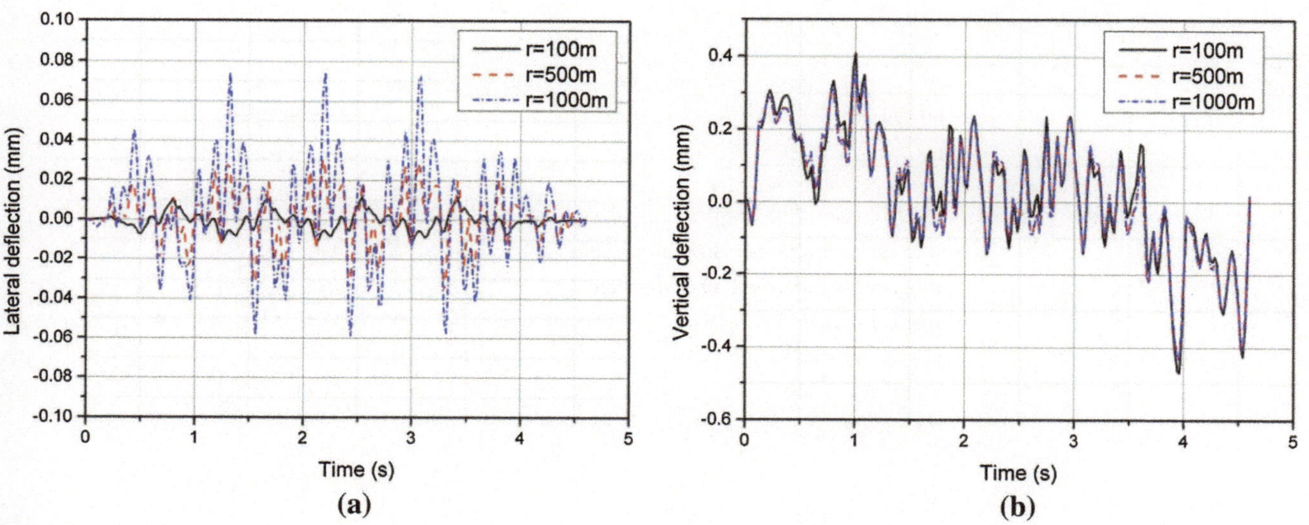

Fig. 7 Record of vibration amplitude at middle span (unit: mm). **a** Lateral, **b** vertical

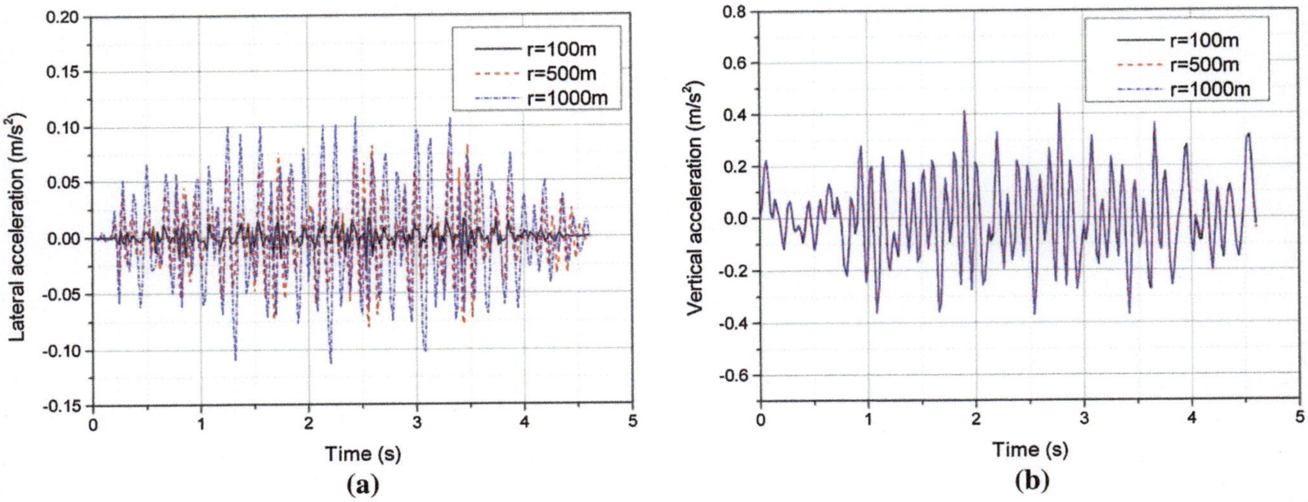

Fig. 8 Observations of acceleration at middle span (unit: m/s^2). **a** Lateral, **b** vertical

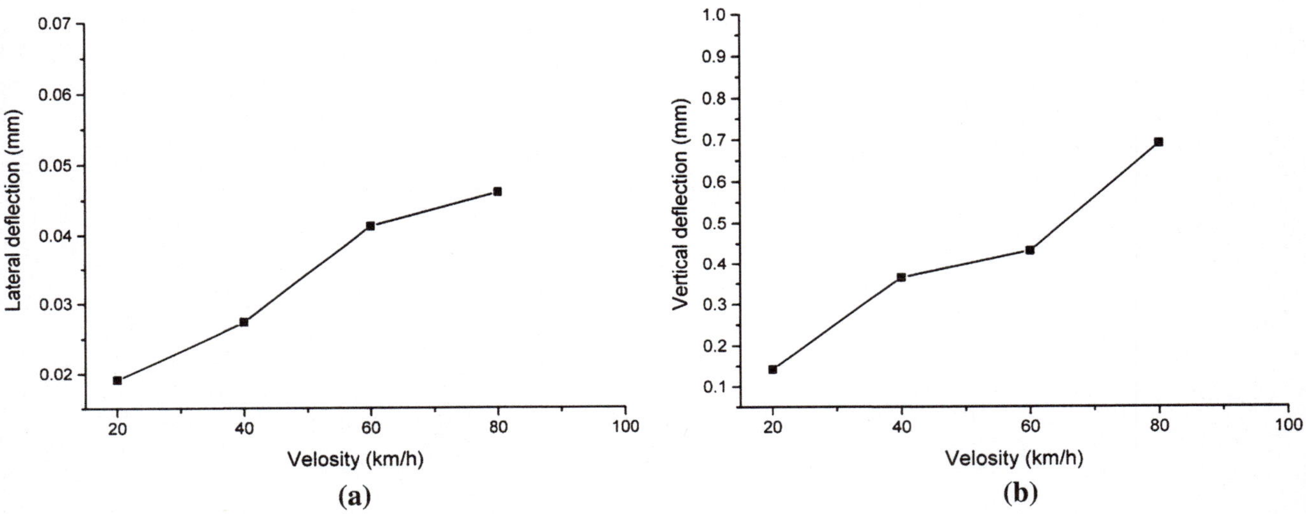

Fig. 9 Maximum of amplitude at middle span (unit: mm). **a** Lateral, **b** vertical

$$U_v^e = \frac{1}{2}\left\{\sum_{i=1}^{nv}\sum_{j=1}^{2}\sum_{m=1}^{2}\left(K_{kijm}^{l}{\Delta_{kijm}^{l}}^{2} + K_{kijm}^{h}{\Delta_{kijm}^{h}}^{2} + K_{kijm}^{v}{\Delta_{kijm}^{v}}^{2}\right) + \sum_{i=1}^{nv}\sum_{j=1}^{2}\sum_{m=1}^{4}\left(K_{zijm}^{v}\Delta_{zijm}^{2} + K_{dijm}\Delta_{dijm}^{2}\right) + \sum_{i=1}^{nv}\sum_{j=1}^{2}\sum_{m=1}^{2}K_{wijm}\Delta_{wijm}^{2}\right\} \quad (16)$$

$$U_v^q = \frac{1}{2}\left\{\sum_{i=1}^{nv}\sum_{j=1}^{2}\sum_{m=1}^{2}\left(C_{kijm}^{l}{\dot{\Delta}_{kijm}^{l}}^{2} + C_{kijm}^{h}{\dot{\Delta}_{kijm}^{h}}^{2} + C_{kijm}^{v}{\dot{\Delta}_{kijm}^{v}}^{2}\right) + \sum_{i=1}^{nv}\sum_{j=1}^{2}\sum_{m=1}^{4}\left(C_{zijm}^{v}{\dot{\Delta}_{zijm}^{v}}^{2} + C_{dijm}{\dot{\Delta}_{dijm}}^{2}\right) + \sum_{i=1}^{nv}\sum_{j=1}^{2}\sum_{m=1}^{2}C_{wijm}{\dot{\Delta}_{wijm}}^{2}\right\} \quad (17)$$

In Eqs. (16) and (17), nv is the number of the vehicles which is running on the beams; subscripts(j and m) are the indexes of the position (shown in Fig. 4). The parameters (Δ_{kijm}^{l}, Δ_{kijm}^{h} and Δ_{kijm}^{v}) indicate deformation of the air suspension in longitudinal, lateral and vertical direction, respectively. The parameters (Δ_{zijm}, Δ_{dijm} and Δ_{wijm}) denote

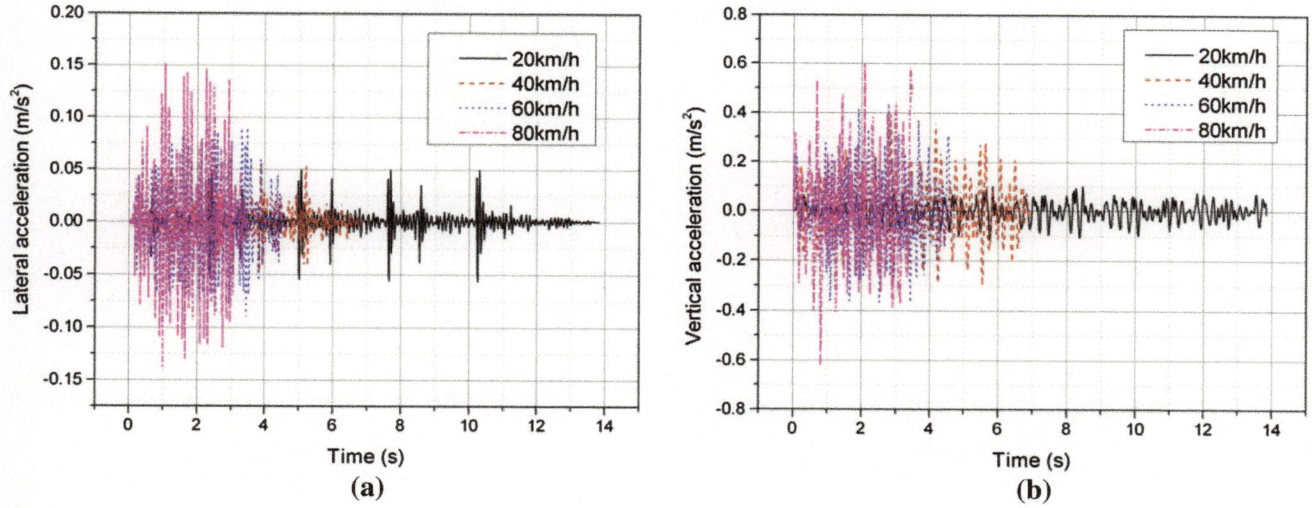

Fig. 10 History of acceleration at middle span (unit: m/s^2). **a** Lateral, **b** vertical

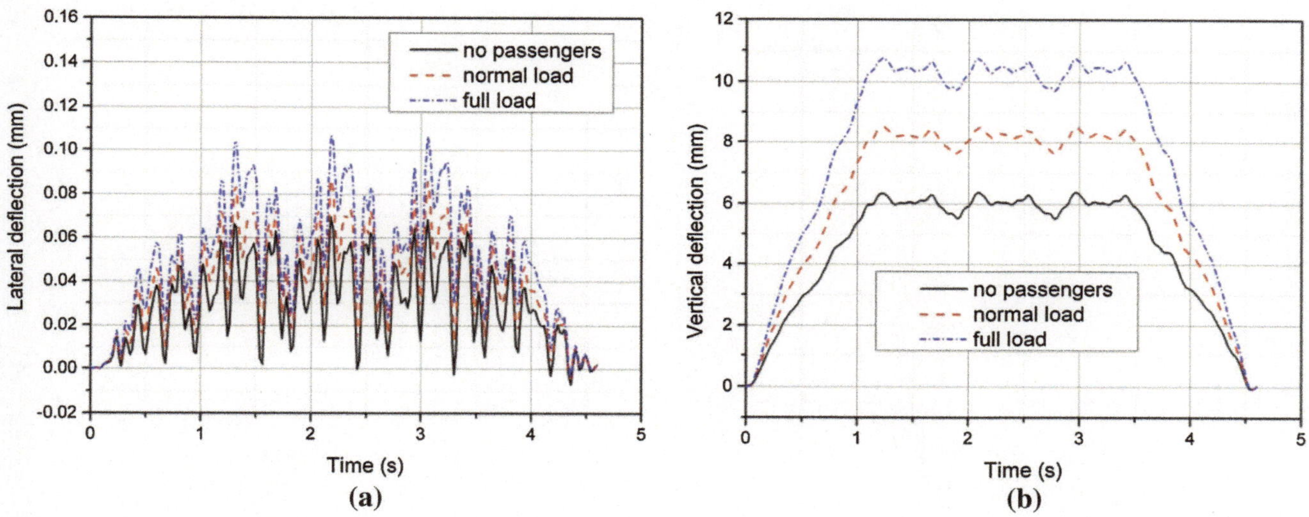

Fig. 11 Record of vibration displacement at middle span (unit: mm). **a** Lateral, **b** vertical

the displacement of the track beam at the position which contacts with the driving wheel, steering wheel and stabilizing wheel, respectively.

4 Case Study

The bridge–vehicle coupling system is a complex time-varying system which makes it impossible to obtain the solutions of motion equations by theoretical calculation. In order to obtain the solutions, numerical simulation is an effective way for the analysis of the bridge–vehicle coupling system. A numerical simulating software was developed on VB programming platform, based on the energy method and the FEM. Then a simply supported monorail track beam in Chongqing monorail transportation line is taken to serve as the dynamic response simulation example (shown in Fig. 5), and the influence of the curve radius, driving speed and vehicle load on the dynamic response of the straddle monorail track beam is studied by using the self-developed dynamics simulation software.

4.1 Influence of Radius of Curvature on Dynamic Responses of Track Beam

The time span in simulation is between the in and out locomotions of the vehicle on the track beam. Through the calculation, the record of vibration displacement and vibration amplitude of the mid-span was obtained, as shown in Figs. 6 and 7. When the track beam has different radius of the curvature, the vibration displacement and the amplitude

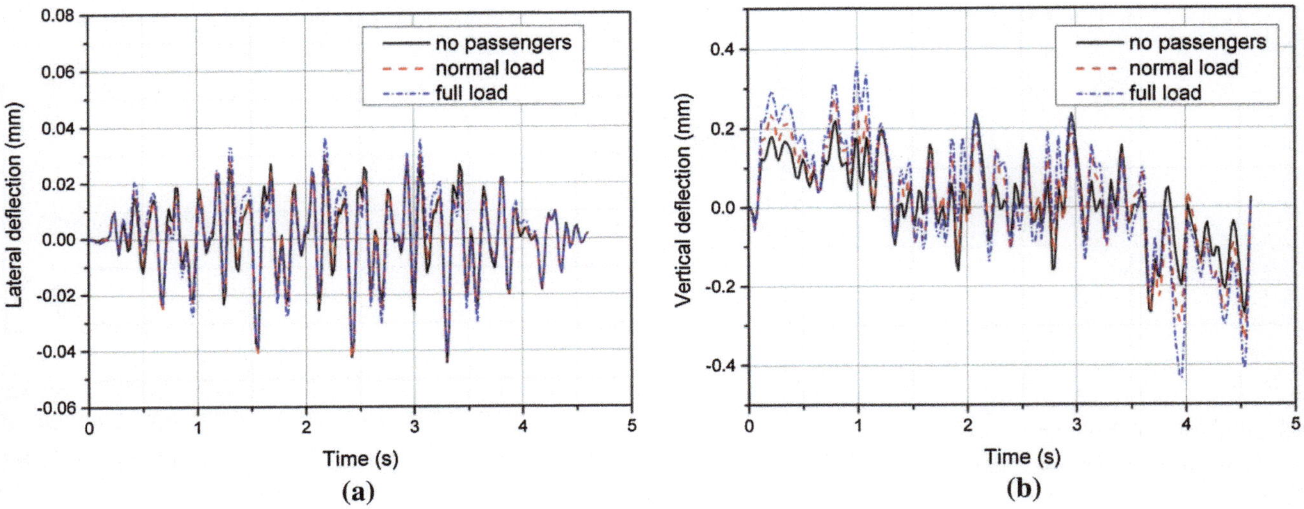

Fig. 12 Record of vibration amplitude at middle span (unit: mm). **a** Lateral, **b** vertical

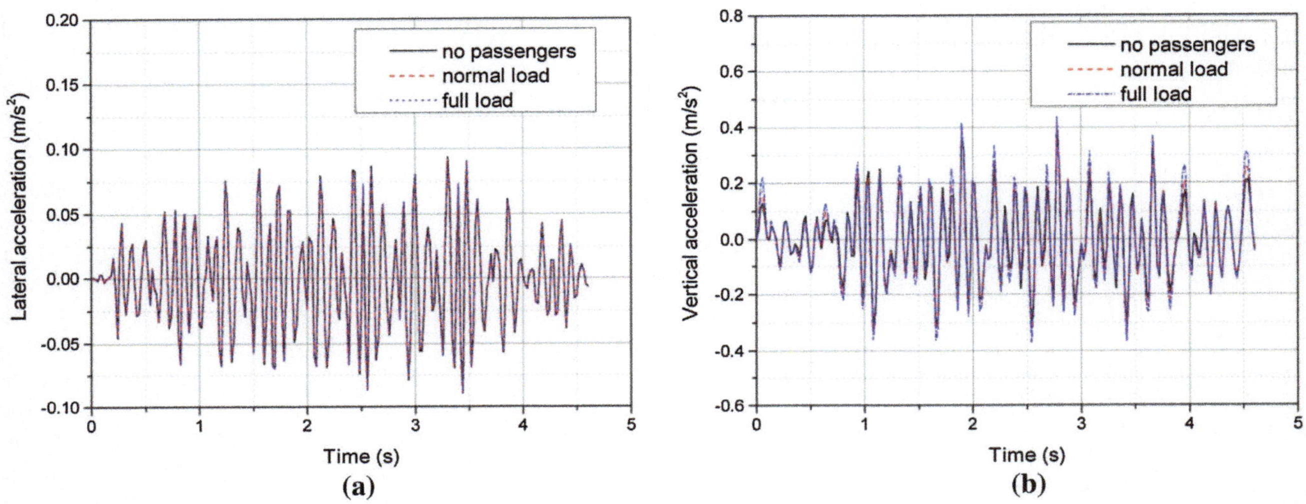

Fig. 13 Record of acceleration at middle span (unit: m/s^2). **a** Lateral, **b** vertical

in the lateral direction are different from those in the vertical direction. It is shown in Figs. 6 and 7 that the vertical vibration displacement and the vibration amplitude of the beam decrease with the increase in the radius of curvature, and the lateral vibration displacement and amplitude increase with the increase in the radius of the curvature. The maximum lateral vibration amplitude is 0.075 mm when the radius of the curvature is fixed at 1000 m, while the maximum vertical vibration amplitude is 0.43 mm when the radius of the curvature is fixed at 100 m.

The observations of acceleration at the middle span with respect to radius of curves is shown in Fig. 8. The valuable information is that the lateral acceleration increases with the increase in radius, and the maximum lateral acceleration is 0.11 m/s^2 in accordance with a radius of 1000 m. But the radius does not influence the vertical acceleration significantly. No significant difference is shown between the curves of acceleration according to different radius is shown in Fig. 8.

4.2 Influence of Vehicle Speed on Dynamic Responses of Track Beam

Vehicle speed is another important factor that influences dynamic responses of the track beam. Through the calculations, dynamic responses of the track beam according to different speeds (20, 40, 60 and 80 km/h) are obtained. It is shown in Fig. 9 that the trends of the maximum amplitude in lateral and vertical direction are similar to each other and the maximum deflection in lateral and vertical direction increases as the driving speed increases. The maximum vertical deflection is 0.69 mm at the speed of 80 km/h, while the maximum lateral amplitude is 0.046 mm at the speed of 80 km/h.

The record of acceleration at the middle span with respect to different speeds (20, 40, 60 and 80 km/h) is shown in Fig. 10. The valuable information is that the lateral and vertical acceleration increases as the speed increases, and the maximum values are 0.15 and 0.62 m/s^2, respectively. The rules investigated in this chapter denote that speed is a very important factor which affects dynamic responses of the track beam; how to control the speed of the straddle-type monorail train will be an important task for the managers.

4.3 Influence of Vehicle Load on Dynamic Responses of Track Beam

As the weights of straddle-type monorail vehicle and track beam are approximately equal to each other, the influence of vehicle load on dynamic responses of track beam should be studied in depth. Figures 11 and 12 show the variation displacement and the vibration amplitude of monorail track beam in three load cases: passenger-free, normal load and full load. The results show that the vibration amplitude and the displacement in lateral and vertical direction increase with the increasing load. When the vehicle is full load, the maximum vibration amplitudes in lateral direction and vertical direction are 0.041 and 0.43 mm, respectively. This calculation denotes that the vehicle load can significantly affect the dynamic response of the track beam.

The record of acceleration at the middle span with respect to different load conditions (no passengers, normal load and full load) is shown in Fig. 13. The valuable information is that the vertical acceleration increases with the increasing load and the maximum value is 0.44 m/s^2. But the lateral acceleration is not sensitive to the variation of load as shown in Fig. 13.

The calculations denote that all the factors (curve radius, driving speed and vehicle load) have important effect on the dynamic responses of the straddle-type monorail track beam under a moving train. These rules with respect to different conditions should be very useful in the procedures of the design or operation of the straddle-type transportation system.

5 Conclusions

A theoretical and mathematical bridge–vehicle interaction dynamics model was set up and analyzed first, and then based on the energy principle and FEM, a bridge–vehicle dynamics numerical simulation software was developed. A time span between the in and out locomotions of the vehicle on the track beam is calculated with the self-developed FEM software. Through the dynamic response analysis of the straddle-type monorail vehicle–bridge coupling system, some valuable conclusions can be obtained.

1. The vertical vibration amplitude of the track beam decreases with the increasing radius of curvature, while the lateral amplitude increases with the increasing radius of the curvature. The maximum lateral vibration amplitude is 0.075 mm with respect to the radius of 1000 m, while the maximum vertical vibration amplitude is 0.43 mm with respect to the radius of 100 m; the lateral acceleration increases with the increasing radius, and the maximum lateral acceleration is 0.11 m/s^2 in accordance with a radius of 1000 m. But the vertical acceleration is not easily affected by the variation of radius.
2. The maximum amplitude in the lateral and vertical direction increases as the driving speed increases, and the maximum vertical deflection is 0.69 mm, while the maximum lateral amplitude is 0.046 mm, at the speed of 80 km/h; the lateral and vertical acceleration increases as the speed increases, and the maximum values are 0.15 and 0.62 m/s^2, respectively.
3. The vibration amplitude (lateral and vertical) increases with the increase in the load, and the maximum vibration amplitudes in lateral and vertical direction are 0.041 and 0.43 mm, respectively; the vertical acceleration increases with the increase in load, and the maximum value is 0.44 m/s^2. But the lateral acceleration is not easily affected by the variation of load.
4. The proposed method for the analysis of straddle-type monorail vehicle–bridge coupling system in this paper can be used to simulate the dynamic responses of the bridge–vehicle interaction system, and the self-developed calculation software is efficient in the analysis of dynamic responses of the complex bridge–train coupling system.

Acknowledgements This study is sponsored by the China Railway Corporation Technology Research and Development Plan (No. 2016G002-L) and Beijing Mass Transit Railway Operation Corporation Research Program (No. 2016000501000005).

References

1. Lee CH, Kawatani M (2006) Dynamic response of a monorail steel bridge under a moving train. J Sound Vib 294:562–579
2. Lee CH, Kim CW et al (2005) Dynamic response analysis of monorail bridges under moving trains and riding comfort of trains. Eng Struct 27:1999–2013
3. Kim CW, Kawatani M (2006) Effect of train dynamics on seismic response of steel monorail bridges under moderate ground motion. Earthq Eng Struct Dyn 35:1225–1245
4. Naeimi M, Tatari M et al (2015) Dynamic interaction of the

monorail-bridge system using a combined finite element multi-body-based model. J Multi-Body Dyn 229(2):132–151
5. Trahair NS (2008) Lateral buckling of monorail beams. Eng Struct 30:3213–3218
6. Kim CW, Kawatani M et al (2013) Seismic behavior of steel monorail bridges under train load during strong earthquakes. J Earthq Tsunami 7(2):1350006
7. Ren L-H, Zhou J-S, Shen G (2004) Dynamics model and simulation study of a straddle type monorail car. China Railw Sci 25(5):26–32
8. Ren L-H, Zhou J-S, Shen G (2003) Hunting stability of straddle-type monorail car based on equation eigenvalue. J Tongji Univ 31(4):469–472
9. Zhao X-L (2008) Simulation research on the dynamic of straddle-type monorail vehicle. Dissertation, Beijing Jiaotong University
10. Du Z-X, Wang X-C (2009) Research on stabilily simulation for straddle-type monorail vehicle. Railw Locomot Car 29(5):56–59
11. Liu Y-Y (2011) Research on dynamic interaction of straddle type monorail vehicle and track beam. Dissertation, Southwest Jiaotong University
12. Liu Y-Y, Ge Y-M, Yang Y-R (2010) The dynamic response analysis of the coupled system of the straddle type monorail train and the track beam. China Railw Sci 31(5):21–26
13. Ma J-B (2008) Research on structural static and dynamic behaviors of straddle-type monorail transportation system. Dissertation, Southwest Jiaotong University
14. Yang X-W, Shan D-S, Li Q (2004) Dynamic property analysis of Yuanjiagang station bridge on Chongqing light rail. China Railw Sci 25(6):77–81

9

Investigating Intercity Rail Transit Scope with Social Economy Accessibility: Case Study of the Pearl River Delta Region in China

Xiaoshu Cao[1] · Linna Li[2] · Heng Wei[3,4,5]

Abstract In response to challenges caused by high gasoline prices, traffic congestion, and greenhouse gas emissions, smart internment on intercity rail transit infrastructures and service suggests a rekindling of many countries' interest in offering a range of benefits over automobile travels. In order to check the suitability of the proposed intercity rail transit system with the local conditions in a region, models for relating the intercity rail transit scope and social economy factors have been developed with the datasets obtained in the Pearl River Delta (PRD) region in Guangdong Province, China. In this paper, the accessibility-based approach is presented to explore the impact of the intercity rail transit system planning on the regional development. The impact of three typical accessibility variables, transportation, population, and economic accessibilities, are considered in the approach. The global rail transit scope planning models are developed by using the regression technique to correlate the lengths of 254 regional intercity rail transit systems in different countries in the world with social economy factors. Those models are used for estimating the intercity rail transit size for the PRD region. The modeled transit scopes are examined with spatial distributions of the defined three accessibilities at each node (or centroid center of a town) in a Geographic Information System environment. The developed method has been proven helpful to understanding the gap between transport supply and potential travel demand and the suitability of each node to alignment of each rail transit route through the PRD region case study.

Keywords Intercity rail transit · Population accessibility · Economy accessibility · Transportation accessibility · Social economic factors

✉ Heng Wei
heng.wei@uc.edu

Xiaoshu Cao
caoxsh@mail.sysu.edu.cn

Linna Li
linnahku@163.com

1 School of Geography and Planning, Sun Yat-Sen University, No. 135, Xingang West Road, Gangzhou 510275, Gangdong Province, People's Republic of China

2 Department of Geography, Beijing Normal University, Room 9719, Jingshi Building, No. 19, Xin Jie Kou Wai Street, Hai Dian District, 100875 Beijing, China

3 Beijing University of Technology, Beijing, China

4 Chang'an University, Xian, China

5 Department of Civil & Architectural Engineering & Construction Management, The University of Cincinnati, 792 Rhodes Hall, Cincinnati, OH 45221-0071, USA

Editors: Haishan Xia and Chun Zhang

1 Introduction

To respond to challenges caused by high gasoline prices, traffic congestion, and greenhouse gas emissions, smart investment on intercity rail transit infrastructures and service suggests a rekindling of many countries' (such as China and USA) interest in offering a range of benefits over automobile travels [1]. Previous studies suggested that the demand for increasing intercity passenger rail transit lies in the region where population density and corridor congestion make intercity rail service a competitive mode with the automobile services and regional air service [2]. However, the expansion of intercity rail transit systems will bring forth more equal regional accessibility but may also incur more uneven development issue [3]. Past experiences in high-speed railways indicate increased imbalances between major cities and their hinterlands [4–6, 3]. A specific example is cities connected through Shinkansen high-speed rail (in Japan) immediately gained location advantages, while non-Shinkansen cities were marginalized [6, 3]. Therefore, it is necessary to investigate the suitable intercity rail transit lengths or scope over a concerned region in the planning stage in order to make investment plans more substantially optimal. The core research question that this study addresses is how to investigate suitable scope of regional intercity rail transit based on social economy factors, and furthermore how the planned intercity rail transit scope may impact the social economy factors.

Besides traditional measures of social economy factors by population and GDP, this study also defines some social economy accessibility variables. According to Geurs and Van Wee [7], accessibility measures adopted in existing research can be classified into four categories: infrastructure-based measures, analyzing the performance or service level of transport infrastructure; location-based measures, analyzing accessibility at locations, typically on macro level; person-based measures, analyzing accessibility at the level of individual; utility-based measures, analyzing the welfare benefits the people derive from access to the spatially distributed activities. It implies that at the regional level, a comprehensive view of the accessibility is needed to well reflect the natures of land use, and demographic and social economic factors. Three categories of factors dictating the demand for intercity passenger rail are usually referred to: population factors (e.g., population growth and densification, increasing urbanization), economic factors (e.g., rising personal transportation costs, economic growth index such as gross domestic product—GDP), and transportation service factors (e.g., congested highways for automobile travels) [2]. These literature findings provide valuable insight to explore the method for estimating the intercity rail transit scope by taking into account the above three categories of impacting factors. In the study as presented in the paper, transportation, population, and economy accessibilities are defined. Transportation accessibility is mathematically expressed as a function of transportation network density and travel time index. Population accessibility is related to population density distribution, and economy accessibility is regarding the local GDP of a concerned area. Thus, this research has combined the infrastructure-based measures and location-based measures to reflect comprehensive accessibility at regional level, among which the transportation accessibility belongs to infrastructure-based measure, while the population and economy accessibilities belong to location-based accessibility.

This study chooses the Pearl River Delta (PRD) region in Guangdong Province, China, as the study case, which is used to test the proposed method. In this PRD case study, three "accessibility" variables, including transportation, population, and economy accessibilities, are defined as impact factors as investigating the intercity rail transit scope (measured by length). The relationship between the intercity rail transit length and the above three categories of factors is firstly estimated based on a set of models that are developed by using the data of 254 intercity rail transit systems in different countries over the world. On the top of it, the accessibility-based approach for spatial suitability analysis of the intercity rail transit planning is developed and tested with data of the PRD region as of 2012. The models involved in the accessibility-based approach are embedded into the ArcGIS system to visualize the result. To make the investment maximally beneficial to the regional transportation and associated economic development, the result from the study is anticipated to provide a valuable reference to other regions in China and even other countries with similar needs.

2 Related Research

Although the impact of intercity rail transit on the regional development has attracted much attention from the transportation studies [4–6, 3], there is little direct research on quantitative investigation about the suitable scope of rail transit, especially at regional level. A few studies have analyzed the social economy conditions for building urban rail systems. For instance, Babalik-Sutcliffe [8] has analyzed eight new urban rail systems in the USA, UK, and Canada by qualitative case studies and identified some important factors for the success of these systems, including urban factors, planning factors, operating policies, urban planning policies, and external factors, among which local economic conditions and population and land use density are factors that need to be considered before the planning stage. Cao and Lin [9] have built a regression model to explore the suitable metro length for cities with different social economy factors, based on data from 121 cities around the world. Loo and Cheng [10] have argued that the cities with a population size of 5 million or above and a GDP of US$11,400 (2008 dollars) per capita are suitable for building metro systems, based on data of 60 metro lines in 21 cities on different continents. However, the discussion about the relationship between suitable rail transit scope and regional and urban social economy condition is still limited. Thus, this study tries to fill the research gap by investigating the suitable intercity rail transit based on regional social economy factors and quantitative analysis.

Traditional studies generally use population, GDP, and land use to measure the social economy factors, while accessibility provides a new measure for regional social economy factors. In geographical transportation field, the accessibility is usually referred to the ease of reaching desired roadway infrastructures or destinations in a transportation network [11, 12]. In a Geographic Information System (GIS) environment, accessibility could be conceptually interpreted in reflecting spatial interactions between transportation services and land use setting as well as facilitation to use the transportation systems in meeting people's social economic needs. Many previous studies have associated the accessibility with potential opportunities of spatial interactions between the transportation activities and land use and social economic factors [13].

The relationship between the accessibility and land use is widely recognized in planning for both urban and rural environments [14–19]. Other social economic factors are also tied with the accessibility, such as poverty, health, and production patterns [20]. Jalan and Ravallion [21] 's study indicates that rural roads are a key factor in explaining why some rural households are able to increase their consumption levels than others. Bryceson et al. [22] interpret the complex and conditional linkages between rural road networks and poverty reduction in terms of mobility enhancement, such as enhanced possibilities to travel for work or services, and accessibility enhancement such as increased attractiveness of rural areas for service infrastructure and staffing.

Hanson [16] argued that the accessibility constitutes of two main parts: proximity and mobility. And the accessibility was estimated as travel time, which depends upon both proximity and mobility. In some studies, the accessibility was modeled by simple measures, such as distances from roads, towns, and other destinations [15, 18]. In other studies, the accessibility was also examined by comparing those variables in different villages with an interview-based approach [23]. Typically, Euclidian distance, measuring the distance to roads, was often used as a simple measurement of the accessibility [15]. While no standards or mature quantitative accessibility models have been found from the literature review, the literature inspired the authors to explore an innovative approach to redevelop a set of regional accessibilities that can be measured and analyzed in a GIS environment.

3 Research Methodology

3.1 Study Case

The PRD region includes nine cities of Guangdong Province in China and is one of the most densely populated regions in China. It has experienced the most rapid urban sprawl over the past decades. In 2008, China announced plans to mesh major cities within the PRD regions, namely Guangzhou, Shenzhen, Dongguan, Zhaoqing, Foshan, Huizhou, Jiangmen, Zhongshan, and Zhuhai, into a single megacity. A series of massive infrastructure projects have been performed to merge transport, energy, water, and telecoms networks across these nine cities. A relentless development has been taken place since then. The World Bank recently named the PRD region as the biggest urban area in the world in terms of population and geographical size [24]. The PRD is hence a good exemplary region to be used as a case study.

3.2 Methodology

First, in order to check the suitability of the proposed intercity rail transit system with the local conditions in a region (e.g., the PRD region in Guangdong Province, China), we first set up the profile of relationship between the intercity rail transit scope and social economy.

The length of a regional intercity rail transit (L) is used to measure the regional intercity rail transit scope. Regional population (P), land use area (S), and GDP are viewed as impact social economic factors to be correlated with the lengths of the corresponding intercity rail transit scope. The datasets of 254 intercity rail transit systems from different countries are collected, including 129 samples in Europe, 60 samples in North America, 38 samples in Asia, 13 samples in South America, 7 samples in Pacific, and 7 samples in Africa, which cover both the developed and developing world. These samples are used to explore the correlation between L and P, S, and GDP. The data are mainly from statistic reports about world urban areas [25–27]. The "Enter" and "Stepwise" methods are then used to develop the regression models that mathematically relate the intercity rail transit length and population, land use, or GDP. Then, the model is applied the PRD region to estimate the suitable scope of intercity rail transit based on the social economy factors in this region. The population, land use area, and GDP data at the city level, the county level, and the town level are from the statistic yearbooks of the cities and counties of PRD.

Second, three accessibility variables were defined to check the modeled rail transit scopes. The transportation accessibility is redefined by integrating the density of transportation network and distribution of shortest travel times from the centroid center of a town (or node) to all other town nodes. The transportation system in the PRD region includes highway network and intercity rail transit system. The highway network consists of freeway, highway, state highway, and county highway. The railway system includes single route and round-route rail transit lines. The density of the transportation network (R_i) is measured by the following equation.

$$R_i = \sum_{i=1}^{n} \frac{m_i l_i}{S} \quad (1)$$

where m_i = land area per unit length of the ith highway or railway which are located within this area; l_i = length of the ith highway or railway within the concerned area; and S = total area of the concerned town.

The values of m_i are obtained from the Highway and Railway Construction Specification and Standards of Guangdong Province, China. However, values for state highway and county highway are not recommended in the Standards. The m_i values are assumed based on the classifications of the highways as the First Class through the Fourth Class, as indicated in the Standards. Although little bias would possibly exist in this way, they could be neglected from the standpoint of the regional spatial distributions. The m_i values for railways are decided by the specifications for the First Class single route electric power railway.

Then, all the shortest travel times from the centroid center of a town (or node) to all other town nodes are estimated. It is defined as transportation accessibility A_i for a specific town node and estimated by using the following equation.

$$A_i = \sum_{j=1}^{n} T_{ij}/n \quad (2)$$

where T_{ij} is the shortest travel time from node i to node j and n is the total destined nodes relative to node i.

Transportation network density is estimated by the ratio of the right-of-way areas of transportation infrastructures over the total land area within which the above infrastructures are located. The density value ranges from 0 to 1. The larger the value is, the higher the transportation network density is. Since A_i is measured in min, it is necessary to convert its measurement unit into a non-measurement unit like R_i by using the following equation. The converted A_i is denoted as UA_i.

$$\mathrm{UA}_i = \frac{A_i - \mathrm{Min}T_{ij}}{\mathrm{Max}T_{ij} - \mathrm{Min}T_{ij}} \quad (3)$$

where $\mathrm{Min}T_{ij}$ = minimum shorted travel time from node i to j; $\mathrm{Max}T_{ij}$ = maximum shorted travel time from node i to j.

Unified transportation network density (R_i) and transportation accessibility (A_i) are averaged to get an integrative transportation accessibility (TA_i) as estimated by the following equation.

$$\mathrm{TA}_i = (R_i + \mathrm{UA}_i)/2 \quad (4)$$

Population and economy accessibility for each node are also defined and quantified. Both of them are defined by aggregating spatial interaction models based on Newton's law of gravitational force to represent the attraction force between two nodes in direct proportion to the attraction size and in inverse proportion to travel resistance. Those two concepts are quantified by the following two equations.

$$\mathrm{PA}_i = \sum_{j=1}^{n} \frac{P_j}{D_{ij}^{a}} \quad (5)$$

$$\mathrm{EA}_i = \sum_{j=1}^{n} \frac{E_j}{D_{ij}^{a}} \quad (6)$$

where PA_i = population accessibility at node i; EA_i = -economy accessibility at node i; P_j = population size at node j; E_j = economic level at node j; D_{ij} = shortest travel time from node i to node j; and α = resistance coefficient (default is 1).

4 Models of Intercity Rail Transit Scope

4.1 General Rail Transit Scope Models Using Global Datasets

Table 1 shows 16 developed models regarding the rail transit length and associated impact factors as a result of the regression modeling with the 254 sample data. The statistical analysis indicates that the intercity rail transit size is most highly correlated with the regional GDP level. Population is the second highest factor in the correlation. Land use is relatively lower in the correlation. Models (a), (d), (g), (i), (l), and (n) are given by using the "Enter" method and other models are given by using "Stepwise" method. There may be other factors influencing the rail transit size, such as policy, investment, technology, and other natural conditions. To address those factors, rail transit length per people versus population density (people per square kilometers) and rail transit length per people versus GDP density (GDP per people) are analyzed.

4.2 Estimating Rail Transit Lengths for the PRD Region

The models as shown in Table 1 can be used to estimate the compatible size of the intercity rail transit systems with the PRD region. Specifically, the regression models (a), (b), (c), (q), and (r) are used to calculate the intercity rail transit sizes. As a result of comparing different model forms, the exponential model is found best fitting the sample data, as expressed by models (q) and (r) (Table 1). The results suggest that the level of intercity rail transit length per people is longer as the level of the GDP/P is higher. The Intercity rail transit density is lower as the population

Table 1 Regression models for rail transit length and social economic factors

Category (# of samples)	Regression model	Equation denotation	R^2	p value
Overall (254)***	$L = 87.205 + 0.168P + 0.024\text{GDP} + 0.059\text{S}$	(a)	0.214	0.000
	$L = 138.001 + 0.104\text{GDP}$	(b)	0.178	0.000
	$\text{L} = 108.086 + 0.066\text{G} + 0.060\text{S}$	(c)	0.202	0.000
Commuting railway (138)***	$L = 151.139 + 0.389P + 0.050\text{S} + 0.010\text{GDP}$	(d)	0.207	0.000
	$L = 177.651 + 0.514P$	(e)	0.182	0.000
	$L = 151.307 + 0.410P + 0.053\text{S}$	(f)	0.206	0.000
High-speed railway (14)**	$L = 247.052 - 0.452P + 0.189\text{S} + 0.107\text{GDP}$	(g)	0.805	0.001
	$L = 81.285 + 0.172\text{S}$	(h)	0.640	0.001
Light rail commuting (240)***	$L = 67.126 + 0.271P + 0.056\text{S} + 0.007\text{GDP}$	(i)	0.166	0.000
	$L = 94.740 + 0.389P$	(j)	0.135	0.000
	$L = 67.441 + 0.284P + 0.058\text{S}$	(k)	0.166	0.000
High-speed light rail (116)***	$L = -18.858 - 0.114P + 0.047\text{S} + 0.092\text{GDP}$	(l)	0.609	0.000
	$L = -6.375 + 0.086\text{GDP}$	(m)	0.584	0.000
High-speed commuting railway (152)***	$L = 200.562 + 0.199P + 0.058\text{S} + 0.021\text{GDP}$	(n)	0.222	0.000
	$L = 230.206 + 0.366P$	(o)	0.187	0.000
	$L = 195.185 + 0.248P + 0.064\text{S}$	(p)	0.221	0.000
Density model	$L/P = 0.228e^{0.230 \times \text{GDP}/P}$	(q)	0.053	0.000
	$L/P = 0.828e^{-1.152 \times P/S}$	(r)	0.137	0.000

** Significance level of 0.01; *** significance level of 0.001; *L* regional intercity rail transit length (km); *P* regional population (10 K); *S* land use area (km^2); and *GDP* gross domestic product index

density is higher. This is possibly because the intercity rail transit systems connect cities with high population densities and areas where the land use has larger open spaces between cities could be less costly utilized.

As shown in Table 2, the total intercity rail transit length is projected as about 2293 km by using the model, 17.5% longer than the planned length of 1890 km (as of 2012). Comparing with the available 254 existing intercity rail transit systems in other countries, the intercity rail transit lengths are relatively shorter in the PRD region. Model (q) is the regression model that is based on per capita economic factor. The estimated length resulted from the model (q) is lower than actually planned length. The estimation gives a room to accommodate the extra demand that may be caused by other unknown factors, thus can be viewed as proactive forecasting in planning.

The PRD region case study focuses on nine cities: Guangzhou, Shenzhen, Zhuhai, Foshan, Jiangmen, Dongwan, Zhongshan, Huizhou, and Zhaoqing. A total of 373 towns (nodes referring to centroid centers of concerned towns) and 259 streets (links) are included in the study area. The intercity rail transit systems are classified into four classes in the planning. Since every county is referred to just one local node over the network, the above models (a), (b), and (c) are inappropriate to check the suitability of the rail transit planning size to each concerned town. The models (q) and (r) are used to estimate the intercity rail transit length for each class. In this study, the data used in the models include population density and per capita GDP of every node, and then the regional intercity rail transit size of every node is obtained by using the model (q) and (r). The average intercity rail transit length of every node ranges from 0.3 to 37.3 km. The node with the largest rail transit size is located in Nanshan, and the node with the smallest rail transit size is located in Gaoyao. Clustering analysis results in four classes of the transit sizes, as shown in Table 3.

As a result of running the models (q) and (r), the first class of the intercity rail transit length, including Guangzhou Tianhe and Shenzhen Nanshan, ranges from 35

Table 2 Comparison between estimated and planned total rail transit lengths

Population size (10,000)	Land use size (km)	GDP Level (hundred estimated million dollars)	Model estimated intercity rail transit size (km)					Model estimated average rail transit size (km)	Planned average rail transit size (km)
			(a)	(b)	(c)	(q)	(r)		
4772	41,515	3342	3418	486	2820	1278	3461	2293	1890

to 38 km in average. The second class includes four nodes located in Guangzhou and Shenzhen, with average length ranging from 9.091 to 31.658 km. The third class, including 20 nodes in Guangzhou, Shenzhen, Zhuhai, Foshan, and Dongwan, is projected with average length of the intercity rail transit routes from 14.183 to 23.094 km. The fourth class includes 347 nodes with smaller average rail transit length.

4.3 Classifying Intercity Rail Transit Lengths at City Level

The estimated average intercity rail transit lengths are further broken down into five classes for each of concerned cities, from shorter to longer length. The result is summarized in Table 4. The result indicates that Guangzhou and Shenzhen obviously need the largest intercity rail transit size, followed by Zhuhai, Foshan and Dongwan. Huizhou, Zhongshan, and Jiangmen also need large intercity rail transit size, even though their need is less than the above cities based the model result. To eliminate the impact of land use area, or just consider the density of the intercity rail transit length per unit area, the transit density is estimated for all the towns and classified into five classes. The first class basically includes the center zone of almost every city in the PDR region such as Guangzhou, Shenzhen, and Dongwan. Foshan, Nanmeng, and Guzhen, as well as part of Guangzhou and Shenzhen, are classified into the second class. The third class mainly covers part of Foshan, Zhongshan, Dongwan, and Shenzhen.

5 Checking Modeled Transit Scopes with Accessibilities

5.1 Spatial Distributions of Population and Economy Accessibilities

The PRD region covers several cities of the most densely populated in China. The PRD regional population density is defined in ten classes with the highest density of 38,458 persons per kilometer and the lowest density of 21 persons per kilometer. To better understand the relationship between population density and transportation development, population density distribution and transport network are overlapped in the ArcGIS platform, as shown by Fig. 1. The current transport network distribution is generally in accordance with the population density distribution. The population accessibility distribution is shown in four classes in Fig. 2. Guangzhou, Dongwan, and Shenzhen have the highest population accessibility. Guangzhou, Foshan, Dongwan, and Shenzhen have the secondly highest population accessibility. Jiangmen, Huizhou, Zhuhai, and Conghua have the lowest population accessibility. It shows that the PRD region is centralized in Guangzhou with the areas of the highest population accessibility, and the population accessibility level becomes lower with farther from the central areas.

Similarly, economy accessibility distribution is calculated by Eq. (6). Guangzhou remains the center position in terms of economy accessibility. Figure 3 shows the economy accessibility distributions in the PRD region. Guangzhou, Dongwan, and Shenzhen are the highest economy accessibility areas. Jiangmen, Huizhou, Zhuhai, and Conghua are the lowest economical accessibility areas. Of the 373 sample nodes in the PRD region, the population accessibility and economy accessibility are pretty correlated and their correlation coefficient is close to 1.

5.2 Transportation Accessibility with Intercity Rail Transit Planning

The transportation infrastructural system provides a "backbone" influencing the transportation accessibility. Accessibility reflects the ability of a location providing people with access opportunities to human life and associated economy activities through connecting land use and transportation facilities. To investigate spatial distribution of transportation accessibility, we use the "natural cutting" method to visualize the defined index in each town included in the PRD region. A total of 373 nodes in the study area are classified into three classes by the ratio of

Table 3 Intercity rail transit size based on society and economy development

Town class	# node	Model	Minimum size (km)	Maximum size (km)	Average size (km)	SD
First class	2	(q)	34.225	35.853	35.039	1.2
		(r)	35.775	40.312	38.044	3.2
Second class	4	(q)	19.772	38.025	31.658	8.1
		(r)	0.740	19.045	9.091	9.2
Third class	20	(q)	7.590	25.427	14.183	5.0
		(r)	15.511	35.856	23.094	4.3
Fourth class	347	(q)	0.142	11.762	2.353	2.0
		(r)	0.376	19.309	5.740	4.1

Table 4 Estimated intercity rail transit size for each city

Estimated intercity rail transit size (km)	Number of nodes									
	Guang zhou	Shen zhen	Zhu hai	Fo shan	Hui zhou	Dong wan	Zhong shan	Jiang men	Zhao qing	Total
≤5	53	5	10	8	51	9	11	72	28	247
5–10	16	10	2	11	5	12	12	4	7	79
10–20	5	7	–	9	1	8	1	2	3	36
20–30	2	2	1	1	–	3	–	–	–	9
>30	1	1	–	–	–	–	–	–	–	2

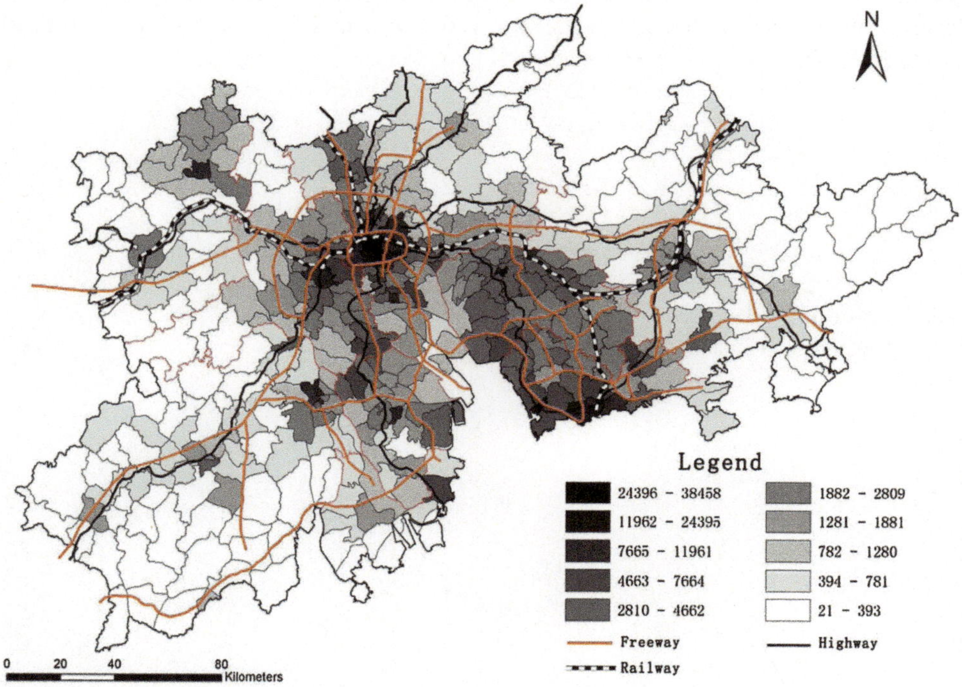

Fig. 1 Overlapped population density and traffic network

transportation land usage to the land area that covers the concerned transportation infrastructures (or transportation land use ratio, R_i). The first class area is defined as the area with the transportation land use ratio ranging from 3.06 to 7.93%. The first class land use covers 29 nodes. The land surrounding Guangzhou Baiyun Songzhou Road has the highest transportation land use ratio (7.93%). For the second class land, transportation land use ratio ranges 1.37–3.06% and 134 nodes are covered, mainly distributed in Guangzhou, Foshan, Dongwan, and Shenzhen. The third class land includes 210 nodes with transportation land use ratio lower than 1.37%. Thus, the layout of the transportation infrastructure development appears like "center + radiation" pattern in the PRD region. The central areas are mainly located in Guangzhou and Foshan, and the second center is in Shenzhen. Those central areas are connected with peripheral areas through freeways. Less highway infrastructures have been developed in the skirting areas. Guangzhou–Shenzhen–Hongkong corridor has been formed in the east area of the PRD region. The transportation land use ratio is much lower in western and northern areas, which is a vulnerable factor in propelling the development inside the PRD region.

The shortest average travel times from each traffic analysis zone (i) to all other zones (j) are calculated, and then A_i values are estimated by using Eq. (2). Of the 373 concerned nodes, the node located in Guangzhou Jinsha Street has the lowest A_i value (i.e., 49 min). The worst accessibility node is located in Huizhou Huidong town with shortest average travel time to other node at 158 min. The overall shortest average travel time for all 373 nodes is 76 min. 21% of the nodes have the shortest average travel

Fig. 2 Population accessibility distribution

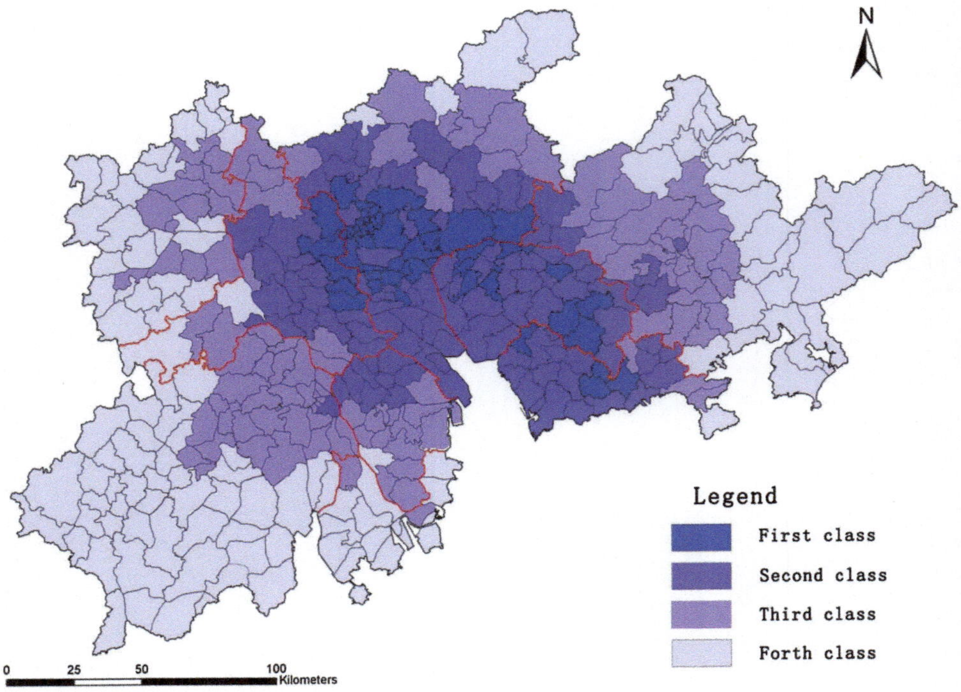

Fig. 3 Economy accessibility distribution

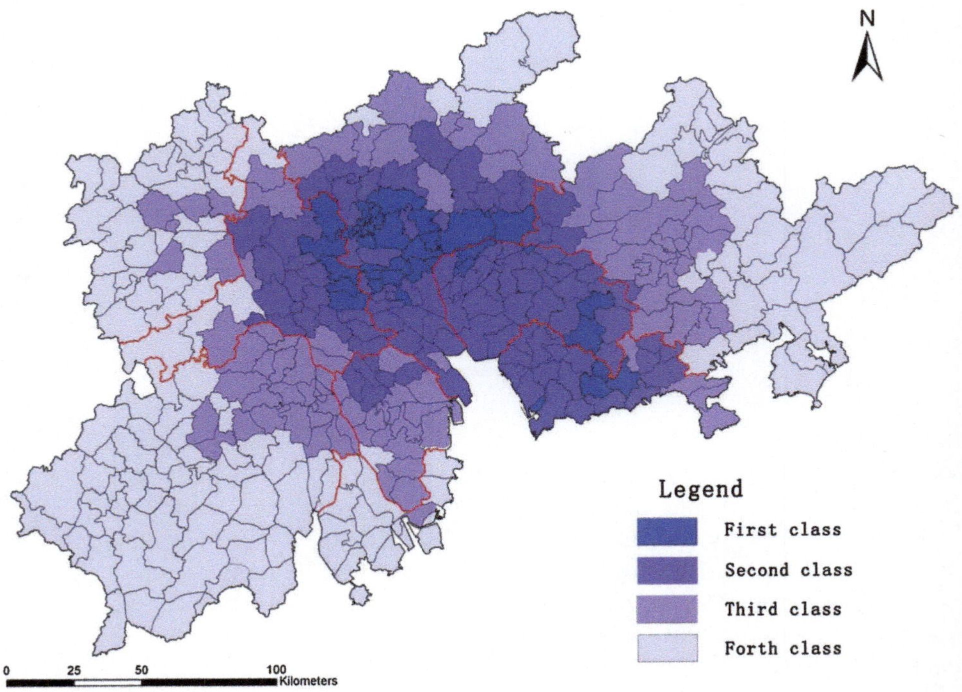

times of less than one hour, and most of them are located in Guangzhou. 80% of the nodes have the shortest average travel time of 90 min, and 3.2% with 120 min. Those nodes mainly located within Huizhou and Jiangmen. With A_i values for all the nodes, transportation accessibility index is then calculated and the result is illustrated by Fig. 4. The result indicates that there are 214 nodes where the accessibility A_i is higher than the average level (76 min) in the PRD region, and they are located in 57.4% of the counties in the region.

The integrated transportation accessibility (TA_i) is classified into six levels by using the "natural cutting" method in the ArcGIS environment (Fig. 4). Sixteen nodes at the first level have the transportation accessibility of

Fig. 4 Transportation accessibility distribution

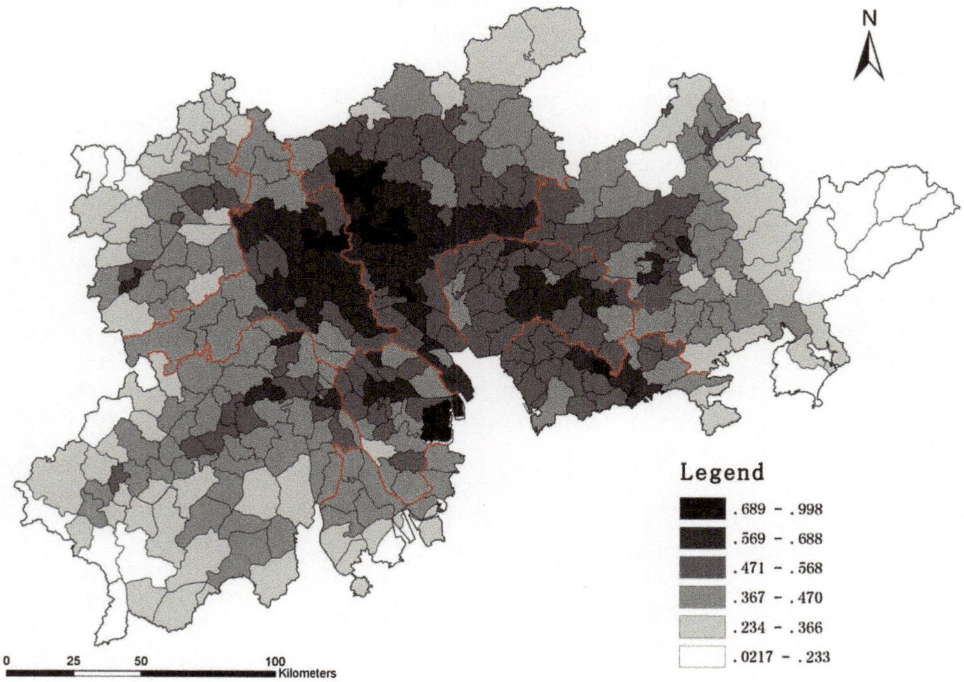

Fig. 5 Transportation accessibility change distribution

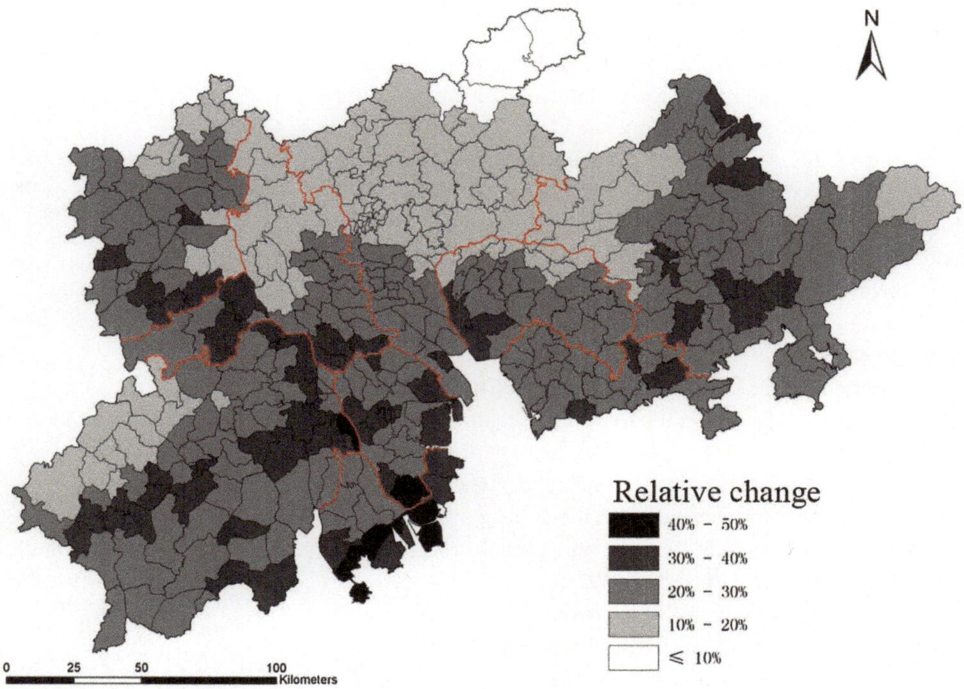

higher than 0.689, most of which are located in Guangzhou and Foshan except for two nodes in Zhongshan and Shenzhen. Sixty-three nodes' transportation accessibility ranges from 0.569 to 0.688 at the second level, of which 26 nodes are located in Guangzhou and 13 located in Foshan. There are 103 nodes that are located in Guangzhou, Dongwan and Shenzhen at the third level. There are 116 nodes in Jiangmen and Huizhou at the fourth level. There are 72 nodes in Huizhou, Jiangmen, Zhaoqing, and Zhuhai. The transportation accessibility distribution appears being centralized in Guangzhou and Foshan. Dongwan and Shenzhen are the second centers. The transportation accessibility strength diffuses around the above centers out-toward skirting areas. The accessibility is higher in east coastal areas than inner western areas of the region.

With incorporated intercity rail transit into the GIS system, the density of the transportation (R_i) would not change much, while the shortest travel time of each town or street (A_i) would change greatly. As shown in Fig. 5, it is found that the intercity rail transit planning can lead 20–30% improvement of A_i in 199 towns/streets and most of them are distributed in the peripheral areas of PRD region. Zhuhai gets the greatest A_i improvement, whereas Guangzhou and Foshan with the highest original transportation accessibility got the least A_i improvement. Thus, the planned rail transit planning can bring more equal regional accessibility in PRD region.

6 Conclusion and Discussion

Since there is no standard for determining the rail transit size, the authors developed a new method by using the regression technique to correlate the sizes of 254 regional intercity rail transit systems in different countries in the world with regional populations, land use areas, and economical (or GDP) factors. The scope models [i.e., models (a), (b), (c), (q), and (r)] resulted from this analysis are used as a reference benchmark for estimating the intercity rail transit size for the PRD region.

Based on the rail transit scope estimation, this research further provided an exemplary framework for the accessibility-based analysis about intercity rail transit planning and visualizing the results in the GIS environment. In this framework, three accessibility measures, including transportation, population, and economy accessibilities measures, were adopted. The application of this framework in PRD suggested that these three accessibilities were distributed with center in Guangzhou and became lower with farther from the central areas. However, with the planned rail transit, the shortest travel time of peripheral areas of PRD got larger improvement than the central areas, such as Guangzhou and Forshan, and the rail transit brought more equal regional accessibility in PRD region.

Based on the case study of PRD region, both the models for estimating intercity rail transit systems scope and the framework for analyzing the accessibility impact of rail transit scope can be applied to other metropolitan areas and even other transport modes. But we have to admit that it has some limitations in applying the accessibility models, especially in terms of only considering population and GDP as the social economic factors influencing location and travel demand. In the future, more factors will be comprehensively considered in the accessibility-based analysis, such as employment, industry, and poverty factors.

Acknowledgements The project is sponsored by Natural Science Foundation of China (#41171139 and #41130747). The results and ideas presented in the paper represent the authors' point of views only.

References

1. Renner M, Gardner G (2010). Global competitiveness in the rail and transit industry. Report for Apollo Transportation Manufacturing Initiative
2. The National Surface Transportation Policy and Revenue Study Commission (NSTPRSC) (2007) Transportation for tomorrow. Report of NSTPRSC
3. Quan H, Li SM (2011) Transport Infrastructure Development and Changing Spatial Accessibility in the Greater Pearl River Delta, China, 1990–2020. Working paper series
4. Gutiérrez J, González R, Gómez G (1996) The European high-speed train network: predicted effects on accessibility patterns. J Transp Geogr 4(4):227–238
5. Jiang H, Xu J, Qi Y (2010) The influence of Beijing–Shanghai high-speed railways on land accessibility of regional center cities. Acta Geogr Sin 65(10):1287–1298
6. Murayama Y (1994) The impact of railways on accessibility in the Japanese urban system. J Transp Geogr 2(2):87–100
7. Geurs KT, van Wee B (2004) Accessibility evaluation of land-use and transport strategies: review and research directions. J Transp Geogr 12:127–140
8. Babalik-Sutcliffe E (2002) Urban rail systems: analysis of the factors behind success. Transp Rev 22(4):415–447
9. Cao X, Lin Q (2008) The evolution of worldwide metro systems: a study on their scales and network indexes. Acta Geogr Sin 63(12):1257–1267 **(in Chinese)**
10. Loo BPY, Cheng AHT (2010) Are there useful yardsticks of population size and income level for building metro systems? Some worldwide evidence. J Transp Geogr 27:299–306
11. Sachs J (2005) The end of poverty. How can we make it happen in our lifetime. Penguin Books, London
12. World Bank (2008) World development report. 2009, reshaping economic geography
13. Reggiani A (1998) Accessibility, trade and location behavior. Ashgate, Aldershot
14. Castella JC, Manh PH, Kam SP, Villano L, Tronche NR (2005) Analysis of village accessibility and its impact on land use dynamics in a mountainous province of northern Vietnam. Appl Geogr 25:308–326
15. Etter A, McAlpine C, Wilson K, Phinn S, Possingham H (2006) Regional patterns of agricultural land use and deforestation in Colombia. Agr Ecosyst Environ 114:369–386
16. Hanson S (1986) Dimensions of the urban transportation problem. Geographic perspectives on urban transportation. Guilford Press, New York, pp 3–23
17. Laurance WF, Albernaz AK, Schroth G, Fearnside PM, Bergen S, Venticinque EM et al (2002) Predictors of deforestation in the Brazilian Amazon. J Biogeogr 29:737–748
18. Nagendra H, Southworth J, Tucker C (2003) Accessibility as a determinant of landscape transformation in western Honduras: linking pattern and process. Landsc Ecol 18:141–158
19. Verburg PH, Overmars KP, Witte N (2004) Accessibility and land-use patterns at the forest fringe in the northeastern part of the Philippines. Geogr J 170:238–255
20. Guagliardo M (2004) Spatial accessibility of primary care: concepts, methods and challenges. Int J Health Geogr 3(1):3. doi:10.1186/1476-072X-3-3

21. Jalan J, Ravallion M (2002) Geographic poverty traps? A micro model of consumption growth in rural China. J Appl Econom 17:329–346
22. Bryceson DF, Bradbury A, Bradbury T (2008) Roads to poverty reduction? Exploring rural roads' impact on mobility in Africa and Asia. Dev Policy Rev 26:459–482
23. Olsson J (2006) Responses to change in accessibility: socio-economic impacts of road investment: the distributive outcomes in two rural peripheral Philippine municipalities. Department of Human and Economic Geography, School of Business, Economics and Law, Göteborg University
24. The Guardian (2016) China's Pearl River Delta overtakes Tokyo as world's largest megacity. The article was drafted on 28 January 2015. Last modified on 15 June 2016. Accessible at https://www.theguardian.com/cities/2015/jan/28/china-pearl-river-delta-overtake-tokyo-world-largest-megacity-urban-area
25. Demographia (2007) Gross Domestic Product (GDP-PPP) Estimates for Metropolitan Regions in Western Europe, North America, Japan and Australasia. Wendell Cox Consultancy, Belleville
26. Demographia (2009) Demographia 2000–2008 metropolitan area population & migration. Wendell Cox Consultancy, Belleville
27. Demographia (2009) Demographia world urban areas: population projections. Wendell Cox Consultancy, Belleville

10

Viability Assessment of Light Rail Line Planning: Case Study of Cincinnati Eastern Corridor

Samuel A. McKinley[1] · Heng Wei[2,3,4]

Abstract With increasing environmental concerns, light rail transit (LRT) has drawn attention for consideration in urban transportation planning by various levels of authorities in the US Government. Advocacy groups need an effective method for viability assessment of the alternatives. Environmentalists may want to support LRT, but assessment of its viability is important to provide judgment of any future transportation project. This paper presents a method for LRT viability assessment through a case study of the "Oasis Line," as part of the Eastern Corridor Major Investment Study (MIS) by the Ohio Kentucky and Indiana Council of Governments (OKI). The study attempts to evaluate chances for "success" of the rail transit component of the MIS. An integrated method is used with a traditional four-step-based demand forecast by OKI, and a development of station-based Light Rail Ridership Regression demand forecast by Pelz. The problem to be solved—whether the line has a good enough chance at success to support and advocate for it—did not demand a full rerunning of the models. A review of appropriate literature—largely assessments of already-built light rail lines in other US cities—is used to characterize the predicted ridership as a success or failure. The predicted ridership falls comfortably above the low end of LRT systems in other US cities. The ridership predictions are found favorable to support the Oasis Line. Extensive literature review suggested that the public's assessments behave in an almost entirely political fashion.

Keywords Viability assessment · Light rail transit (LRT) · Diesel Multi-Unit (DMU) · Light Rail Ridership Regression (LRRR) · Ridership · Boarding · LRT station

✉ Heng Wei
heng.wei@uc.edu

Samuel A. McKinley
sam_mckinley@yahoo.com

1 Price Hill Will, Sustainable Places Studio, LLC, Cincinnati Area, KY, USA

2 Beijing University of Technology, Beijing, China

3 Chang'an University, Xian, China

4 Department of Civil and Architectural Engineering and Construction Management, The University of Cincinnati, 792 Rhodes Hall, 2850 Campus Way, Cincinnati, OH 45221-0071, USA

Editors: Haishan Xia and Chun Zhang

1 Introduction and Background

While rail transportation in the USA is primarily comprised of freight shipments, rail passenger transportation is available but playing a limited role as compared to transportation patterns in many other countries [1]. According to the Bureau of Economic Analysis [2] of US Depart of Commerce, however, consumer using public transportation outgrew spending on new automobiles between 2000 and 2015. In 2015, consumer spending on public transportation exceeded $100 billion in the USA for the first time ever [3]. Use of mass transit is usually concentrated in highly urban areas. According to the American Public Transportation Association [4], the three largest public transit agencies by ridership are in New York City, Chicago, and Los Angeles. New York City's Metropolitan Transit Authority serves about one-third of those who use mass transit in the USA [3]. With such increasing concerns with congestion and associated adverse impact on air quality and hence public health, light rail transit (LRT) has drawn attention for reexamination and consideration in urban transportation planning at various levels of transportation planning authorities in the USA [3, 5–9].

Government and advocacy groups have called for an effective method for viability assessment of the alternatives. It is the major goal of the rail lines to be supportive of transit, bicycle, and walking-based travel and can reduce reliance on cars as a primary mode of local transportation. It can also help stimulate community enhancement, revitalization, and economic growth opportunities through ongoing service and transit-oriented developments (TODs) that could grow around the line's rail stations. TODs are walkable, mixed-used community spaces that typically include office, retail, residential, and social gathering facilities [10].

As an American environmental organization, Sierra Club has hundreds of thousands of members in chapters located throughout the USA. The Greater Cincinnati Chapter of the Sierra Club (or called the Miami Group) and other environmental groups have worked for years to forestall a major widening and realignment of Ohio Route 32, in eastern Hamilton County. The presented study was part of a Major Investment Study (MIS) by the Ohio Kentucky and Indiana Council of Governments (OKI), the metropolitan planning organization (MPO) for the Greater Cincinnati area which covers four counties located in State of Ohio, three counties in Kentucky and one county in Indian: The Eastern Corridor MIS [11, 12]. It was handled by the Hamilton County Engineer's Office as a Transportation Improvement District (TID) (6–7). In this project, the LRT component of the MIS is termed the "Oasis Line" as it uses existing rail right-of-way with that name.

The Oasis Line rail transit plan had followed the proposed road over a new bridge across the Little Miami. However, the existing railroad tracks that had made much of the rail transit plan relatively cost-effective already have a bridge in place. If the light rail component were truly divorced from the road component, environmentalists would stand a much better chance of stalling the road component and getting rail transit to use the existing bridge. Also, while rail transit still represents increased capacity through the Eastern Corridor, it tends to induce much more compact, less environmentally costly patterns of development [13]. While environmentalists may want to line up strongly in support of the light rail component of the Eastern Corridor Plan, assessment of the viability of the Oasis Line is highly desired to provide convincing or solid analysis to judge success of the future transportation project..

Sierra Club, an environmental organization in the USA that was founded on May 28, 1892, in San Francisco, California, is generally supportive of transportation alternatives to the personal automobile. But they would not want to win this major commitment to rail transit only to have it viewed as a failure, giving a bad name to all fixed-guideway modes of mass transit [14]. To minimize this addressed concern, the viability assessment is focused on evaluation of the magnitude or degree of the rail transit line's positive impact on the existing job connectivity with expanding or new non-retail employment, alongside the estimated rider shares potentially attracted by the rail transit plan and its environmental impact. On the other hand, we need to ensure whether the Oasis Line has a reasonable-enough chance of success that Sierra Club can support it without worrying that it will preclude future rail transit projects in the Greater Cincinnati area. From the long run, the Oasis Rail Transit line is desired to be a foundation upon which Greater Cincinnati's envisioned regional rail system could ultimately be built to better connect our region—from Hamilton, Clermont, Butler, and Warren counties in Ohio to Campbell and Kenton counties in northern Kentucky and locations in southern Indiana [10].

The study presented in the paper looks at several results of the Eastern Corridor MIS by OKI [11, 12]. Specifically, the study is an attempt to evaluate the chances for "success" of the rail transit component of the recommendations that resulted from the MIS. The study uses an integrated method with a more traditional "four-step" demand forecast by OKI, and station-based Light Rail Ridership Regression (LRRR) demand forecast for the proposed Oasis Line by Pelz [15]. While it would be preferable to produce original assumptions and modeling, the problem to be solved by this study—whether the Oasis Line has a good enough chance at success for Sierra Club to support and advocate for it—does not demand it. Additionally, a review of appropriate literature—largely assessments of already-built light rail lines in other US cities—will be used to categorize the predicted ridership as representing a line that is a success or a probable failure.

This paper is organized as follows: the introduction is followed by scope of the study. Then, the methodology section presents the core of the methods used for the study and the results in predicting success of the concerned LRT line and other findings, as well as the analysis of national ranking by ridership as a reference to weigh a possible success or failure of a LRT project. Conclusions are at the end.

2 Scope of the Study

The study is motivated to (1) examine the appropriateness and validity of demand forecasting done by traditional "four-step" versus "station-based" LRRR modeling; (2) compare the demand forecasts generated by each and explain why they differ, and what that means for each set's defensibility in public debate; (3) compare both sets to what have proved to be successful or unsuccessful lines in other US cities, suggesting various measures; and (4) determine whether demand forecasting predicts the proposed line will likely be popularly judged a "success" or a failure that should not be supported or pursued, based on comparisons with other cities' riderships and how they were judged.

Figure 1 shows the study area. The proposed route of the Oasis Line light rail begins at the transit center at the southern end of downtown Cincinnati, essentially following the Ohio River up until the Little Miami River's confluence with the Ohio. Thence, it follows fairly close to the Little Miami, until crossing the river, either at the Cincinnati neighborhood of Linwood or at the southern middle of the Village of Mariemont. If it crosses at Linwood, it follows State Route 32. If it crosses at Mariemont, it drops down to the southern part of the Village of Newtown. The two alternatives rejoin in the eastern portion of Newtown and then head northeast, to the southern bank of the Little Miami, south of the Village of Terrace Park. The reunited alternatives follow that river until about halfway past the southern border of Milford. From there, the line heads almost due east, to terminate near the junction of US Route 50 and I-275.

Of all the sources in the present study's bibliography that define light rail [15–21], each of them mentions electric power for the rolling stock. The proposed rolling stock for the Oasis Line is actually diesel-powered, called "DMU." DMU stands for Diesel Multi-Unit, and the proposed stock is made by Siemens Company [22]. DMU is used in many applications that would otherwise be described as light rail [23]. In the case of the Oasis Line the frequency of proposed stops, the fact that it serves mostly urban short haul and the morphology of the proposed rolling stock means that it is logical to evaluate it against light rail standards. Siemens' DMUs are available in three gauges, including the standard gauge of the existing Oasis rails. The advantages of DMU are thought in this case to include the fact that it obviates the need for the installation of overhead power and the fact that—with time segregation or perhaps advanced switching and scheduling—it could be certified by the Federal Rail Administration (FRA) to run on freight rails, so the local rail workers' unions might sign on in support.

3 Methodology

3.1 Demand Forecasting Modeling

In this study, two demand forecasts for the Oasis Line are studied and compared: results using OKI's traditional travel demand model that are released in the *Eastern Corridor Tier One Draft EIS Report* (DEIS) [12], and results by Pelz's study [15] in which the station-based LRRR model was used. Both models included station locations. OKI's model basically used public input combined with professional planning expertise, depending upon or referencing no numerical models for station placement. Pelz' study employed a spatial and statistical methodology to determine optimal station locations.

OKI's forecasting models appear to be something of a "black box." The DEIS mentions the input factors first and then gives the outputs of successive model runs. The travel demand modeling adapts the "well-established" four-step modeling process, namely trip generation, trip distribution, mode choice, and trip assignment. The input statistics for the Cincinnati Eastern Corridor were taken from the corridor only, rather than the super-region, as had been done in previous versions [12]. It does not share the actual calculations, which would generally involve some adjustment factors, calibrated to the region [24]. A feedback loop to model the effects of loading roadway segments to the point of congestion was employed.

The black box is the fact that the inputs were fed through a computerized model, OKI/MVRPC RTDM 6.0 (Ohio, Kentucky, and Indiana/Miami Valley Regional Planning Commission Regional Travel Demand Model 6.0), which just means it is a computerized model calibrated to the OKI/MVRPC super-region. This model was used in a "building" process [12], where pieces of the final recommended alternative are added in one at a time. The model is then rerun. These pieces include:

1. Existing + Committed—What facilities are already there, plus those that have been committed for implementation;
2. Regional Baseline—Adds foreseeable facilities to the Existing + Committed model;
3. Transportation System Management—Various management techniques are assumed to be implemented upon the system, and added onto the Regional Baseline model;
4. Expanded Bus—Recommended bus system improvements are added onto the Transportation System Management model;
5. Rail Transit—Built up from the Expanded Bus model, with the Expanded Bus model taken to act as a feeder system for the Oasis Line, plus five different scenarios run with different levels of additional rail lines, including a line along I-71 and the Wasson Line;
6. Highway Multi-Modal Plan—Adds modeling onto the Rail Transit that includes the Rte. 32 expansion and realignment, as well as other roadway improvements;
7. Multi-Modal Plan with Land Use Vision Plan—Adds onto the Highway Multi-Modal Plan model the "Vision Plan" developed for land uses in the corridor.

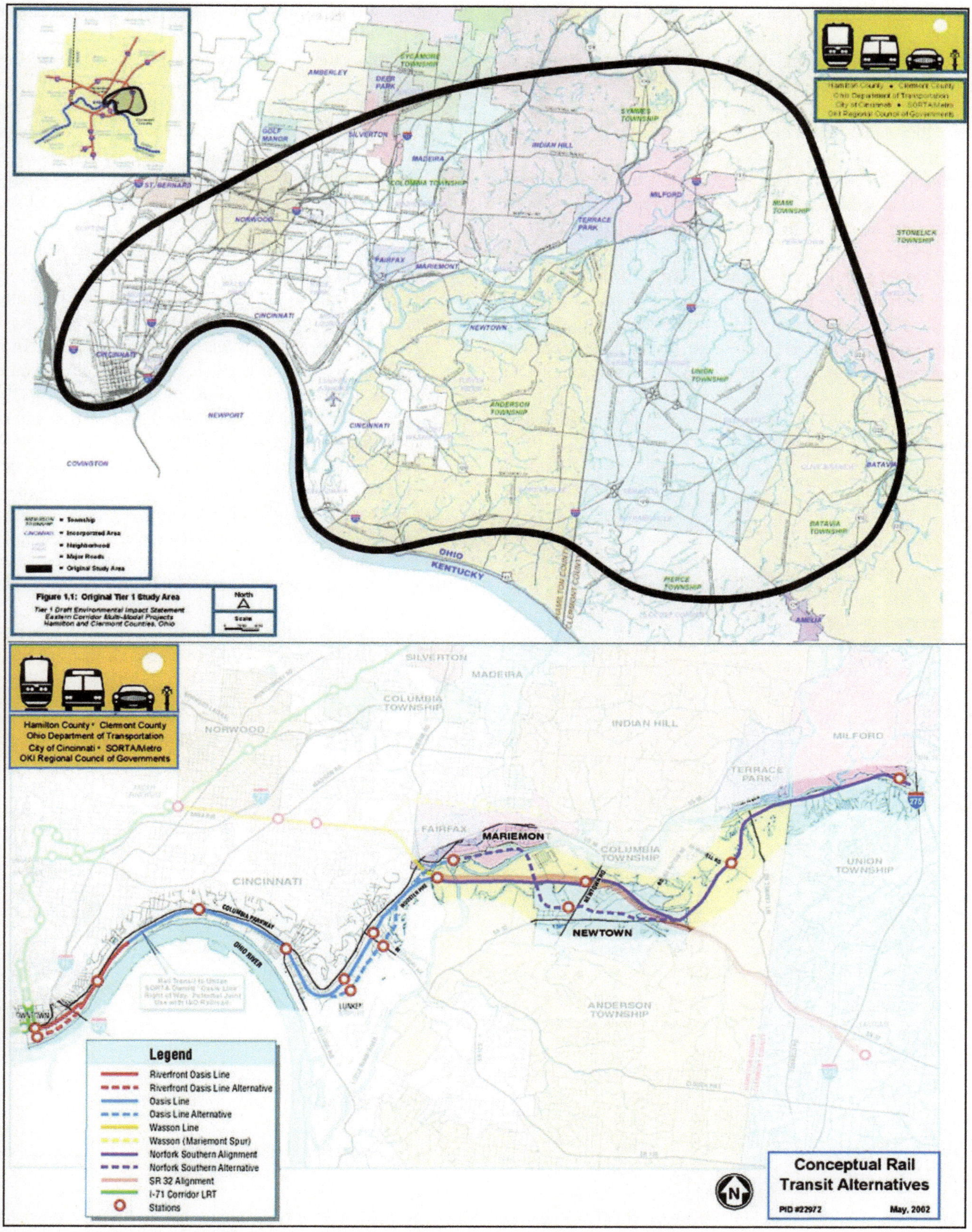

Fig. 1 The Cincinnati Eastern corridor study area and the Oasis *Line* study area [12]

All these were modeled for the year 2030, the target year of the authority's next Long-Range Transportation Plan (LRTP). At the end of all the runs, the Oasis Line was projected to have approximately 8000 daily riders, as part of 36,000 daily rail ridership for the whole OKI region including the Wasson and I-71 lines.

Pelz's study [15] was developed from two sources: Kuby et al. [18] and Upchurch [20]. Kuby et al.'s [18] study developed a station-based ridership model, specifically targeted at light rail. They tested 17 variables, through regression analysis, over 268 stations in nine cities with existing light rail systems. Only 12 variables are correlated with the observed boardings. Typical variables are shown in Table 1, along with whether they were hypothesized to be negatively or positively correlated with boardings and whether or not they proved to be significant after regression.

The variables shown in Table 1 are briefly explained as follows:

- Employment within walking distance—The job locations within ½ mile of the station;
- Population within walking distance—Population living within ½ mile;
- Airport—A dummy variable, one if the station served airport terminals, zero if not;
- International border—Also a dummy variable, reflecting whether the station serves and international border;
- College enrollments—Whether a station had a college within walking distance;
- CBD dummy variable—Whether a station was located within a Central Business District;
- Park-and-ride spaces—The number of parking spaces available at stations;
- Bus connections—The number of intersecting bus lines;
- Other rail lines—A 0–1 dummy variable, reflecting only whether the station was colocated with one for another type of rail transit;
- Heating and cooling degree-days—The number of days above or below 65 °F, times the absolute value of the degrees from 65;
- PMSA population—The population of the Primary Metropolitan Statistical Area;
- Terminal station—A dummy variable, reflecting whether a given station was the last one on the line;
- Station spacing—The distance between a the edge of a ½ mile buffer around a given station and the ½ mile buffer around the next station;
- Designated transfer station—Whether a station offers transfer to another light rail line;
- Normalized accessibility (centrality)—The average travel time for each station, divided by the highest average travel time for its system;
- Percentage of PMSA employment covered by system—What percentage of the PMSA's employment lies within ½ mile walking distance of any station in the system; and.
- Percent renters within walking distance—The percentage of the population within ½ mile of stations who rent their homes.

In this modeling, statistics were used for areas within a ½ mile "buffer" around each station. In some cases,

Table 1 Variables [18]

Independent variable	Hypoth. correlation	Significant
Employment within walking distance	+	Yes
Population within walking distance	+	Yes
Airport	+	Yes
International border	+	Yes
College enrollments	+	No
CBD dummy variable	+	No
Park-and-ride spaces	+	Yes
Bus connections	+	Yes
Other rail lines	+	No
Heating and cooling degree-days	−	Yes
PMSA population	+	No
Terminal station	+	Yes
Station spacing	+	No
Designated transfer station	+	Yes
Normalized accessibility	−	Yes
Percentage of PMSA employment covered by system	+	Yes
Percent renters within walking distance	+	Yes

statistics were only available by census block and a simple calculation of the percentage of the census block covered by the buffer was conducted. After performing the regressions and calibrations, Kuby et al. [18] developed the equation:

$$\begin{aligned}\text{Boardings} = {} & 1584 + (0.023\ \text{Employment}) + (0.092\ \text{Population}) \\ & + (915\ \text{Airport}) + (12{,}055\ \text{Border}) \\ & + (0.0774\ \text{Park-and-ride}) + (123\ \text{Bus}) \\ & - (1.52\ \text{Degree-days}) + (660\ \text{Terminal}) \\ & + (5735\ \text{Transfer}) - (1872\ \text{Centrality}) \\ & + (1301\ \text{Employment Coverage}) \\ & + (624\ \text{Percent Renters}) + \text{Error}.\end{aligned} \quad (1)$$

The border variable was left off as zero in every case, as the nearest international border to Cincinnati is about 5 h drive by automobile. The error variable only makes sense when trying to calibrate existing systems to each other.

Based on the method resulting from Upchurch's study [20], a geographical information system (GIS)-based model was developed, which resulted in a surface whose high points suggest the ideal station locations. Six criteria were used with this model: employment density, population density, feeder bus service, location on the alignment, location at a street intersection, and a minimum of ½ mile distance from the next station.

A composite of employment and population densities, plus feeder service within ½ mile of the rail alignment, was made. Statistics for population density were only available for full census blocks, so simple percentages of the blocks were used according to how much of their area fell within the buffer. Likewise, employment density was available by transportation analysis zone (TAZ), so the same clipping procedure was carried out on those. While this is not perfectly fine-grained, it is much more precise to the alignment than is OKI's full-corridor methodology.

Next, these areas' scores were turned into "heights" within the GIS model. This could then be overlayed on the street map, with the highest points at street intersections, but not within ½ mile of another station, being chosen as stations. Finally, Pelz' models produced boardings (which can be equated with unlinked trips) of 9133 for the Oasis Line in project year 2030 [15]. What is noteworthy is that his models produced about 17% higher ridership predictions for the western 8 miles of the Oasis Line than OKI's models for its entire 17-mile length. Population and especially employment densities start to drop off just east of Pelz' study area, so perhaps that helps to explain why his station-based model in that area predicts higher numbers than OKI's whole-corridor model.

3.2 Success Assessment Through Ridership Forecasting

At the heart of the present study is a basic decision: would the Oasis Line likely be perceived as successful enough that it would not damage the chances for further rail transit in the Greater Cincinnati area? Sierra Club has found that the fully recognized and allocated outcomes of rail transit are, in general, vastly better for the environment than are the outcomes of road projects [13]. So, if the public in the Greater Cincinnati area would likely perceive the Oasis Line as a success, and therefore demand and support more rail transit for the area, then Sierra Club should probably support the Oasis Line. Conversely, if it is predictable that the public would see the Oasis Line as a failure, a "boondoggle," or a giant "pork" project, then Sierra Club should probably not be associated with that. Moreover, Sierra Club should probably work actively against it in that case, as it would obviously reduce the chances to build any further rail transit.

The original model of success for the present study was a simple one: hold the ridership predictions of the two models up against the observed ridership of already-built systems whose constituents seemed to perceive them as successful. Some discussion of how likely other cities' results were to predict perceptions in Greater Cincinnati would be necessary, but it was hoped that a review of articles in those cities would reveal public perceptions.

As it turned out, this model provides only scant, almost anecdotal evidence. The American Public Transit Association (APTA) lists current ridership statistics for only 26 existing US light rail systems for the first quarter of 2008 [16]. One of those, in Charlotte, NC, was not developed enough to have a full-quarter's statistics. Of the rest, most are too old to have a body of readily available newspaper articles representing public reaction to them. The organization "Light Rail Now Project" [19] features articles on just about every system, but they are all attributed to the organization, and its name might suggest bias.

A series of five articles were uncovered in the *Houston Chronicle* [17], wherein the author traveled to five cities to evaluate their systems. The cities were San Diego, Denver, Cleveland, Portland, OR, and Dallas, TX. However, he wrote these articles in the run-up to a vote on light rail for Houston, TX. Furthermore, they appeared on the Web site of the organization Bicycle Austin [17], which was promoting efforts in Austin, TX, to build a light rail system. Again, one could predict which way their argument would land.

Many, many hours of research gave an impression that light rail is so politically charged that local sources—and even most national sources—are unlikely to give much sense of the "public" reaction to the projects. Rather, they show their authors' biases, with supporting statistics interpreted in predictable directions.

In the end, the best evaluation is probably to use the most basic—least prone to bias—statistics and apply good sense and logic to them. Therefore, the following section will show where OKI's and Pelz' predictions rank the Oasis Line versus the first quarter 2008 ridership APTA reports [16]. The "good sense and logic" part will come in with various attempts to equalize and discuss them.

4 Evaluation of the Proposed Oasis Line

At the risk of repetitiveness, the real standard of evaluation of the present study is whether the Oasis Line has a reasonable-enough chance of success that Sierra Club can support it without worrying that it will preclude future rail transit projects in the Greater Cincinnati area. Given that, we will evaluate first its chances of being built at all. Next we will look at its rank—using both OKI's and Pelz' ridership predictions—against other whole light rail systems in the USA

The project is listed for funding in the "Fiscally-Constrained Detailed Project Scoring" section of OKI's *2030 Long-Range Transportation Plan* [25]. Tables summarizing OKI's scoring criteria are available in that report. The project scores 49 out of a possible 105 points in this list. The cutoff point is 47. That is, anything scoring 47 points or less will not be submitted for funding, while the projects that score 47.4 and higher will. The scoring criteria are certainly vulnerable to charges of subjectivity and pro-road bias, but the Oasis Line made it, in any case. On the other hand, the I-71 rail transit alignment and study for the Wasson Line did not make the cut, which could have an effect on the ridership models of both Pelz and OKI.

In a strict ranking by ridership versus APTA's 1st Quarter 2008 Light Rail Agency Statistics as shown in Table 2, both Pelz' and OKI's predictions rank fairly well. They are well down the list, but by no means low outliers. As discussed earlier, though, the systems listed vary widely by many possibly relevant criteria, such as age of system, city characteristics, system track-mile length, and system network topology.

It is probably best to use Pelz' and OKI's 2030 projections and just keep in mind that they are predictions for a fully mature line. The fact that they fall midway between Cleveland's and New Orleans' ridership is then perhaps appropriate. Table 2 immediately begs the question of how comparable the Oasis Line might really be to these other systems. For instance, while the Kuby regression found that PMSA population did not matter that was for *station* boardings. The APTA statistics are for entire systems, and common sense tells us that PMSA population probably does matter over the entire system. Speaking of stations, might that variable make a difference? As shown in Table 3, the Oasis Line is presented on a boardings-per-mile basis and the Pelz and OKI forecasts really diverge. Even the OKI numbers stay out of the bottom two quintiles, while Pelz' forecast puts the Oasis Line halfway down the second quintile. As stated before, both population and employment start to get more sparse in the area around the Oasis Line, just about at the eastern limit of Pelz analysis, so it is quite possible that forecasts for an alignment the length of the one Pelz used would not drop off much.

Finally, since the Kuby model focused on stations, perhaps boardings per station might offer some insights, as shown in Table 4. Interestingly, this gives very similar rankings for the Oasis Line. The Pelz forecast is still in the second quintile, although a bit farther down, while OKI falls in the same 11th from the bottom position, or just into the third quintile. One important factor to note in the other systems is that the San Francisco Muni number is skewed very badly. A footnote to APTA's table says that streetside stops may not be counted as stations. Obviously, Muni relies heavily on these, considering they show only nine stops for 72.9 track miles.

Two sets of demand forecasts and evaluations against national light rail ridership statistics predict that ridership on the Oasis Line would be strong enough to support the Oasis Line. The above results also imply a sign of success for the proposed Oasis Line based on the ridership forecasting. The ridership numbers say the Oasis Line would be a credible light rail project and that Sierra Club can support it as long as it is not bound with the Rte. 32 expansion. Results shown in Tables 2 through 4 give a sense that the Oasis Line would be well into the thick of the national rankings, which is what is meant by "qualitative." Perhaps a truly quantitative route to evaluation would be through Kuby's model. Rather than just using it to predict ridership, perhaps given a forecast, its parameters could be used to normalize all the cities with light rail and give a "true" ranking that way.

5 Discussions and Conclusions

To date, there exist no real quantitative measures of success for a light rail system. Paul M. Weyrich and William S. Lind in "Does Transit Work? A Conservative Reappraisal" [21] pointed out one very good reason (among many) for this—light rail never turns a profit. They go on to point out that no other transportation systems do either (nobody complains that sidewalks don't turn a profit), but could take it a step farther. Whether or not it is fair for light rail to be held to a profitability standard, the fact is that

Table 2 Pelz and OKI Forecasts Ranked versus Existing Systems [12], [15, 16]

Non-linked avg. daily ridership 1st quarter 2008			
Agency	State	City	AVG daily boardings
Massachusetts Bay TR Auth	MA	Boston	226,186
San Francisco Muni Rwy	CA	San Francisco	118,534
Los Angeles County MTA	CA	Los Angeles	114,473
Tri-County Metro Trp Dist	OR	Portland	92,053
Southeastern Penn TA	PA	Philadelphia	90,552
San Diego Trolley, Inc.	CA	San Diego	88,368
Bi-State Dev Agency	MO	Saint Louis	61,532
New Jersey Transit Corp	NJ	Newark	55,163
Regional Trp District	CO	Denver	53,307
Dallas Area Rapid Transit	TX	Dallas	50,275
Sacramento Reg Tr Dist	CA	Sacramento	40,815
Metro Tr Auth of Harris Co	TX	Houston	34,568
Santa Clara Valley Trp Auth	CA	San Jose	26,534
Metro Transit	MN	Minneapolis	22,674
Maryland Transit Admin	MD	Baltimore	20,301
Port Auth of Allegheny Co	PA	Pittsburgh	18,769
Niagara Frontier Trp Auth	NY	Buffalo	18,249
Regional Transit Auth	LA	New Orleans	10,997
Oasis Line	**OH**	**Cincinnati**	**9133 (Pelz 2030 Proj)**
Oasis Line	**OH**	**Cincinnati**	**8000 (OKI 2030 Proj)**
Greater Cleveland Reg TA	OH	Cleveland	6963
Sound Transit	WA	Seattle	2436
Memphis Area Transit Auth	TN	Memphis	2173
Hillsborough Area Reg TA	FL	Tampa	1352
King County Dept of Trp	WA	Seattle	874
Kenosha Transit	WI	Kenosha	73
City of Galveston/Island Tr	TX	Galveston	69

Bold indicates if the concerned items are statistically signifincant at 95% level or not

Americans simply are not familiar enough with it to appraise it by other standards. That is, in America "the market" is the fallback standard and often the primary standard. Where nobody can agree on another, it is supposed that the market will sort it out. At the very least, it is supposed that light rail's benefit/cost ratio should be competitive with roads.

However, as Weyrich and Lind go on to point out, this is not very productive, either. Rail transit and roads serve different purposes and operate in different cost allocation and geographical universes. The main thesis of their study is that rail should only be held responsible for "rail-competitive" trips, where it turns out to do quite well. When the focus is only on those who are served by convenient, high-quality rail, and for trips which make sense on rail, they find that Chicago's system, for instance, captures as many as 50% of the trips. Since the system operates primarily in areas where automobile congestion is worst, and eminent domain and parking costs are the highest, it is highly cost-effective in its own universe.

Another author, on the Light Rail Now Web site [19] (17), points out that, while light rail is often maligned for not having much impact on a *region's* transportation problems, it can be proved to have a dramatic impact within its own corridor, especially on signalized arterial roads within the corridor. The author further points out that no single transportation project has much impact on an entire region's transportation, and in fact increased roadways often exacerbate the problems they are meant to solve through induced demand.

The present study attempts to hold up forecast numbers for comparison to systems in other cities. It even went further and tried to equalize the data by comparing trips per track mile and trips per station. While these procedures looked quantitative, they probably work best in a qualitative sense. With a universe of only 25 systems, how valid

Table 3 Pelz and OKI 2030 forecasts ranked against national light rail systems, boardings per track mile [12], [15, 16]

Q1 2008 non-linked avg. daily ridership ranked by boardings/track mile

Agency	State	City	AVG daily boardings	Track miles	Boardings/track mile
Massachusetts Bay Tr Auth	MA	Boston	226,186	78.0	2899.8
Metro Tr Auth of Harris Co	TX	Houston	34,568	20.0	1728.4
Regional Trp District	CO	Denver	53,307	32.1	1660.6
San Francisco muni Rwy	CA	San Francisco	118,534	72.9	1626.0
Sound Transit	WA	Seattle	2436	1.8	1353.5
Niagara Frontier Trp Auth	NY	Buffalo	18,249	14.1	1294.3
Oasis Line	**OH**	**Cincinnati**	**9133**	**8.0**	**1141.6 (Pelz 2030 Proj)**
Los Angeles County MTA	CA	Los Angeles	114,473	116.3	984.3
Tri-County Metro Trp Dist	OR	Portland	92,053	97.9	940.3
Metro Transit	MN	Minneapolis	22,674	24.2	936.9
San Diego Trolley, Inc.	CA	San Diego	88,368	97.0	911.0
New Jersey Transit Corp	NJ	Newark	55,163	67.1	822.1
Bi-State Dev Agency	MO	Saint Louis	61,532	81.0	759.7
Sacramento Reg Tr Dist	CA	Sacramento	40,815	62.6	652.0
Southeastern Penn TA	PA	Philadelphia	90,552	171.0	529.5
Dallas Area Rapid Transit	TX	Dallas	50,275	101.2	496.8
Oasis Line	**OH**	**Cincinnati**	**8000**	**17.0**	**470.6 (OKI 2030 Proj)**
Regional Transit Auth	LA	New Orleans	10,997	26.0	423.0
Hillsborough Area Reg TA	FL	Tampa	1352	3.2	422.4
Port Auth of Allegheny Co	PA	Pittsburgh	18,769	44.8	419.0
King County Dept of Trp	WA	Seattle	874	2.1	416.0
Maryland Transit admin	MD	Baltimore	20,301	54.0	375.9
Santa Clara valley Trp Auth	CA	San Jose	26,534	71.5	371.1
Greater Cleveland Reg TA	OH	Cleveland	6963	33.0	211.0
Memphis Area Transit Auth	TN	Memphis	2173	10.5	206.9
Kenosha Transit	WI	Kenosha	73	1.9	38.2
City of Galveston/Island Tr	TX	Galveston	69	5.0	13.8

Bold indicates if the concerned items are statistically signifincant at 95% level or not

are city-to-city comparisons? For instance, San Francisco has had continuous rail transit service for several generations, and its light rail operates in coordination with multiple layers that include the BART heavy rail subway, CalTrains commuter rail, cable cars, and a huge bus system. Its continuous emphasis on public transit and its highly constrained geography give it density that would be expected to yield high ridership. Another example, Cleveland's RTA, does not show well on the national rankings. However, it probably has a good chance of survival, since it serves some of the wealthiest suburbs in the nation—suburbs which have grown up and formed, dependent on light rail.

At this point, if a debate opens up, the ridership may be only one among many factors argued. Somebody may assert that the line could give enough accessibility to foreclose any desire for the Rte. 32 expansion. Another member may question whether it would present similar problems in crossing the Little Miami River. Finally, one hopes that a politically sensitive member would speak to the attitude of the community and how the project is likely to be received and what other battles it may spawn. And that's just for a subcommittee of one organization, at a local level! So the final judgments are not only qualitative, but highly politicized, as with any question involving large expenditures of public monies.

The answer to the central question of the present study is that the Oasis Line will probably have enough riders that Sierra Club won't get hurt on that front, but there is a lot more to it than that. While the ridership data cannot be available because the Oasis Line Corridor has not been fully completed and put into operation yet, the trend will be monitored and then analyzed in the future report. Also, the study and associated assessment method is also expected to provide an experience as a reference for future similar projects in other areas in the USA [26].

Table 4 1st Quarter 2008 Light rail agency ridership, boardings per station [12], [15, 16]

Q1 2008 non-linked avg. daily ridership ranked by boardings/station					
Agency	State	City	AVG daily boardings	Number of stations	Boardings/station
San Francisco Muni Rwy	CA	San Francisco	118,534	9	13,170.5
Massachusetts Bay Tr Auth	MA	Boston	226,186	70	2231.2
Los Angeles County MTA	CA	Los Angeles	114,473	49	2336.2
Regional Trp District	CO	Denver	53,307	23	2317.7
Bi-State Dev Agency	MO	Saint Louis	61,532	28	2197.6
Metro Transit	MN	Minneapolis	22,674	16	2160.5
Southeastern Penn TA	PA	Philadelphia	90,552	46	1968.5
San Diego Trolley, Inc.	CA	San Diego	88,368	49	1803.4
SORTA	**OH**	**Cincinnati**	**9133**	**6**	**1522.2 (Pelz 2030 Proj)**
Tri-County Metro Trp Dist	OR	Portland	92,053	62	1484.7
Dallas Area Rapid Transit	TX	Dallas	50,275	34	1478.7
Metro Transit	MN	Minneapolis	22,674	17	1333.7
Regional Transit Auth	LA	New Orleans	10,997	9	1221.9
Niagara Frontier Trp Auth	NY	Buffalo	18,249	15	1216.6
New Jersey Transit Corp	NJ	Newark	55,163	49	1125.8
Sacramento Reg Tr Dist	CA	Sacramento	40,815	41	995.5
SORTA	**OH**	**Cincinnati**	**8000**	**9**	**888.90 (OKI 2030 Proj)**
Port Auth of Allegheny Co	PA	Pittsburgh	18,769	25	750.8
Maryland Transit Admin	MD	Baltimore	20,301	32	634.4
Santa Clara Valley Trp Auth	CA	San Jose	26,534	57	465.5
Sound Transit	WA	Seattle	2436	6	406.0
Memphis Area Transit Auth	TN	Memphis	2173	7	310.4
Greater Cleveland Reg TA	OH	Cleveland	6963	34	204.8
Hillsborough Area Reg TA	FL	Tampa	1352	8	169.0
King County Dept of Trp	WA	Seattle	874	9	97.1
Kenosha Transit	WI	Kenosha	73	2	36.3
City of Galveston/Island Tr	TX	Galveston	69	3	23.1

Bold indicates if the concerned items are statistically signifincant at 95% level or not

Acknowledgements This research was supported by the University of Cincinnati, Cincinnati, Ohio. The authors also express gratitude to the Miami Group of the Sierra Club for exposing many of the issues that gave rise to this study. The results and ideas presented in the paper represent the authors' points of view only.

References

1. Wikipedia (2017) "Rail transportation in the United States." Available at https://en.wikipedia.org/wiki/Rail_transportation_in_the_United_States. Accessed 9 March 2017
2. Bureau of Economic Analysis (BEA) (2017) National data. Accessible at https://www.bea.gov/regional/index.htm. Accessed 9 March 2017
3. Hickman L (2016) "The rising demand for public transport in the US". Accessible at http://blog.marketresearch.com/the-rising-demand-for-public-transport-in-the-us. Accessed 9 March 2017
4. American Public Transportation Association (APTA) (2017) Public transportation fact book. Accessible at http://www.apta.com/resources/statistics/Pages/transitstats.aspx. Accessed 7 March 2017
5. Brown JR, Neog D (2008) "Urban structure and transit ridership: a reexamination of the relationship in the United States." In: Compendium of papers CD-ROM, 87th transportation research board annual meeting, January 13–17, 2008, in Washington, DC
6. Federal Highway Administration (FHWA) (2005) Traffic congestion and reliability: trends and advanced strategies for congestion mitigation. Available at http://www.ops.fhwa.dot.gov/congestion_report/index.htm
7. Pages ER, Lombardozzi B, Woolsey L (2016) The emerging U.S. rail industry: opportunities to support American manufacturing and spur regional development. Available at https://www.nist.gov/sites/default/files/documents/mep/Rail-Report.pdf. Accessed 9 March 2017
8. Schrank D, Lomax T (2007) The 2007 urban mobility report. Available at http://mobility.tamu.edu/
9. Weisbrod G, Fitzroy S (2008) "Defining the range of urban congestion impacts on freight and their consequences for business activity." In: compendium of papers cd-rom, 87th transportation research board annual meeting, 13–17 Jan 2008, in Washington, DC
10. Eastern Corridor Program (ECP) (2017) "Oasis rail transit overview." Available at http://easterncorridor.org/projects/oasis-

rail-transit/oasis-rail-transit-project-overview/. Accessed 9 March 2017
11. Ohio, Kentucky, and Indiana Council of Governments (OKI) (1999) 1999—eastern corridor major investment study (MIS). Available at http://www.easterncorridor.org. Accessed 30 May 2008
12. Ohio, Kentucky, and Indiana Council of Governments (OKI) (2005) Eastern corridor tier one draft EIS. Available at http://www.easterncorridor.org. Accessed 29 May 2008
13. Sierra Club (2008) "Stop sprawl." Available at www.sierraclub.org/sprawl. Accessed 8 June 2008
14. McKinley SA (2008) A feasibility analysis of hamilton county's proposed oasis light rail line through traditional and advanced demand forecasting vs. best practices. In: Course project report for CEE605 travel demand forecasting offered at University of Cincinnati, Spring Quarter, 2008
15. Pelz ZL (2007) A station level analysis of competing light rail alternatives in cincinnati's eastern corridor. Master's thesis in community planning approved by the University of Cincinnati, June 2007, University of Cincinnati: Cincinnati
16. American Public Transit Association (APTA) (2008) Light rail agency ridership for Q1 2008. Available at www.apta.com/research/stats/rail/lrmiles.cfm. Accessed 1 June 2008
17. Hoffman K (2000) "It's a smooth ride in mile high city," Houston Chronicle June 6, 2000. Available at http://bicycleaustin.info/rail/Houston-chronicle-06-06-00.html. Accessed 9 June 2008
18. Kuby M, Barranda A, Upchurch CD (2004) Factors influencing light-rail station boardings in the United States. Transp Res Part A 38:223–247
19. Light Rail Now (LRN) (2017) "Light rail now! light rail progress." Available at www.lightrailnow.org. Accessed 3 June 2008; re-accessed 20 Jan 2017
20. Upchurch CD (2005) A spatial decision support system for predicting light-rail transit ridership in phoenix. Master's thesis approved by Arizona State University, September 2005
21. Weyrich PM, William OSL (1999) "Does transit work? A conservative reappraisal." Available at http://www.apta.com/research/info/online/weyrich2new2.cfm. Accessed 9 June 2008
22. Portune T (2008) April 10, 2008 interview by Author, with other members of Sierra Club
23. Siemens AG (2008) Light rail system references (Internet database). Available at http://references.transportation.siemens.com/refdb
24. Ortuzar JdD, Willumsen LG (2001) Modelling Transport Third Edition 2001. Wiley, Chichester
25. Ohio, Kentucky, and Indiana Council of Governments (OKI) (2008) DRAFT 2030 OKI regional transportation plan update. Available at http://www.oki.org. Accessed 9 May 2008
26. HDR (2016) OASIS rail conceptual alternative solutions HAM/CLE—OASIS rail corridor. In: PID No. 86463 final report for Ohio Department of Transportation. 8 Feb 2016

11

Tribological Aspects of Wheel–Rail Contact: A Review of Wear Mechanisms and Effective Factors on Rolling Contact Fatigue

Hesam Soleimani[1] · **Majid Moavenian**[1]

Abstract Nowadays, the railway is considered as a means of fast, aggregate and secure transportation. The issue of the rail and wheel contact is one of the most fundamental aspects in the railway system since inappropriate interactions create problems such as wear and a negative effect in the dynamic functioning of the train. In this research, we analyze one of the most important phenomenon in a railway system which is the contact of the rail and wheel. This review is done in the scope of the most fundamental deteriorating aspect of the rail and wheel system, the wear. The wear in the contact of objects along with its occurrence between the wheel and rail is also analyzed.

Keywords Wear · Wheel–rail contact · Railway system · Surface roughness · Transportation · Inappropriate interaction

1 Introduction

In the present century, transit and transportation in terms of loads and passengers are essential issues of human life and a large volume of air, territorial, sea and rail transportation is being performed every day. The railway system has been considered as a means of fast, aggregate and secure transportation as well as a serious rival for air transportation in the fields of security and speed. The most significant advantages of railway transportation are lower energy consumption, aggregate transportation, higher security and environment preservation. The issue of the rail and wheel contact is one of the most fundamental aspects in the railway system, and the interaction of the wheel and rail in the railway transportation system has always been taken into consideration for many years. Studying the wear phenomenon between the rail and wheel has led to a design of the suspension system and wheel profile which is proportionate to the wear process because using the optimum wheel profile will lead to dynamic stability, comfort in expedition and security against emersion from the rail especially in arched paths [1–4].

The interaction of the rail and wheel and the resulting phenomena such as wear and noise are considered as very important issues in railway transportation which allocates considerable research activities in advanced railways in the world yearly since the inappropriate interaction of the rail and wheel has always brought about costly items such as extreme structure of the instruments and fractures of pavement and road bed of the lines and also automotive transportation and most importantly causes irreparable accidents. By analyzing the repair and maintenance railway costs worldwide, we found that a huge expedition is done in the field of purchase, change and installation of the rail and wheel in the railway industry every year [5, 6]. The wheel has a special geometric form, and the geometric incompatibility of a wheel's profile with the geometric characteristics of the rail line could have negative circumstances such as extreme wear of the rail and wheel and dynamic instability of the railway transportation appliance, since the stability of the railway appliance and the passengers comfort is dependent on the wear of the rail and

✉ Hesam Soleimani
Hesamsoleimani@gmail.com

Majid Moavenian
Moaven@um.ac.ir

[1] Department of Mechanical Engineering, Ferdowsi University of Mashhad, Mashhad, Iran

Editor: Gary Barber

wheel [7]. Thus, it could be inferred that the compatibility of the rail and wheel's profile is the most vital factor in controlling the wear. The rolling contact of the rail and wheel will cause wearing of the rail and wheel profile and in return there will be phenomena such as negative effects in dynamic function of the train and comfort in traveling, the possibility of emersion from the line, permanent misshapes, reduction in the wheel longevity, reduction in security, reduction in the repairing period of the line and, as a result, a reduction in efficiency and an increase in costs [8]. Recognition and reduction in the mentioned phenomena require sufficient knowledge of the wheel and rail wear. Due to the solidity of the contact surfaces of the rail and wheel, small contact area and the heavy weight applied in railway cars, there is high stress that would lead to phenomena such as wear and fatigue of the rail and wheel.

The importance of the wearing issue in terms of security and financial aspects has made wear a significant issue in the railway industry. The studies regarding wearing of the rail and wheel in the rail industry began a half century ago because in the late 1960s there had been an increasing wheel wear. Moreover, the wear problem was taken into consideration from 1970 to 1996 as an important issue by railway researchers [9]. In these studies, speed and the load base of the railway appliances were introduced as the most important factors of wear in railway appliance wheels [10]. Friction and wear between two surfaces when these surfaces have rolling movements, while being loaded, are an inevitable consequence. Friction is an important factor in energy loss and wear is an important factor in material loss; thus, by reducing or controlling wear you can economize the material and energy use. To control and reduce the friction and the wear rate of two surfaces, there are preventive and controlling methods such as using different materials in the production of rail and pavement materials of the line, effective methods of lubrication, speed regulation and radial load used in different railways [5]. It must be noted that using the railway system in non-ideal conditions would cause the rail and the wheel to be worn and would reduce the functionality and security of the system. Studies that have been done using field measurements, using the byline materials, show that the wear between the wheel and rail especially in the flange area relies to a large extent on the bogie condition (axle) and its functionality [11].

The introduction of the rail and wheel could be separated into two periods. In the first period, the speed of the trains was relatively low and the rail and wheel contact was analyzed statically. Many of the research studies in this era led to a series of theories that were all based on the Hertz contact theory and would hypothesize that the contact area had an oval shape [7]. In 1926, the two-dimensional rolling theory with friction was introduced by Karter [12] to be used in railway appliances dynamics. Johnson and Vermoulen [13] have extended their theory to include an ideal surface with creep and without rotation. In this theory, the contact surface of the rolling parts was divided into two areas (slip and adhesion). Haling et al. [14] have introduced improvements for the approximate theory of oval contact with the net linear tripping. Kalker demonstrated a new approximate method that resembled Karters's model [15]. Moreover, Kalker [16] used numerical theories and methods based on the finite element method. Xia and Gave [17] demonstrated a model to analyze the side stability in the arch, using the recoil rate and the damping of the rail and its infrastructure. Zhandel [11] worked on the wear forecasting of the wheel in comparison with the results of field measurements. Telliskivi and Olofsson [18] used the Metallurgical point of view and the concept of wear number to obtain the wear amount of the line and rail. Diana and Brownie [19] analyzed the effects of the wear of the rail and wheel on dynamic characteristics. Braghin and Louis [20] investigated the wear amount of the wheel profile using a mathematical model. Ranger [21] analyzed the side stability of the wagon with high speed in the arches. Turner [22] used a disk on disk instrument to determine the effect of speed and the percentage of trip on the wear of rail and wheel.

The first level of wearing includes surface destruction, but in the following steps it could lead to friction, noise, unwanted heat and dimensional changes in different parts. These phenomena would interfere with the functionality of the products and could possibly lead to a disastrous failure. In this study, we will analyze the most important existing phenomenon which is the contact of the rail and wheel. This consultation is done in the scope of the most fundamental deteriorating aspect in the rail and wheel system, the wear. In this research, we will analyze the wear phenomenon in the contact of objects along with its interaction between the wheel and rail, and then, the effective factors of the wheel and rail are introduced.

2 The Contact Surface of the Rail and Wheel

An ideal sample design for the rail and wheel profiles is demonstrated in Fig. 1. There are three contact areas. The first area is for the motion of straight lines and the conical coefficient, which is defined as the ratio of half the difference between left and right wheel rolling radius of an axis to movement of transverse axis, is often designed from 0.15 to 0.2 for passenger trains. Higher values depending on the structure of bogies would cause dynamic instability, and lower levels will cause higher number of flange hits and faster increasing wear. The second area is for passing through the normal arcs and needs to have a conical

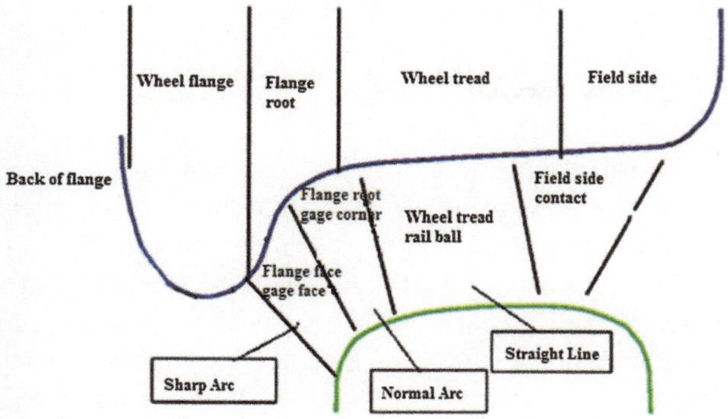

Fig. 1 Contact points of the rail and wheel in a variety of paths [23]

coefficient of 0.2–0.6 in the region, while in the third area higher values are required [23].

In the absence of sidelong stimulations in straight lines, wheel and rail contact areas are shown in Fig. 1 and will generally be focused in the center of the wheel and the rail. In passing arch paths, the outer wheel of the front axle of a bogie is diverted to a greater extent to the flange rather than outer wheel of the back axle one.

Keeping all of the above general information, contact areas of the wheel and rail (Fig. 2) can be divided into three areas [24, 25].

2.1 Contact Area A (The Central Rail and Profiles Crown Center)

More frequently in straight lines and gentle arcs, the type of contact is rolling, as we have a little slip, and the wear is called vertical abrasion wear. The contact area can be characterized as follows [24]:

- Contact stress between wheel and rail is lower than in other areas.
- Longitudinal creep forces are larger compared to the side creep forces.
- Crown wheel depression occurs due to wear.

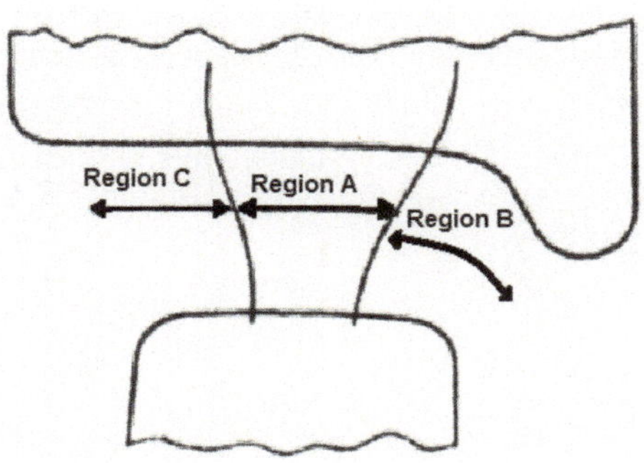

Fig. 2 Contact areas of rail and wheel [25]

Exfoliated skin of rail, rot, deformation and transverse and longitudinal cracks are the consequences of vertical wear in region A.

2.2 Contact in Area B (Contact Between the Inner Edge of the Track and Wheel Flange Arc)

In sharp curves due to centrifugal forces, the contact is pure slip and abrasion wear is called side wear. Contact characteristics in this area can be mentioned as follows [24, 25]:

- High contact stresses.
- Wear of the inner surface of the outer arc rails.
- High side creep and sharp flange.
- This area is most important in determining the life of the wheels and rails.

In the contact area B, there are three contact parts.

2.2.1 Single Contact Point

This type of contact is the worst contact condition of wheels and rails (Fig. 3). High contact stresses, due to focus of load on a small area of contact with high creep in this area, cause fatigue in the rolling contact of the inner side of the rail. This type of exposure can lead to wheel flange wear, too. This disaster results in an asymmetric grinding wheel in straight lines [24].

2.2.2 Two Contact Point

In this contact mode, the wear of flange is severe enough that the lubricant layer is eliminated in the arcs and metal-to-metal contact occurs (Fig. 4). The internal side rail in this contact is worn intensively [24].

2.2.3 Conformal Contact

This type of contact can be seen with the simultaneous wear of the wheel flange and inner side of the rail and contact of the two profiles in curved lines (Fig. 5). In this

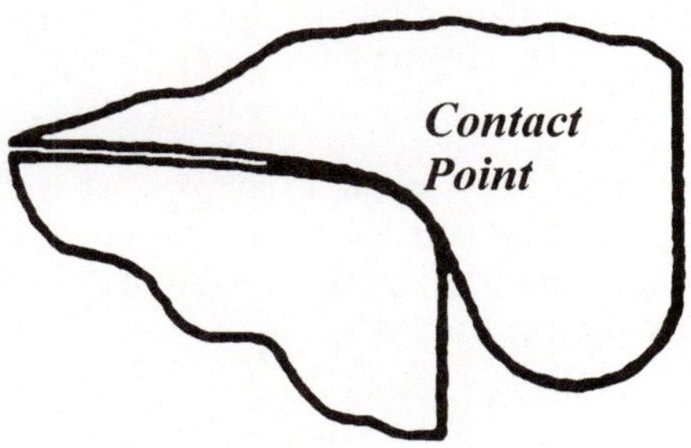

Fig. 3 Single contact point schematic [24]

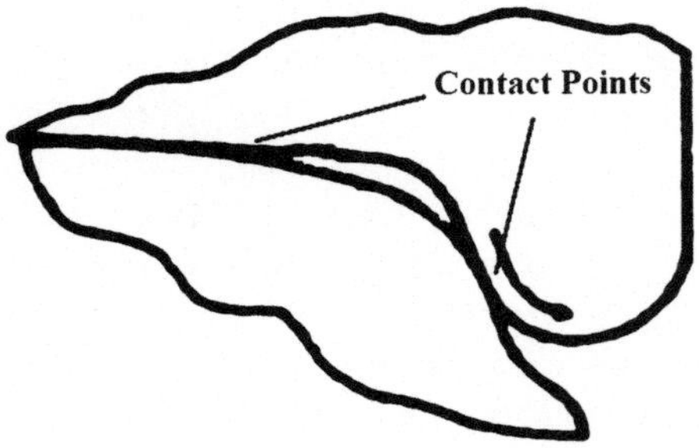

Fig. 4 Schematic of two contact points [24]

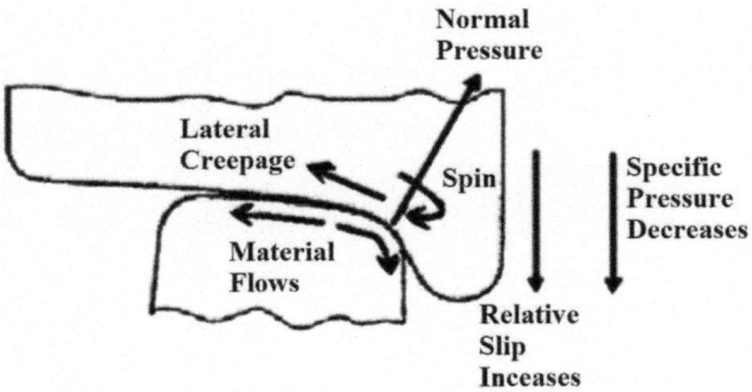

Fig. 5 Wear model of flange contact

case, the inner side wear of the rail is not considerable. Due to the reduced contact stress, there is always the possibility of a thin layer of lubricant to prevent direct contact between wheel and rail that is a contact of metal with metal. In this contact, the wear rate remains constant throughout the service life. In addition, lubricating this contact mode reduces the creep force between the flange and the medial collateral to reduce the wear rate profile. Moreover, this contact type is more resistant to rolling contact fatigue of rail and wheel contact. This type of exposure is more resistant to indentation of rail and reduction in under surface rolling contact fatigue destructions in rail and wheel contact [24, 26].

2.3 Contact Zone C (Between the End of the Crown Wheel and the Crown of the Rail)

The contact may occur due to poor design of structures and in this case arise against the wheel flange contact with the internal side rails and removal from the line occurs. In this area, one of the two modes is possible [24]:

- High contact stresses due to the contact of the end of the crown wheel with the rails.
- Completion of the contact before the virtual wheel flange meets the end of the crown.

3 Wear in a Variety of Rail Lines

3.1 Wear in Straight Lines

Rolling friction of the wheels and the rails on the side of the cap causes wear in rails in straight lines and can make the most of this type of wear vertical resulting in reduction in rail height. In this case, the amount of wear on rail transit burden depends on the weight of the vehicle, while passing the rail and the weight of the load at the contact of wheels and rails is transferred to the rails. The large amount of stress that exceeds the strength of steel rails results in the deformation and plastic strain to a limited depth below the top surface of the rails [5].

3.2 Wear in Curved Lines

The most important part of the wheel to be affected by wear is the flange area, which is used to maintain dynamic stability of the vehicle and prevent its removal from the line. According to published reports from wear situation on the London Underground, the factors affecting the deterioration flange, the most effective agent, are moving in severe arches [27]. The severity of wear is to the extent that the flange wear is reduced 16–21 times only by using appropriate lubrication in arches.

In the arches, particularly in the outer rail of the arch and other than vertical abrasion, sidelong wear can also be seen that depends on the amount of traffic and the radius of the arch in addition to the width and speed of trains. This type of wear is usually seen in less than 500-m-radius arches. In examining the causes of this type of wear, it can be said that the wheels of the train input forces when they pass through an arch on a line, which includes the centrifugal force and the forces of the arch itself, and it must be taken into consideration that the centrifugal forces are caused by the transverse slope. If the transverse slope of curved lines is less and line width is more than a limit value, the wear rate decreases [9]. In some lines, transverse slope is practically considered more than the balanced amount which intends to help move trains pass faster. In this case, the internal rail has a greater burden to bear, causing more wear and a greater chance of breaking, as it appears with a vertical slit in the crown. If the slope width is less than the balance, the outer rails bear greater loads and rail rolling risk or the wheel coming off the rail is increased [5].

4 Wear Mechanisms

4.1 Abrasive Wear (Scratch)

Abrasive wear occurs when a hard surface is rubbed against a soft surface and by penetration, causing a track on the soft surface. The special symbol to separate this from other types of wear is that there is no wear regarding tiny particles that cause the severe losses [28]. This type of wear is divided into two-object and three-object wear. In the two-object wear, there is a scratch on the surface and it only happens when a hard surface has slipping movement against a softer one and causes penetration and tracks, and leads to weight decreases in the softer one. The three-object type occurs when an abrasive matter separates some particles from the interface between the slipping parts [29]. The presence of sand utilized to improve the adhesion and increase traction of locomotives can be seen as the main factor leading to grazing of locomotive wheels [30] (Fig. 6).

Abrasive wear occurs as follows by different mechanisms (Fig. 7).

The microplowing phenomenon is primarily seen on very soft surfaces. The key indicator for this mechanism is a strain in a relatively large area around the surface tracks. In this case, scratches on the soft body would not scrape off the material, and the material only move on the surface and is regularly stored as a bulge on both sides of the created groove. Transfer of microplowing mechanism to microcutting takes place when the hardness of the material is increased. If the hardness of the sliding surface increases more and more, the abrasive wear can transform to microcracking. In this mechanism, the particles are rubbed off due to the formation and growth phenomenon of a crack embedded in the grooves. Particles accumulated in slots around the abrasive surface move constantly that would lead to a microfatigue phenomenon [31].

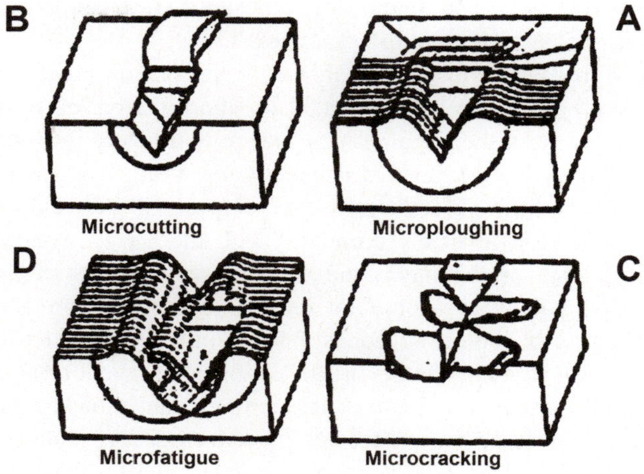

Fig. 7 Abrasive wear mechanisms: **a** fine plowings; **b** fine cuttings; **c** fine cracks; **d** microfatigue [32]

The rate of hardening, ductility of metal, homogeneous strain distribution and mechanical instability are the most important factors in the abrasive wear mechanism.

4.2 Adhesive Wear

Surfaces of engineering parts which are produced by different processes are not completely flat and have different non-homogeneities. As a result, contact of the engineering parts on the surface occurs at some scattered points which creates high stress around these points. This incident causes plastic deformation and local contacts on the surface. Adhesive wear occurs if there is local slip between two surfaces in the joints and, eventually, causes failure by the transfer of material from one surface to the other (Fig. 8). Adhesive wear mechanism is divided into two forms of moderate and severe decline in the rate of wear of the metal from the body and particle size [33]. The moderate form appears in straight lines and gentle arches, in which lubrication is not used. In this mode, a form of severe wear and lubrication in dry flanges occurs.

The tendency to form adhesive connections depends on the chemical and physical properties of the materials

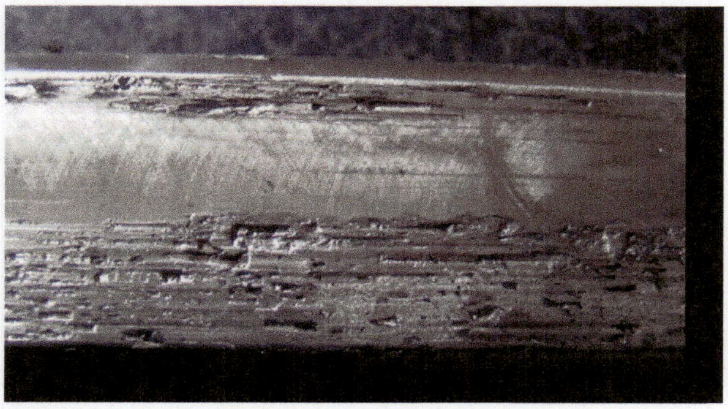

Fig. 6 Wear scratches on rails [30]

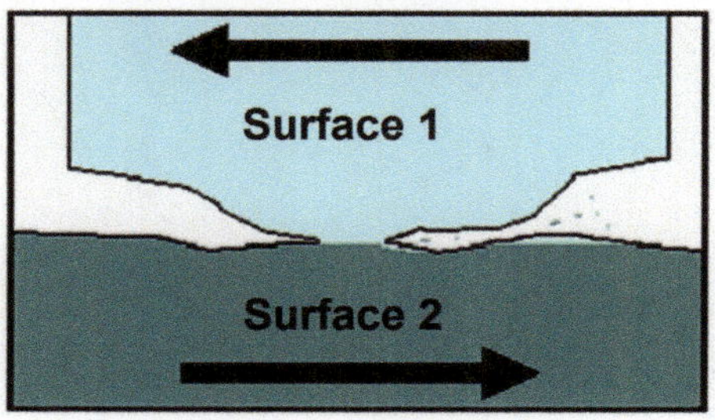

Fig. 8 Adhesive wear mechanism [34]

involved, type and amount of load and surface properties such as pollution and the roughness [34].

Since the surface is rough and wavelike, the points of contact time are short-lived. This short period of time characterizes the slipping of a system and determines the thickness of the layer that is undergoing wear. Component life correlates directly with the contact of the sliding velocity and inversely correlates with the surface roughness. One of the main sources of energy dissipation in slipping is what should be done in the formation and breaking of molecular bonds that are in contact with two levels of slipping contact (Fig. 9). Molecular reactions lead to the formation of more or less strong connections to the so-called microseizure. Bodies involved rigidity, modulus of elasticity and other factors that determine the level of contact determine the type of stop.

The Archard wear equation to examine the phenomenon of adhesion is presented as follows:

$$W_{ad} = \frac{V}{L} = K\frac{F_N}{H} \quad (1)$$

In this relation, W_{ad} is wear rate (volume per unit of distance traveled worn sliding), K coefficient of wear, V worn volume, L sliding distance, F_N vertical load and H the hardness of the softer material. In this equation, hardness is considered as the only reason of wear as the feature of the material K depends on the contact pressure levels and sliding speed of the two objects.

4.3 Delamination Wear

Delamination wear can cause microscopic wear. If the plastic deformation surface layer is prevented, the wear rate is greatly reduced. In this type of wear, the surface of material is thought to be separated by the process of wear [31]. Delamination wear occurs in three steps by plastic deformation, germination of cracks and crack propagation [28]. According to the delamination wear theory, the deformation of a plastic shape, crack formation and growth are created in a short depression of the surface that ultimately caused separation of sheets of wear particles.

When the sliding surfaces are in contact with each other due to repeated loading, smoother surface is easily deformed. When this heterogeneity of the surfaces is modified, the surface is relatively become smooth [35].

Due to alternate loading of harder heterogeneity on the softer surface, plastic shear deformation is accumulated and some points are deformed at the surface and below the surface. This causes the creation of cracks below the surface. The plastic deformation causes it to expand and grows to the adjacent cracks and cavities. In locations with low strength, these cracks cut the surface from depth and wear particles appear as thin and long sheets. The thickness of wear particle formed depends on the growth of cracks in the surface and controlled vertical and tangential forces (friction). When a hard metal with a high elastic modulus contacts a soft metal with a lower elastic modulus, surface dislocations in the soft metal is driven into the interior, so hardworking is created under the layer of softer metal. Figure 10 shows an example of this type of wear on the wheel [30].

4.4 Tribochemical Wear

This type of wear on contact surfaces results from reactions with the environment. The surrounding environment can be in the state of gas or liquid. Wear layers are produced through a continuous separation process, and the reaction products of the reaction layer at the interface is seen in this type of wear. This means that the process of the abrasion of the surface layers of the products in chemical reactions begins and continues with the separation of them [28]. Tribochemical wear is greatly influenced by the surface layers and their properties that determine the resistance to separation. These properties include flexibility, strength and adhesion to the substrate. The tribochemical wear mechanism is shown in Fig. 11.

On one hand, abrasion-resistant materials used as additives for lubricants can lower metal to metal contact friction and therefore reduce adhesive wear; but on the other hand, the use of these substances leads to the formation of surface layers. As a result, the use of lubricants and

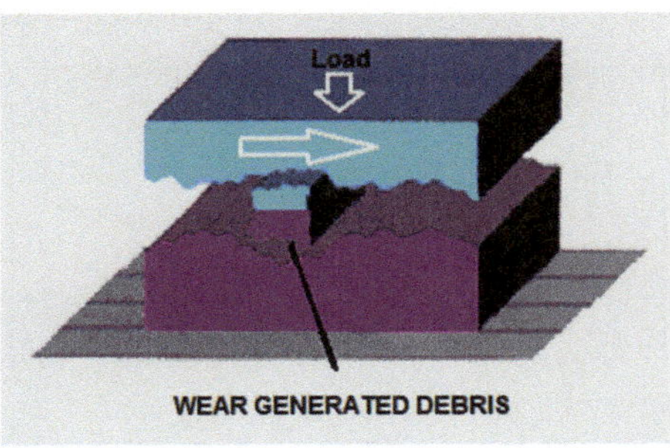

Fig. 9 Particles created from adhesive wear [34]

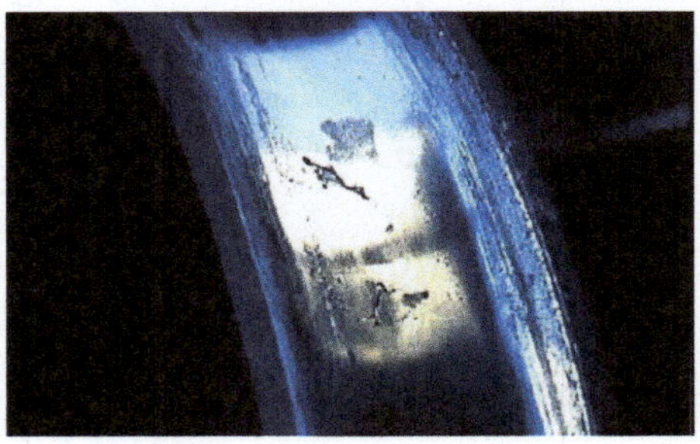

Fig. 10 Delamination wear of wheel [30]

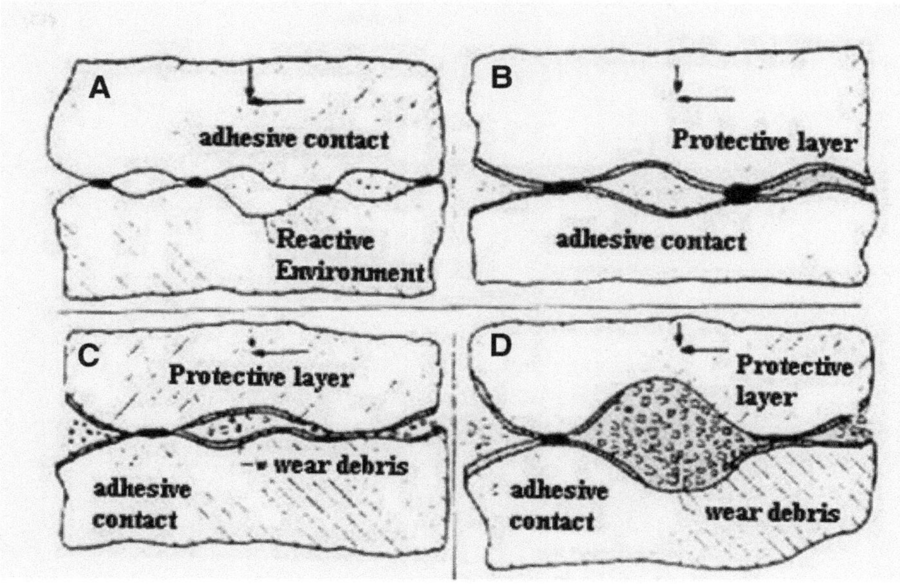

Fig. 11 Tribochemical wear mechanism [34]

additives prevents a sharp deterioration due to the increased risk of adhesion [34].

Increased chemical activity levels happen by increasing the thickness of the protective layer. Therefore, the level of adhesive contact and eventually adhesive wear will be reduced. Wear of a thin surface layer leads to a wear reduction, by increasing the thickness of the brittle surface layers, which increases the tendency to crack. Chemical activity to minimize the amount of wear depends on the severity of the slip, contact pressure, temperature and surface quality. Tribochemical reactions which produce a harder surface layer can reduce wear, but on the other hand increases abrasive wear.

4.5 Fretting Wear

Fretting wear is caused when two surfaces would have tangential and oscillating movements under the applied loads with low relative amplitudes and the slip is caused by vibratory or cyclic stresses [36, 37]. This phenomenon is often associated with corrosion and oxidation. As shown in Fig. 12, observed by electron microscopy, fretting wear mechanisms are evident in four stages [38]:

A. Adhesion and metal wear particles (Adhesive wear).
B. Creation of wear particles by mechanical–chemical effect. Mechanical action causes breaks off the oxide film and cleans the surface and strains of the metal, which is due to be active in the next half cycle, and the presence of this level in atmosphere causes oxidation (tribochemical wear).
C. Production of a uniform particle abrasion by fatigue (fatigue surface wear).
D. Wear oxide particles are created from the process as an abrasive powder and result in the ongoing destruction of the surface (abrasive wear).

Among the factors affecting the deterioration due to fretting are the number of cycles, the range of motion, force, frequency, temperature, hardness, lubrication and atmospheric environment.

4.6 Surface Fatigue Wear

Surface fatigue is a phenomenon caused as a result of the effects of stress fluctuations under the condition which two solid materials have sliding contact [38]. This kind of wear is created with the formation of cracks and separation of material from the surface due to use of the repetitive alternating forces [36, 37]. The rotating contact or slipping and also alternative impact can cause cycle stresses on the surface. Despite the pollution-free lubricant, adhesive and abrasive wear is not created, but surface fatigue causes destruction of parts due to the alternative loading caused by the reaction of the components, and the destruction may begin with a defect or surface cracks [39]. Figure 13 shows the schematic of the crack formation process.

This kind of wear is observed in smooth lines and arches with proper lubrication. The wear particles stick to the surface of the wheel and the rail and they don't separate until they are removed like sheets of contact surfaces (Fig. 14). The reason of this kind of wear is the separation of particles which were from the plastically deformed. One of the most useful ways to overcome this mode is suitable grinding of rails [40].

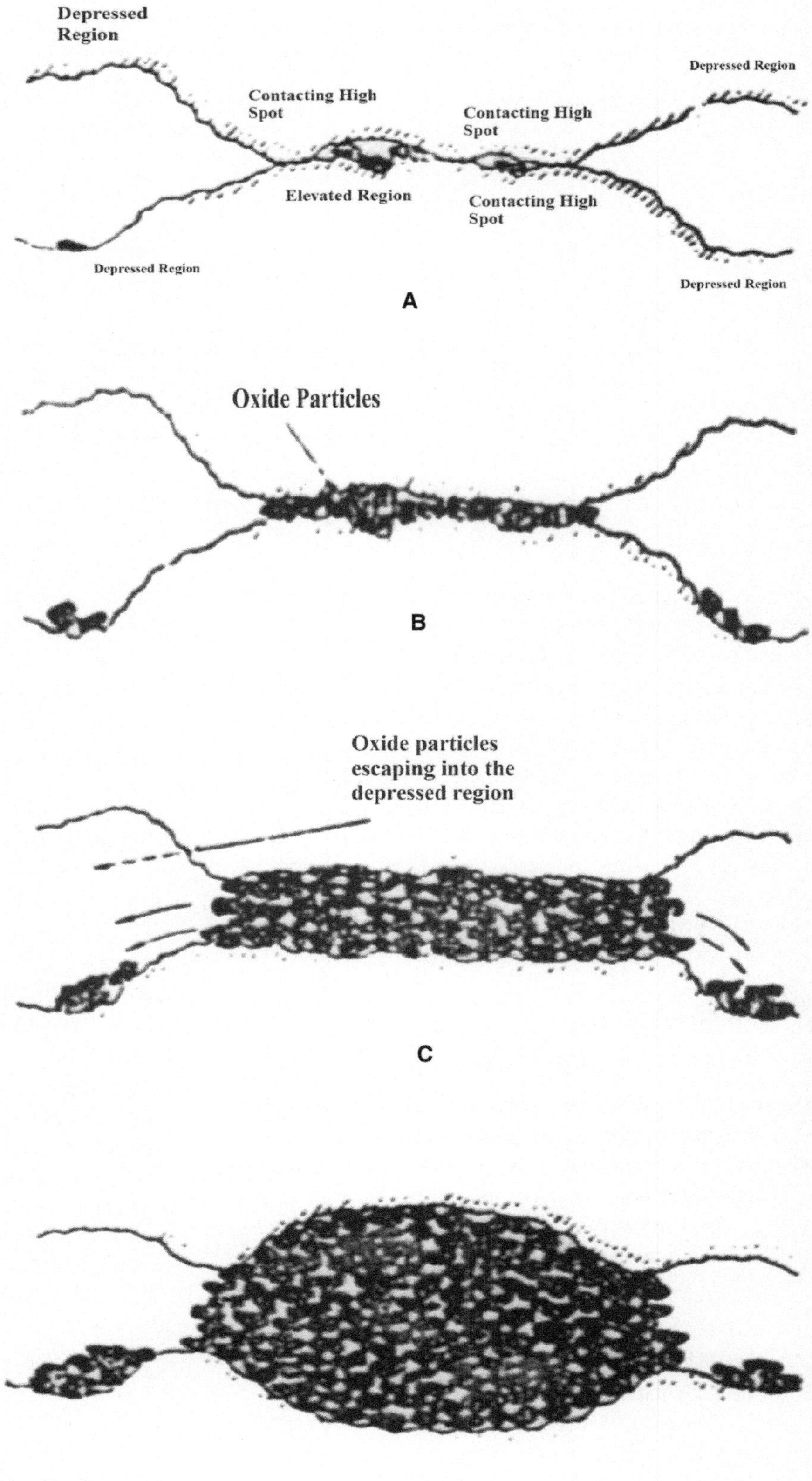

Fig. 12 Various stages of fretting wear, **a** accumulation of trapped particles in the space between high spots, **b** integration of contact areas into a united area, **c** spalling of particles into the adjoining depressed regions, **d** formation of the curved zone in the middle of the contact area caused by the intense wear process

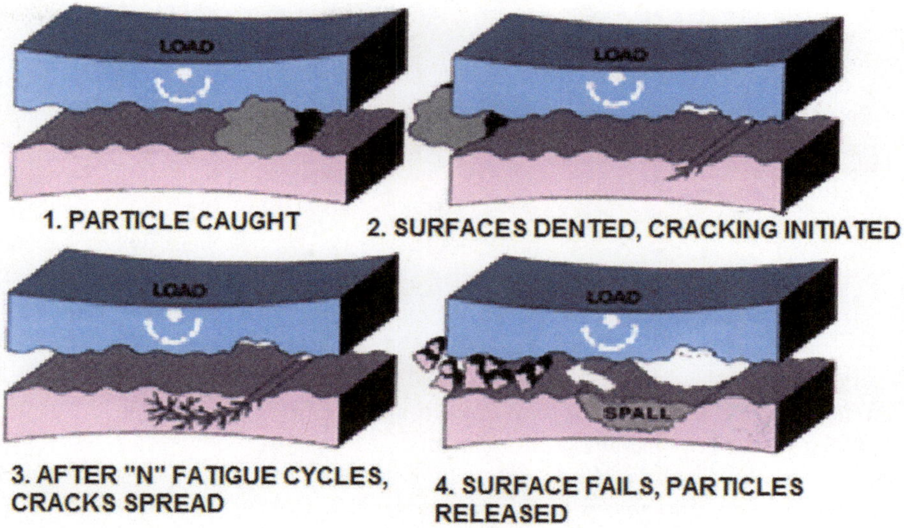

Fig. 13 Formation and propagation of cracks in the surface fatigue mechanism [39]

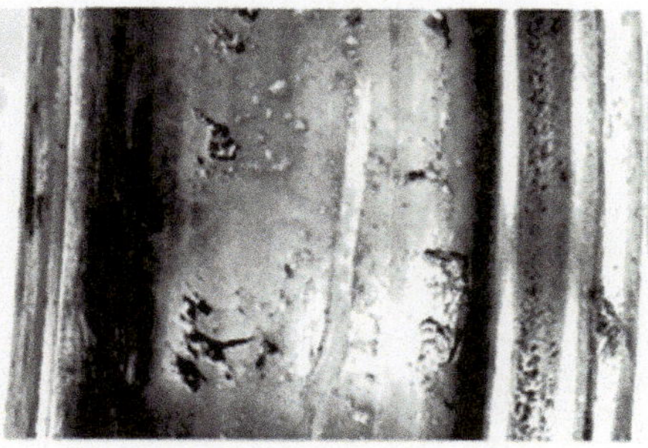

Fig. 14 Surface fatigue wear on the wheel [40]

4.7 Impact Wear

Impact wear can be produced when a solid surface is continuously in contact with another solid surface. The wear mechanism are related to [28]:

- Material used: for example, thermal wear is likely to occur in polymers.
- Contact tension range: such as contact fatigue in comparison with corrosive fatigue
- Speed: increasing speed makes corrosive wear with low stress to the adhesive wear.
- Type of loading: for example, tangential load causes a slip or vertical causes impact.

5 Parameters on the Wheel and Rail Wear

Theory and practice of management in terms of wheel and rail contact during the past 20 years are to increase the life of wheel and track and improve their adaptability and the use of greater axial loads. Wheels combined with high-strength steel-polished rail profiles lower wear and improve lubrication. Various effects such as difficulty of body contact, coefficient of friction, changes in lateral and longitudinal creep and wear volume are effective. Due to the lack of accurate predictions of actions and reactions created between rail and wheel (adhesion between the rail and the wheels, the alignment of rails and wheels together, rail level pollution, ambient temperature, humidity, hard surfaces, railways, the quality of the natural and physical infrastructure and other factors), determination of the level of each factor on the force between the rail and wheel wear rate is difficult [5].

5.1 Effects of Friction Coefficient

The coefficient of friction depends on factors such as the microstructure of materials, load, quality levels and how it is the measured. Contact zone temperature is also significantly affected by friction. By increasing temperature, the mechanical properties of the lubricant have changed, the speed of oxidation increased and there is even the possibility of fuzzy degradation. So, all of these affect the amount of friction [31].

Rail and wheel and glide wheel and the locomotive traction are reduced, to prevent the infiltration of lubricant when rolling solid lubricants can be used that have high viscosity [41, 42] (Fig. 15).

5.2 Effects of Adhesion Coefficient

The maximum force applied by the wheel and rail mechanism depends on two parameters of the coefficient of friction at the contact surface and axial load. Adhesion or adhesion coefficient is defined as the maximum tensile

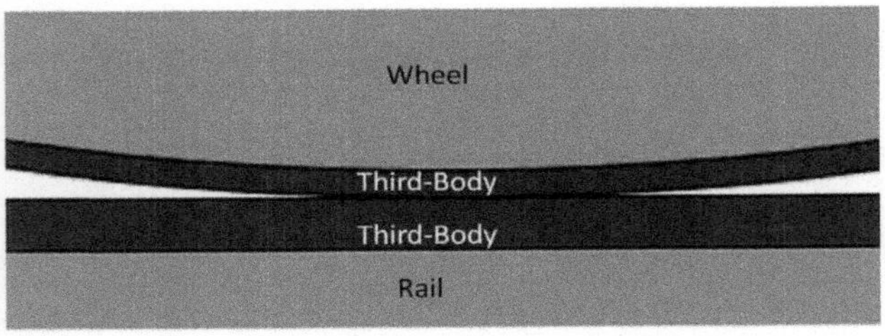

Fig. 15 Lubricating rails to reduce surface friction coefficient [42]

strength in the wheel rim to the vertical load of the wheel and is defined in the contact area at the slip moment [33].

When the wheel spins under the influence of torque to move the tub (Traction effort), stagnation or slip phenomenon occurs. The goal is to achieve maximum coefficient of adhesion at high-speed insertion. This is usually done through the use of special materials in the contact area between the wheel and rail in addition to increase in the coefficient of friction [40].

The coefficient of adhesion depends on several factors such as creep in the contact area of wheels and rails, the surface profile of wheels and rails, wheel and vehicle speed rail, rail vehicle dynamics and characteristics of the traction system. Surface conditions of the line such as the presence of oil and water can be effective in reducing the coefficient of friction and cause a decrease in the coefficient of adhesion. For this purpose, sand particles are used to overcome this defect. General parameters that affect the coefficient of adhesion are [40]:

- Locomotive weight and axle load distribution on the drive shaft
- Speed
- Dimensional changes due to abrasion wheel
- Surface rail
- Curved line (straight line or arc)
- Steel wheels and rails

In addition, the negative effects of excessive adhesion factor include [43]:

- Abnormal and uncontrollable movement of the wheels in the arches
- Creation and development of cracks in rails
- Increase wheel wear
- Unwelcome noise and high arches

Different materials are now used to increase or decrease the friction to control the adhesion in rail transport, but there should be a lot of care about the use of these materials because some of them (especially materials that are used to reduce friction) have a relatively permanent effect that will only be eliminated by machining the surface of the rails.

6 Conclusion

One of the most critical and the most important problems of the railway industry is wear. In some cases, the wheel and rail wear is severe enough to spread to the adjacent rail profiles and damage the adjacent rail lines. It is noteworthy that the most destructive factor causing wheel and rail wear, which can be significant and have negative consequences, is dynamic instability which is common to rail vehicles. Hence, different types of wear mechanisms, causes and the importance of preventing the emergence of any of them are fully examined. Identification of and reduction in the phenomenon, which can create the permanent deformation of the rails, reduce wheel life, reduce safety, increase maintenance time of the vehicle and thus reduce productivity and increase costs, require a good understanding of the wear mechanisms of the wheel and rail. This study investigated the phenomenon of the interaction between wheel and rail wear, including influencing factors such as friction, adhesion, abrasive and ways to reduce these phenomena in contact such as lubrication were also introduced.

References

1. Jendel T (2002) Prediction of wheel profile wear-comparison with field measurements. Wear 253:89–99
2. Kumar S (2007) The investigation of derailment. Indian Railways Institute of Civil Engineering, Pune
3. Soleimani H (2015) A study and comparison of different methods for wheel–rail wear measurement and proposing an applicable local method. MS Thesis, Ferdowsi University of Mashhad, School of Mechanical Engineering, Mashhad, Iran
4. Soleimani H, Nejad RM, Moavenian M (2016) Common failures in wheel and rail and different methods of measuring their profiles. In: Proceedings of the 1st international conference on mechanical and aerospace engineering, Iran
5. Zakeri JA, Ghorbani V (2011) Investigation on dynamic behavior of railway track in transition zone. J Mech Sci Technol 25(2):287–292
6. Mohammadzadeh S, Ghahremani S (2012) Estimation of train derailment probability using rail profile alterations. Struct Infrastruct Eng Mainten Manag Life Cycle Des Perform 8(11):1034–1053
7. Mahlooji V, Ghazavi MR (2014) Railway vehicle derailment in curve due to wear. Modares Mech Eng 14(8):199–208
8. Mir Majidi SMH, Ohadi A, Rezvani MA. Wear of wheel flange and its effects on Tehran's subway trains, Tehran. In: 10th conference and 3rd exhibition of railway transportation achievements
9. D'Acunto M (2003) Wear and diffusive processes. Tribol Int 36:553–558
10. Zakharov SM, Zharov I, Komarovski I (1999) Tribological aspects of rail/wheel interface. In: International heavy haul railway conference, St. Louis
11. Zakharov SM, Zharov IA (2005) Criteria of bogie performance and wheel/rail wear prediction based on wayside measurements. Wear 258(7):1135–1141

12. Carter FW (1926) On the action of a locomotive driving wheel. Proc R Soc Lond A Math Phys Eng Sci 112(760):151–157
13. Vermeulen PJ, Johnson KL (1964) Contact of nonspherical bodies transmitting tangential forces. J Appl Mech 31:338–340
14. Garg VK, Dukkipati RV (1984) Dynamics of railway vehicle systems. Academic Press, Ontorio
15. Kalker JJ (1967) On the rolling contact of two elastic bodies in the presence of dry friction. Doctoral Thesis, TU Delft, Delft
16. Kalker JJ (1987) Wheel–rail wear calculations with the program contact. Delft University of Technology, Delft
17. Zhai WM, Cai CB, Guo SZ (1996) Coupling model of vertical and lateral vehicle/track interaction. Veh Syst Dyn Int J Veh Mech Mobil 26(1):61–79
18. Telliskivi T, Olofsson U (2004) Wheel–rail wear simulation. Wear 257(11):1145–1153
19. Diana G, Bruni S, Braghin F (2005) Wheel–rail contact: wear effects on vehicle dynamic behavior. Dipartimento di Meccanica, Politecnico di Milano, Milan
20. Braghin F, Lewis R (2006) A mathematical model to predict railway wheel profile evolution due to wear. Wear 261:1253–1264
21. Ranjbar M, Ghazavi MR (2013) Bifurcation analysis of high-speed railway vehicle in a curve. AMAE Int J Prod Ind Eng 4(1):13–17
22. Pombo J (2012) Application of a computational tool to study the influence of worn wheels on railway vehicle dynamics. J Softw Eng Appl 5:51–61
23. Molatefi H, Firouzabadi Z (2011) Analyzing the interaction between wheel profile S1002 and Iran rail profiles to investigate wheel–rail wear and wheelset behavior. J Transp Eng 2(48):363–374
24. Serajian R (2005) Comparison of contact theories of wheel and rail. The School of Railway Engineering, Iran University of Science & Technology, Tehran
25. Salehi M, Farahi G (2012) Stress analysis worn profiles based on Hertz theory. In: Twentieth annual conference of mechanical engineering, Tehran
26. Tournay H, Giani J (1995) Rail/wheel interaction: multidisciplinary practices developed in South Africa. In: Railway engineering conference, Melbourne, Australia
27. B. G. Engineering, The impact of MMU research on the optimisation of railway vehicle–track interaction. Manchester Metropolitan University
28. Mollasalmani M, Rezvani MA, Shahverdi H (2013) Filed study on the effect of lubrication on controlling wheel flange wear in 6-axle locomotive (type DF8BI). In: The 3rd international conference on recent advances in railway engineering, Tehran
29. Olofssona U, Zhua Y, Abbasia S, Lewi R, Lewis S (2013) Tribology of the wheel–rail contact—aspects of wear, particle emission and adhesion. Veh Syst Dyn Int J Veh Mech Mobil **(Special Issue: State of Art Papers of the 23nd IAVSD)**
30. Kaneharaa H, Fujioka T (2002) Measuring rail/wheel contact points of running railway vehicles. Wear 253(1–2):275–283
31. Jin X, Zhang J (2001) A complementary principle of elastic bodies of arbitrary geometry in rolling contact. Comput Struct 79:2635–2644
32. Chevaliera L, Cloupet S, Quillien M (2006) Friction and wear during twin-disc experiments under ambient and cryogenic conditions. Tribol Int 39(11):1376–1387
33. Jahed H, Nasr A, Eshraghi A (2004) Parameters influencing wear in Iranian state railways. In: Proceedings of the 7th international conference of rail ways, Sharif University, Tehran
34. Jamison WE (1981) Wear test results on eight samples of rail from FAST. Tribology Materials Development Corporation
35. Archard JF (1953) Contact and rubbing of flat surfaces. J Appl Phys 24(8):981–988
36. Zheng J, Luo J, Mo J, Peng J, Jin X (2010) Fretting wear behaviors of a railway axle steel. Tribol Int 43(5–6):906–911
37. Waterhouse R (1984) Fretting wear. Wear 100(1–3):107–118
38. Bhushan B (1980) Review of experimental techniques in EHD-lubricated contacts. DOE Tech. Rep. No. AT80DO69, SKF Industries, INC., King of Prussia, PA
39. Matsumoto A, Sato Y, Ono H, Wang Y, Yamamoto M, Tanimoto M, Oka Y (2002) Creep force characteristics between rail and when on scaled model. Wear 253(1–2):199–203
40. Ahlstrom J, Karlsson B (1999) Microstructural evaluation and interpretation of the mechanically and thermally affected zone under railway wheel flats. Wear 232:1–14
41. Cantizano A, Carnicero A, Zavarise G (2002) Numerical simulation of wear mechanism maps. Comput Mater Sci 25:54–60
42. Zhu Y (2013) Adhesion in the wheel–rail contact. Doctoral Thesis, Department of Machine Design, Royal Institute of Technology, Stockholm, 2013
43. Yazykov V, Pogorelov D, Mikhalchenko G (2004) Railway vehicle simulation using non-elliptical wheel–rail contact model. In: Mechanics of 21st century—ICTAM04 proceedings, Warsaw, Poland

12

Innovative Interior Designs for Urban Freight Distribution Using Light Rail Systems

James Kelly[1] **· Marin Marinov**[1]

Abstract High levels of pollution and congestion in urban centres are an increasing concern for local councils in the UK. Heavy goods vehicles delivering urban freight to city centres are a leading cause of this problem. This paper discusses the concept of using light rail networks to deliver freight to city centres from surrounding businesses. Specifically, various innovative designs are considered for the interior of the metro carriage and developed into visual models using Autodesk Inventor software. A full evaluation of all the designs developed is completed, resulting in a proposed design for consideration. The conclusion reached is that the proposed interior design is viable and coincides with the future metro fleet designs and concepts.

Keywords Metro · Urban freight · Interior design · Autodesk Inventor · Tyne and Wear metro · Light rail · Urban environment

1 Introduction

Freight distribution is an important part of economic and national development in the UK [1]. The problem of increasing congestion and pollution in urban centres is a growing concern for urban centres in the UK. In 2015, 1.65 billion tonnes of goods were lifted by HGVs in the UK; this represented an increase of 11% compared with previous years; during that time, a distance of 18.4 million kilometres was travelled by domestic freight in the UK. Every year, businesses in the UK lose £10 billion per year [2] due to congestion. If HGV usage can be reduced by using more sustainable forms and means to distribute goods to city centres, then this could generate major economic and environmental benefits.

Most air pollution in the UK comes from road transport and residential emissions [3]. In 2014, the EU announced it was taking legal action against the UK; this was due to pollution levels persistently exceeding the safe limit. In 2003, 61% of particulate emissions in London were produced by freight transportation, causing an impact on the health of those working in the city as well as the productivity of the firms located in urban zones [4]. Currently, the main method of freight distribution in the UK is via heavy goods vehicles (HGVs). The annual report from the European Environmental Agency (EEA) claims that HGV pollution costs EU countries £40 billion per year, making up half of all transport air pollution costs. The EEA also stated that HGVs should be charged with an increased amount in certain European countries due to the higher levels of pollution [5]. The UK Government is funding various projects which are researching the use of gas and dual-fuelled HGVs under the low-carbon truck trail [6]. However, even if HGVs were emission-regulated, the problem of congestion caused by HGV freight delivery to local businesses and shops still exists in city centres.

As stated by the European Commission [7], the combined effects of congestion, safety, noise and environmental issues are both economic and societal issues. These issues not only reduce efficiency and effectiveness of urban freight logistics but also impact the well-being of society due to the negative effects on health. One way to reduce urban pollution is to introduce low-emission zones (LEZ), which aim to promote cleaner urban centres through the

✉ Marin Marinov
marin.marinov@ncl.ac.uk

1 NewRail, Mechanical and Systems Engineering School, Newcastle University, Newcastle upon Tyne, UK

Editor: Xuesong Zhou

use of cleaner vehicles and a reduction in the use of older, highly polluting vehicles [8]. In Europe, a modal shift towards environmentally friendly freight distribution is promoted by the European Commission, governments and local authorities. However, as the financial engineering of the projects is complex, a long period of time will be required before implementing the modal shift [9].

A multimodal system using light rail is an important concept and needs serious considerations. For the case of Newcastle upon Tyne, the interested reader is referred to [10]. It is evident that a modal shift towards more sustainable transport methods is needed. Such a project could generate employment, environmental benefits and business, although large amounts of public and private investments in the infrastructure and construction would be needed in order to implement such a system.

The Tyne and Wear metro service in Newcastle upon Tyne is chosen to act as a case study, outlining the potential for light rail services to become well-adapted and suitable for a complementary freight delivery service. We focus on how the interior of the metro vehicle can be adapted in order to operate as a freight service. Various designs are developed, discussed and analysed along with a thorough review of the safety and security considerations. A full evaluation of the designs is completed, while discussing the levels of investment needed with the benefits and weaknesses of each design.

Autodesk Inventor simulation package is used to develop visual models, demostrating the way the interior of the metro carriage could be adapted without major changes to the existing rolling stock construction. The advantages and disadvantages of such a system are highlighted, conclusions about the effectiveness of this innovative idea are derived and consequently whether this freight system could be practical for the Tyne and Wear metro and other light rail systems.

2 Current Systems

A HGV refers to a large goods vehicle; it is the term used by countries within the European Union (EU) for vehicles with a total mass of over three and a half tonnes [11]. HGVs are currently the most common form of freight distribution in the UK. In 2014, road freight was the main method of freight transportation in the UK, with 75% of goods transported by HGV over rail and water [2]. HGV use peaked in the UK in 2004, before the recession and has been on the steady increase since 2012 [12].

The advantage of using HGVs for transportation of goods is that they can deliver goods directly to the door of the customer, often eliminating the need for intermodal transport. However, there are restrictions related to the use of HGVs in the UK. The maximum vehicle weight in the UK is 44 t (truck, fuel and load) [13]; this government-enforced law means that freight distributed by HGV often cannot compete with the high volume of goods transported by rail or water. HGVs are generally seen as being environmentally unfriendly and are a major cause of inner city pollution. HGV transport by road produces 76% more CO_2, 90% more particulates and up to fifteen times more NO_X than the equivalent journey by rail [14].

Dense inner city zones require smaller logistics vehicles in the modern environment. City freighters are vehicles with a relatively small capacity that can perform distribution activities to locations along narrow and crowed streets in city centres [15]. They are often significantly more environmentally friendly (electric or hydrogen based).

Electric vehicles (EVs) are a cleaner alternative to HGVs which use lithium-ion batteries that can be charged at different locations such as at home or the workplace. The UK Government has offered various benefits to those who use EVs at home or for business such as a grant of up to 20% of the cost of the vehicle (up to £8000) [16]. This scheme aims to improve the air quality in city centres by reducing the carbon emissions from freight vehicles. Using EVs for freight distribution in city centres also reduces noise pollution due to the near silent state at running speed.

However, the case for solely using EVs for freight and logistics services has its downfalls. More is required to make a feasible business case than just replacing vehicles with electric cars. The current logistics systems in many cities are set up for conventionally powered vehicles. The number of charging points in urban areas would have to be significantly increased to deal with the demand. A new freight network would have to be devised among freight companies, resulting in high initial costs but low operational costs.

Many large cities in the UK, Europe and around the world benefit from light rail transit systems such as metro or tramways. The potential for freight distribution by light rail is a topic that has been researched, though it has not experienced a desirable implementation. Amsterdam City Cargo started running in 2008 following a successful pilot scheme, testing the capability on using the existing tram tracks within the city. City Cargo did not affect passenger tram services as it did not use passenger stops.

For the first phase of testing, the trams travelled without carrying any cargo, to assess the traffic flows along the routes. Delivery points outside the city were used to distribute goods to the corresponding consolidation hubs in the city centre, where small electric vehicles (as shown in Fig. 1) were used to transport the goods to the final destination. Cargo transportation in Amsterdam took 15 min more than direct HGV transportation; however, this is counterbalanced by the fifteen percent reduction in expenses compared to conventional delivery [17].

Fig. 1 Amsterdam City Cargo concept

Urban consolidation centres (UCCs) have drawn attention in recent years as a means of promoting existing logistics systems. The appropriate application of a UCC has the ability to improve supply chain performance and reduce societal and environmental impacts of freight transport activity [18].

Automated freight delivery has the potential to be a game changer in the UK logistics industry. Technologies such as radio frequency identification (RFID) and automated guided vehicles (AGVs) are an expanding sector within the freight service market, from which light rail freight systems could largely benefit. Such a fundamental change to the industry could produce benefits and improvements for congestion, emissions and road safety [19].

3 Current Metro Interior Designs

Interior designs of metro vehicles are not suitable for the transport of most type of freight, especially not suitable for the transport of hand luggage and passengers' bags. Worldwide light rail and metro systems use various different seating layouts. There are currently two fundamental seat configurations when it comes to metro carriage interiors. The Tyne and Wear metro (as shown in Fig. 2 [20]) uses a transverse (facing) arrangement, which is often used in wider trains. Evidence suggests that passengers prefer sitting in the direction of travel rather than sideways [21]; therefore, this seating arrangement is usually used for services with longer journey times, as the tolerance for discomfort is lower than that in short journeys.

The transverse layout is said to be the least effective at increasing passenger flow on services with multiple stops [21].

This transverse arrangement involves a higher seating density than the alternative of [17] using longitudinal seating, as shown in the London underground (shown in Fig. 3 [22]). Neither layout shows clear potential for freight storage; therefore, alterations would need to be made. The Institute of Transport Research Studies at Monash University conducted a study [23], and researched on how seating arrangements can affect dwell times on metro services. The Stockholm metro was used as a case study, in which the seating arrangement was changed from transverse to longitudinal and tested during an eight-week period. The study found that there was a 2.4 s reduction in dwell time caused by the new seating arrangement. These marginal gains would accumulate over the service to contribute to a better compliance with the timetable.

Through a review of alternative freight transport methods, large benefits related to switching to an unconventional mode of freight transport such as the reductions in CO_2 and particulate emissions could be maximized. HGVs are the most common form of freight transport in the UK today, yet they cause many of the environment problems due to their inefficiency. Certain transport modes are clearly not suitable for some applications; however, combining modes of transport would be a good approach towards the goal of increasing efficiency while reducing emissions within urban centres. Cities are unique and vary from one to another, and thus, one freight system may not be suitable to all urban areas.

Fig. 2 Tyne and Wear metro interior

Fig. 3 London underground interior

According to [10], for the case of Tyne and Wear region, the utilisation of the Tyne and Wear metro service for freight distribution would allow for a greater usage of the light rail network.

4 Design Software

Two computer-based packages have been identified for the visualisation of interior designs: SolidWorks and Autodesk Inventor.

SolidWorks is the most common 3D modelling software package used in industry for detailed CAD models of complex components and assemblies. This software has successfully been used for similar design projects previously. However, the software does not provide the same easy user interface as other software packages provide.

The Autodesk Inventor's easy and user-friendly interface makes it a great package for the use of visual models. After reviewing various software packages that could have been used for the project design work, Autodesk Inventor was chosen due to its availability and easy user interface.

5 Tyne and Wear Metro

5.1 Infrastructure

The Tyne and Wear metro has been in operation since 1980 and is considered to be one of the world's first electric urban railways, alongside London and Glasgow. The light rail system serves over 60 stations, spanning from Newcastle to Sunderland. Since its introduction, the rolling stock interior design has remained largely unchanged.

The service carries 38 million passengers a year, making a vital contribution to the region, socially and economically. Nexus, the owner of Tyne and Wear metro, claims that every pound invested in the metro yields a return of eight pounds [24]. This shows that if a freight service was introduced on the metro, such a scheme would very likely be profitable.

There is a clear potential for the establishment of a metro-based freight system in Newcastle upon Tyne. In order for such a system to operate, a micro-consolidation centre would be developed at Palmersville, near the metro station [10]. All goods from surrounding businesses would be loaded onto the modified metro train. The train would depart from Palmersville metro station and arrive at a track situated between Jesmond and Manors metro stations (as shown in Fig. 4 [25]), currently only used at night to return trains back to the depot. This section of the track is where a distribution centre would be constructed. Goods would be sorted and transported at the final stage of the journey by bicycle couriers or electric vehicles.

5.2 Rolling Stock

The Tyne and Wear metro has one of the largest seat numbers per train when compared to other light rail services in Europe. This high number of seats does not reflect the usage statistics of the metro, suggesting that most passengers travel on the metro for a total of 10 min [26]. This shows that there is potential for removal and reconfigurations of seats within the carriage.

The current metro trains are usually comprised of two carriages, each made up of two cars [25]. Refurbishments took place in 2010, extending the life of the carriages by 10–15 years before being replaced [27]. The capacity of the carriage is 232 standing and 68 seated passengers [24]. Each carriage has eight doors and a swivel bogie which connects the two cars. The carriage can be split up into three sections (as shown in Fig. 5 [25]). Sections A and C show potential for freight storage without the need to change the interior due to the larger floor space and accessibility to the doors at either side of the carriage. To increase the freight capacity, section B would need to be reconsidered.

6 Methodology for Improving Existing Design

Research into current methods of freight distribution in urban centres was considered when developing designs for a metro carriage capable of freight transport. There were three design scenarios to be explored. The first being a design that utilised the current metro interior, requiring very low investment and could work as a short-term option. The second scenario was to design a reconfigurable carriage that would be able to provide freight and passenger services alongside one another, and will possess adaptable seating to free up space within the carriage. The final scenario was to design a fully automated freight system that operated using the metro system. Various designs were considered before the final designs were chosen.

An initial CAD model was produced to obtain an accurate visual aid of the current interior layout; this is shown in Fig. 6. This model acted as a foundation for the development of design work. The designs were developed from initial sketches to 3D CAD models using Autodesk Inventor. These models helped to present the designs in a visual and realistic way, compared to the hand sketches. An evaluation of the final designs was completed using a decision matrix, comparing scenarios against a set criterion. The potential for such a system is discussed and conclusions are made, discussing the benefits and limitations of implementing such a concept.

Fig. 4 Metro map, showing the track between Jesmond and Manors

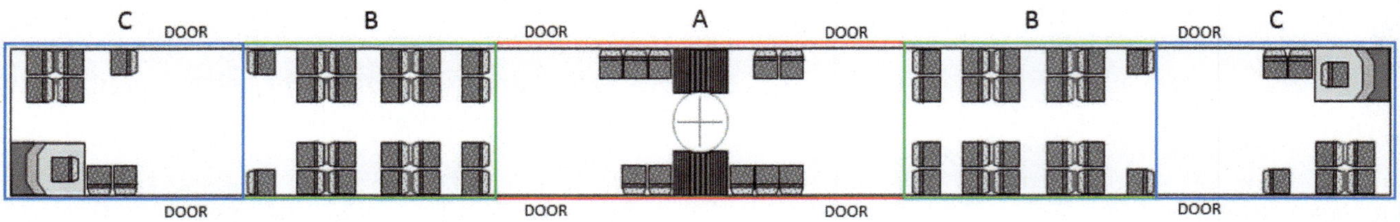

Fig. 5 Current metro configuration, with highlighted areas A, B and C

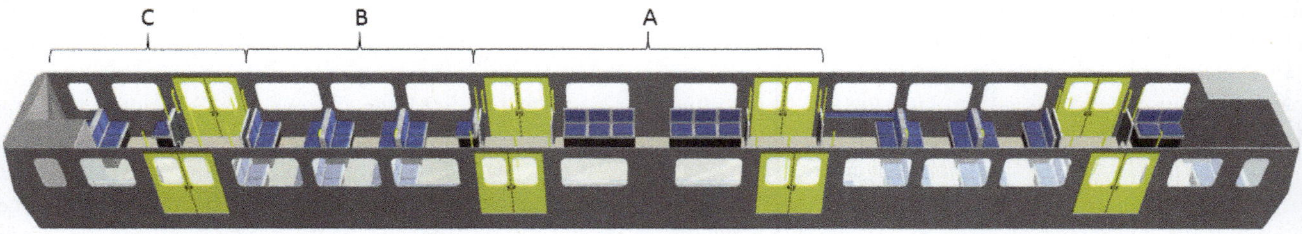

Fig. 6 Autodesk Inventor model of current design

7 Application

7.1 Design 1: Utilising the Existing Design

Designs 1.1 and 1.2 have been developed using the current metro design and have utilised it for freight storage. These designs would involve the lowest investment due to the minimal amount of changes made to the carriage interior design. As shown in Fig. 7, there is significant space in the doorway and longitudinal seating area (this area is depicted by section 'A' in Fig. 5). In order to optimise the use of space in this part of the carriage, the production of bespoke cargo boxes would be needed.

7.1.1 Design 1.1

Design 1.1 makes efficient use of the open space by using wheeled boxes to transport the goods on and off the carriage. A full depiction of the design in full operation is given in Fig. 8. The boxes would be stored uniformly within the carriage, fitting two boxes side by side in the longitudinal area and three boxes in the door area. The boxes would have to be carefully arranged around the pivoting bogie in the stationary section of the carriage, making sure all four wheels are positioned securely, preventing unsafe movement. The boxes would be locked into position using a footbrake. This method of freight storage on the metro could be applied to a passenger service by

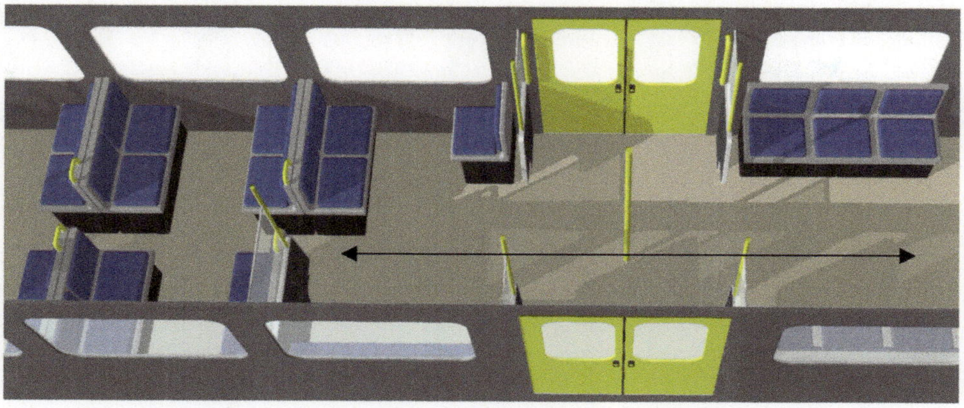

Fig. 7 Inventor model showing the potential for goods in the middle of the carriage

Fig. 8 Display of Design 1.1 when in operation

closing off one of the carriages; however, this arises the problem of increased dwell time (the time taken at each station). For this method to be efficient enough to work alongside the passenger service, the boxes would need to be removed from the carriage by workers extremely quickly. This design would work best as a service outside of passenger service hours. In order for the freight service to act alongside the passenger service during the day, the dwell time of loading and unloading cargo would need to be significantly reduced (see Design 3).

7.1.2 Equipment and Features

Design 1.1 involves bespoke cargo boxes, specifically dimensioned to fit within the constraints of the current metro. The boxes would have detachable shelving that can be rearranged as needed, to suit different freight sizes. Using shelving allows different goods to be stored in one box while being easily differentiated, acting similar to a bento box [10]. The height of the box is just under that of the average human height, to allow the user to see ahead when manoeuvring the box. The boxes will be mounted with swivel wheels possessing brakes to allow them to be moved onto the metro by hand. This application eliminates the need for forklifts as these would not be able to load the cargo onto the metro effectively.

7.1.3 Design 1.2

Design 1.2 (as shown in Fig. 9) also utilises the metro design without modifying the carriage. The transverse seating could be used to accommodate platforms in which freight could be stored above. This design is suited towards smaller freight volumes, as the packages would have to be individually placed in the storage area. A sack truck or lifting trolley could be used to move the goods on and off the carriage. These platforms could simply be made from a sheet of plywood or other rigid material. This design is easily set up and could be very effective for smaller freight volumes at less busy times.

7.2 Design 2: Reconfigurable Carriage

Design 2 focuses on modifying the metro design to act as a reconfigurable carriage, allowing the parts of the train to be quickly switched between passenger and freight service.

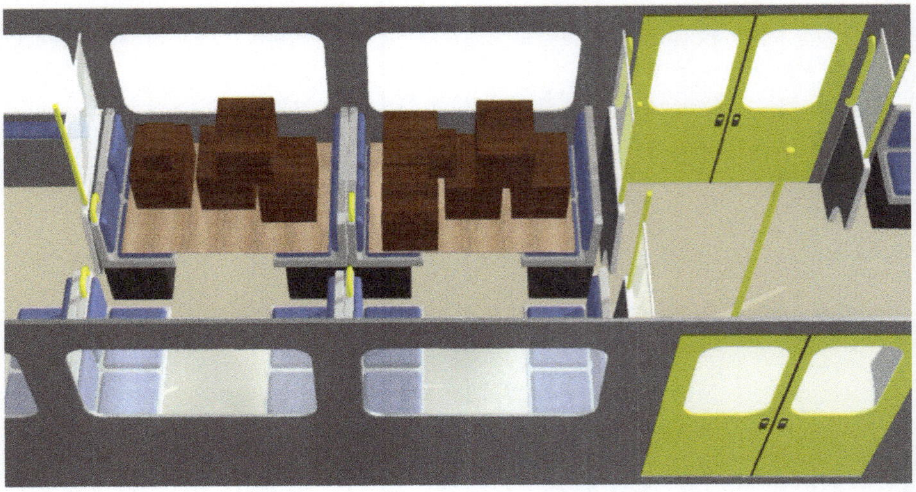

Fig. 9 Design 1.2 using platforms over the current seating

7.2.1 Design 2.1

After considering results from the research on rolling stock design by Monash University [21], an alternative design using longitudinal seating along the carriage length was developed. Design 2.1 makes use of the more efficient seating arrangement, creating more floor space for freight storage. Standing space on board the train would be increased; however, this configuration would reduce the number of seats available in the carriage. Due to the average journey time on the Tyne and Wear metro being only 10 min [26], this arrangement would unlikely affect or increase passenger discomfort. The introduction of foldable seats would make the carriage more adaptable for a quick change between freight and passenger services. This folding seat design drastically increases the space inside the carriage, as shown in Fig. 10.

The panels (as shown in Fig. 11) act as a barrier from the doors, causing a reduction in floor space due to the distance from the side wall (roughly a seats width). To address this problem and create more space, these panels could be made foldable (as shown in Fig. 12) by using a lock and turn mechanism with access via a key. There are limited options to remove the panels completely as they act as a safety barrier from the automatic moving parts that operate the carriage doors.

Removing the central handrail and adding more grab rails at head height increases space without compromising passenger comfort. An addition to the design would be to add portable shelving to the part of the carriage between the doors and pivoting bogie. The shelving would be mounted on wheels with brakes to allow easy entry and removal from the carriage. This shelving would be used for smaller, individual packages that would not be transported using the cargo boxes.

Figure 13 shows the organisation of the shelving and cargo boxes inside the newly designed carriage. The removal of the door panels creates an open plan space in which a larger freight volume can be stored. In order for the freight section of the carriage to be kept separated from the passenger section, barriers would be needed to deter passengers from entering the unauthorised area. Currently, when metro drivers are under instruction, an elasticated sheet is raised by connecting the handrails to section off that part of the carriage. However, this method is only applicable to the areas of the carriage with an abundance of handrails. In the middle section of the carriage, another method would need to be used. Through the research of

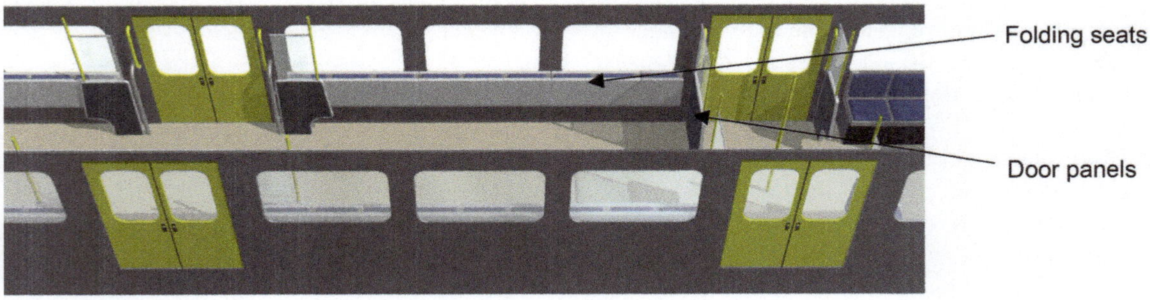

Fig. 10 Folding seat design

Fig. 11 Folding panel design

Fig. 12 Door panels

various barrier methods, it was concluded that the best solution would be to use extendable fibre barriers. This method would create a quick and easy way of sealing off one part of the carriage from passengers. These fixtures could be mounted to the carriage wall, eliminating the need to bring any equipment on and off the carriage. Figure 14 shows how these barriers could be fitted to the metro.

7.2.2 Design 2.2

Design 2.1 uses an innovative transverse seating design (as shown in Fig. 15), allowing seating to be folded away during freight service. The design uses a central bar in which both parts of the seat are mounted. This support bar is able to rotate around a pivot, connected to the metro wall. This pivot would be locked in and out of place using a simple spring lock mechanism, similar to the mechanism commonly used in gym equipments.

This design would have the same effect on floor space as the longitudinal folding seating arrangement presented in Design 2.1. Using this design would increase the number of seats available in peak times. However, the time taken to fold in all the seats would be longer than that of a simpler longitudinal seating arrangement shown in Fig. 10. Figure 16 shows how the folding design would work when in full operation.

This design would allow the metro to be utilised more effectively during peak times when capacity is much greater than usual, such as after a concert or football match. Therefore, the option of being able to adapt the carriage in this way would positively affect both the freight and passenger services.

Fig. 13 Model showing the positioning of portable shelving on board the carriage

Fig. 14 Extendable barriers in the mid-section of the carriage

Fig. 15 Transverse folding seat design

Fig. 16 Transverse seat design in operation

7.2.3 Design 2.3

Design 2.3 has been developed from Design 1.2, by creating a horizontal platform in which smaller packages would be stored during low-freight volume periods. The design utilises the longitudinal seats next to the pivoting bogie, by making the seats foldable into a flat platform. Figure 17 shows how the seats would be refitted with a locking mechanism that would allow movement into the horizontal position by simply applying a force to one part of the seat. This mechanism is similar to those used in caravans to convert a sofa into a bed [28].

When the seat is in the horizontal position, smaller packages could be stored above the surface. The position of the mechanism at the side of the seat eliminates the need to modify the electric components underneath the seat. This design allows for easy configuration between passenger and freight service. However, this design would not operate with the same freight volume as Design 2.1 or Design 2.1,

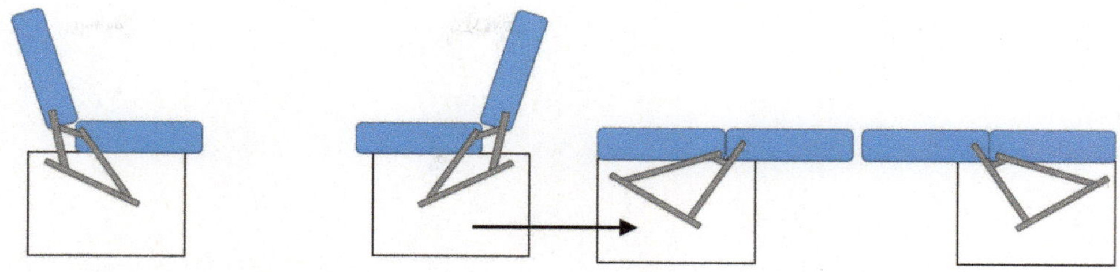

Fig. 17 Folding seat mechanism diagram

and therefore, it would be more applicable to a quieter service.

7.2.4 *Motored Seating Concept*

B/E Aerospace [29] is a company that designs interiors and utility systems for commercial aircraft. The company has designed a seat that would adjust for taller passengers via small motors mounted in the floor. The design arose to address the problem of passenger discomfort on board long flights [30]. The movement of the seats would be controlled by the staff on board the plane via a single control unit (see Fig. 18 [29]). Mechanically controlled seating is a concept that could be applied to a metro car. This seating design would allow quick and easy movement of seats when positioned in a transverse seating layout. However, this design could be said to be overengineered and inapplicable to the metro, as the complexity of the electronics needed would require a full redesign of the metro rolling stock.

7.3 Design 3: Full Automation

Design 3 focuses on the use of automation within the carriage. The concept includes the use of autonomous guided vehicles (AGVs) to distribute the freight from the consolidation centre directly to the customer. Full automation means that the vehicle would be capable of safely completing full journeys without the need for an operator, while encountering normal traffic and weather conditions. This method would involve a high level of private and public investment to become fully integrated into the system, but it could be the most efficient and effective way to distribute freight in the long run.

7.3.1 *Automated Guided Vehicle (AGV)*

The AGV would act as the cargo container, distributing the goods directly to the customer. AGVs are commonly used in modern warehouses, making use of systems such as radio frequency identification (RFID) [31]. AGVs can help reduce product damage, eliminate lost loads and simplify equipment maintenance in warehouses [32]. The use of AGVs can deliver year after year cost reductions. Figure 19 [32] shows various applications of AGVs for moving goods.

7.3.1.1 Radio Frequency Identification (RFID) RFID utilises radios waves to identify goods by using chips [31]. Electromagnetic fields are used to identify and track chips attached to the goods. The RFID device serves a similar purpose to that of a bar code or magnetic strip on a credit card [33]. Active chips have a local power source such as a battery and can operate far away from the RFID reader. RFID has many applications such as access management, tracking goods and airport baggage logistics.

Using RFID in an automated system on the metro would benefit the operational efficiency of the system by eliminating the need to check the goods manually. An RFID reader could simply be mounted on the carriage door, which would monitor the inward and outward flow of goods on the metro. A computer would store this information, allowing all freight to be tracked effectively.

7.3.1.2 Navigation In order for the AGV to navigate to the correct destination, an advanced navigation system would need to be applied to the system. Visual guidance would allow AGVs to operate by using cameras to record features along the route. Specially designed stereo cameras with 360 degree images would be used to build up a 3D map, allowing the AGV to follow the route without human assistance. However, geo-guidance are considered to be a more accurate method of navigation. A geo-guided AGV would recognise its environment to establish its position. Obstacles would be detected and identified, allowing the AGV to avoid obstructions.

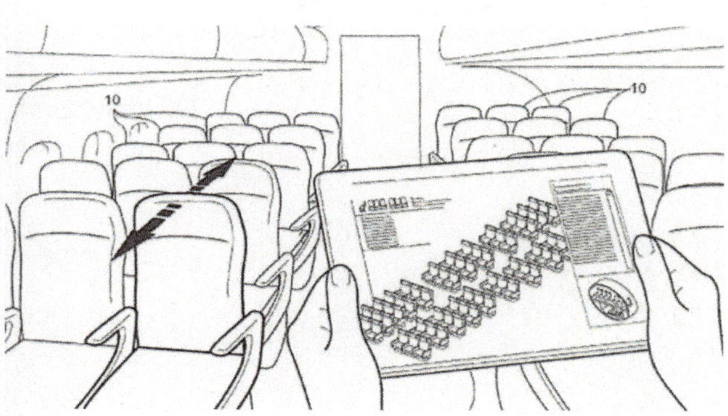

Fig. 18 Motored seating concept on board a commercial aircraft

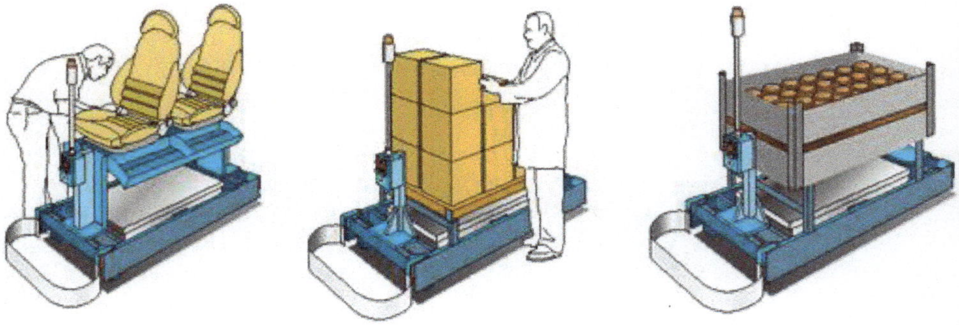

Fig. 19 Automated guided vehicle applications

7.3.1.3 The Metro AGV The metro AGV would operate using geo-guidance navigation technology and would be powered by a lithium-ion battery, located underneath the goods in order to lower the centre of gravity. Lowering the centre of gravity of the AGV is a key to inclined transit, helping to prevent the AGV from tipping over. The AGV would travel at average walking speed (1.4 m/s) [34] in busy pedestrian areas, with the ability to speed up in open areas. The AGV would be stored at the unused area at the end of the metro platform, before positioning itself in front of the yellow line at the end of the platform specifically reserved and cornered off for freight use. The AGV would have automated sliding doors at either end allowing easy access to the goods inside. Figure 20 shows the Inventor model of the AGV design and outlines the main features. The AGV would be roughly 1 m wide, 1.5 m long and 1 m tall. These dimensions would allow two AGVs to fit side by side within the modified metro carriage with sufficient room to walk in between. A key feature of the AGV is the double wheel. The double wheel allows the AGV to turn at very sharp or tight corners without wearing down the rubber tyres. The double wheel would simply rotate around 90 degrees and the AGV would move in the perpendicular direction. This feature allows the AGV to move within the carriage and along the platform with ease.

For the purpose of this design, the interior design of the carriage has been extended or carried forward from Design 2.1. The AGVs would have the ability to arrange themselves within the carriage during transit; this would allow the correct load to be by the exit door when it arrives at its destination and needs to leave the train. To do this, one space would need to be kept free at all time to allow the AGVs to move in a way similar to a block puzzle (see Fig. 21). This feature greatly improves the efficiency of the freight service, reducing dwell time at each station. The current dwell time at metro stations is roughly 20 s; this

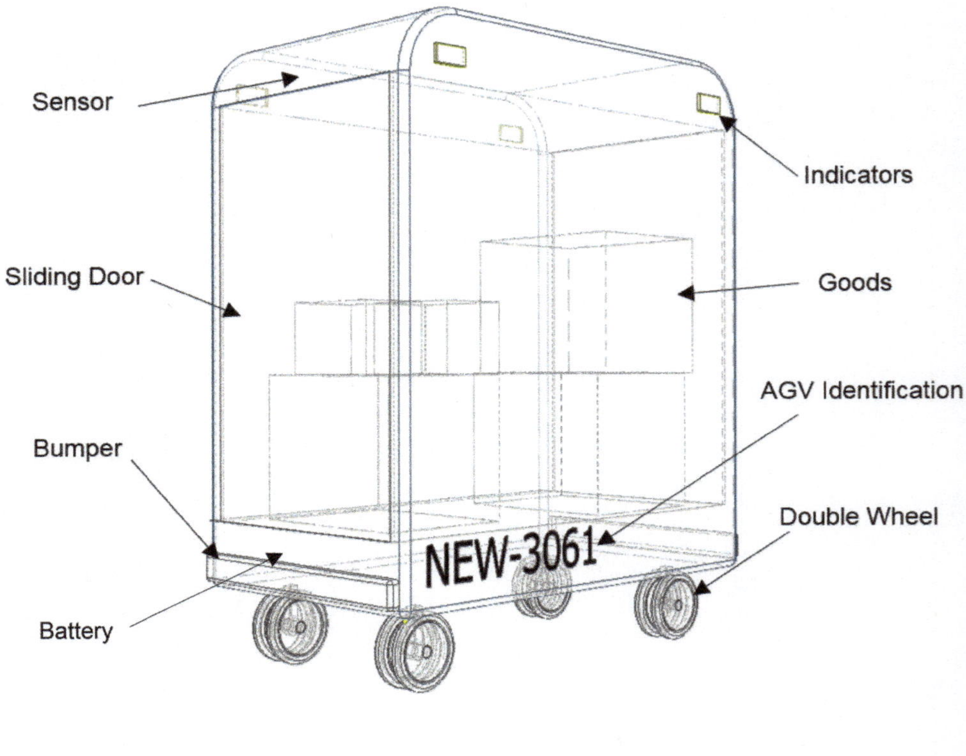

Fig. 20 Metro automatic guided vehicle

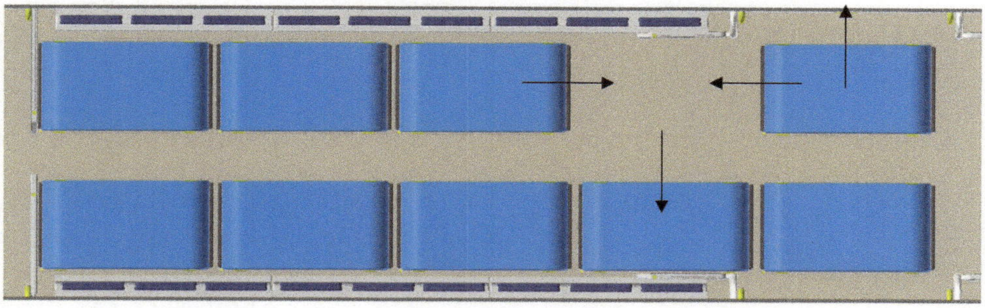

Fig. 21 Plan view of AGVs on metro

time would allow two or three AGVs to leave or board the train.

Figure 22 shows an Inventor model of the AGVs in full operation on the metro, displaying how the AGVs could position themselves on the station platform, ready to enter and leave the carriage as the doors are opened.

7.3.2 Greggs: Four Lane Ends Case Study

Greggs PLC is a Newcastle-based company, operating with 29 outlets in Newcastle, 9 of which are located in the city centre. In 2011, Greggs opened a new £16.5 million site in Longbenton [35], 4.5 miles from Newcastle city centre. The site is located one mile from Four Lane Ends metro station; this shows that there is large potential for a metro-based delivery system to supply its outlets in the city centre. Figures 23 and 24 show the location of the Greggs manufacturing site in relation to Newcastle city centre and Four Lane Ends metro station.

The route from Greggs in Benton to the metro station consists of predominantly wide, open pedestrian areas and cycle lanes, making it an ideal environment for the use of AGVs. The Greggs site would be able to load AGVs with bakery goods, which would be taken on the metro to the city centre outlets, many of which are located next to metro stations. The potential for such a system is large, with many positive economic and environmental impacts that could incur. The AGV would be modified to take long trays

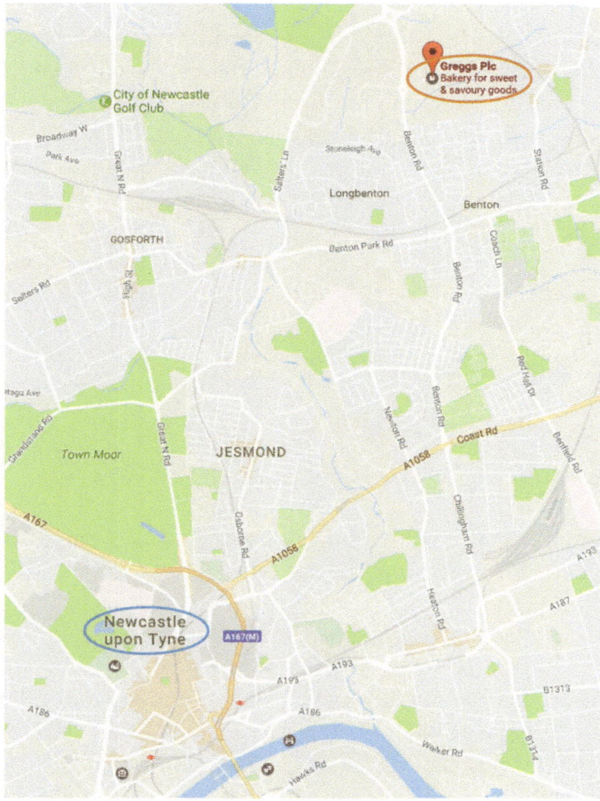

Fig. 23 Map showing location of Greggs PLC

Fig. 22 Metro AGVs in full operation

Fig. 24 Map showing location of Four Lane Ends metro station

of goods that could simply be slotted in and out of the vehicle due to its ideal dimensions.

7.3.3 Potential for Mainline Application

The AGV system has potential for application on mainline train systems as well as light rail systems. New developments in rail such as the HS2 line could be adapted for national AGV freight transport, although the freight volume per train might not meet national demands. A problem that arises is that the train is not levelled with the platform in mainline services, unlike light rail services. This would cause a problem for the AGVs entering the carriage. An automatic ramp could be mounted to the train, allowing easy access for the AGVs. Automatic freight services are unlikely to be introduced on mainline services before light rail services due to the increased logistical complexity and investment needed, although it is a point of discussion for governments in the future.

7.3.4 Safety and Security Measures

In order for the AGVs to be safe and secure, various design features need to be addressed. The AGV would have an obstacle sensor at the front and rear of the vehicle to detect obstacles such as pedestrians, preventing unwanted contact. Two areas will be demarcated in front of the AGV: the first distance being a warning area and the second being an emergency area. If an object is in the emergency area then the AGV will stop immediately. All AGVs will be equipped with flashing lights and sounds as to warn people if the AGV is too near. The AGV will only be accessible to the receiver of the goods via an access card or code, preventing any unauthorised access.

Regulations for the development and use of autonomous vehicles are yet to be confirmed by global regulatory boards, and the potential legal issues surrounding autonomous vehicles are undefined [36]. Issues such as assessing liability in the results of accidents involving AGV need to be discussed. The majority of new vehicles in Europe are subject to an approval process and requires that standards are met [19]. However, specialist vehicles intended for limited road use are exempt from this regulatory process, for example, forklift trucks, airport baggage hauling vehicles. The metro AGV would be likely to be exempt to this process due to the nature of its operation.

8 Evaluation

8.1 Methods for Evaluating Designs

In order to evaluate the designs and propose the final solution, two evaluation methods were completed. A decision matrix uses specified criteria in which the designs are compared and scored against each other, allowing the designs to be ranked from highest to lowest scoring. The designs in this study have been evaluated against a set of key design criteria. The main criterion was system performance as freight storage within the carriage is a vital part of the design. The level of investment needed was another key criterion for the final design; this was an important factor contributing to the overall success of the project proposal. Feasibility, safety, security and versatility are other criteria in which the designs were compared against. For this evaluation, the three key design scenarios were considered as a whole. Design 1 involved utilising the current metro, Design 2 introduced the concept of a reconfigurable carriage and Design 3 developed the idea of fully automated freight system. Another evaluation method is to compare the proposed design to existing freight systems. This comparison would provide a clear comparison of the performance of alternative freight distribution methods.

Table 1 Design decision matrix

Criteria	Weight (%)	Design 1		Design 2		Design 3	
		Rating/5	Score	Rating/5	Score	Rating/5	Score
Performance	20	2.5	0.5	4	0.8	5	1
Cost/investment	18	5	0.9	4	0.72	2	0.36
Feasibility	15	5	0.75	4	0.6	3.5	0.525
Ease of implementation	15	5	0.75	4	0.6	2	0.3
Safety	10	3.5	0.35	4	0.4	4	0.4
Security	8	3	0.24	5	0.4	4.5	0.36
Versatility	8	3.5	0.28	5	0.4	5	0.4
User-friendly	6	4	0.24	5	0.3	4	0.24
Total weighted score		–	4.01	–	4.22	–	3.585
Rank		2		1		3	

Table 2 Comparison of Design 2.1 to existing systems

Criteria	Weight	City Cargo [17]	HGVs [11]	Baseline Design 2.1	EV [37]	Monoprix [38]
Capacity	4	0	− 4	0	− 4	0
Environment	3	0	− 3	0	0	0
Versatility	2	− 2	+ 2	0	+ 2	− 2
Congestion	1	0	− 1	0	− 1	0
Total	10	− 2	− 6	0	− 3	− 2

8.1.1 Decision Matrix Results

The design evaluation can be summarised in the form of a decision matrix as shown in Table 1. Design 2 scored overall the highest as it would provide the best solution in the short and medium term. The reconfigurable carriage design scores highly in many of the given criterion, making it the most suitable design for final proposal. The design utilises the space within the carriage by making simple yet very effective changes, allowing for a large increase in the floor space without compromising the passenger comfort. Other benefits are also created by this design, such as the increase in passenger capacity on the train and a reduction in dwell times [21], thus benefitting both the passenger and freight services. With regard to performance, though, Design 2 did not score the highest. Design 3 using full automation scored the highest as this would provide the fastest and most efficient freight solution. Design 1 scored the highest in terms of ease of implementation with the current metro as no changes would need to be made to the carriage itself, but failed to score highly with regard to performance as the freight volume would not be as high as other designs.

As for long-term investment, Design 3 may be the best option. Design 3 shows great potential for future considerations, but this table highlights the weak points in the design, mainly being the high initial investment costs and implementation planning. For Design 3 to be successfully implemented into the Tyne and Wear metro network, further considerations would need to be made to make the scheme more compliant with the current service. This makes this design a long-term choice, as it would not be a viable solution in the short term.

8.1.2 Comparison of System Performance with Existing Systems

In order to further evaluate the proposed design, a comparison to existing freight systems has been completed. Design 2, specifically Design 2.1, has been chosen as the proposed design for comparison as this design scored the highest using a decision matrix (see Table 1). The comparison table is given in Table 2.

Each system was compared to Design 2.1 using the criteria as given in Table 2. Capacity of the service was a key criterion, in which no other service scored higher than Design 2.1. The comparison shows that the new design provides similar performance to the already successful Monoprix scheme. The capacity of Cargo Tram and Monoprix services would be similar to that of the reconfigurable metro carriage. HGVs were proved to be more harmful to the environment than other methods of transport [14], causing a negative score in the comparison table. In terms of versatility, heavy goods vehicles (HGVs) and electric vehicles (EVs) scored higher due to their ability to access more destinations than the metro, which is restricted to the metro stations. However, the metro could be seen to be more versatile in the sense that it can shift between freight and passenger services easily. The table shows that all other systems have a negative score, showing that they do not theoretically perform as well as Design 2.1 overall.

9 Conclusion

This paper has presented solutions for the design work needed to implement a metro freight system in the future. Conclusions about the possibility of such a system being carried out are discussed in this section.

There are many challenges that need to be considered in order to achieve a functional freight service operating on the Tyne and Wear metro. This paper outlined the potential for a freight service on board the metro, without the need for large changes to existing infrastructure or redevelopment of the network.

In order to validate the designs, it must be confirmed that the metro would be able to withstand the extra weight added by the freight packages. The metro trains are designed to withstand a capacity of six passengers standing per square metre. The heaviest equipment on board the carriage would be the loaded AGVs. An AGV would take up 1.08 m^2 of floor space when on board the metro and would be the equivalent of just over six people. Six people weighing 80 kg would be a mass of 480 kg per square metre. Therefore, the AGV with freight would be able to represent up to 500 kg in weight, assuming that the majority of the space on board is taken up with AGVs. However, due to the nature of the AGVs on board the carriage, not all space will be taken up (see Fig. 23). It can therefore be said that the axle loading caused by the additional freight in the carriage is acceptable. However, this would need to be confirmed by completing engineering analysis on the existing and new rolling stock.

There were three potential design scenarios researched in this study. The first being a design which utilised the current metro design by using bespoke cargo boxes in the open region of the carriage. This space provided enough room for freight equipments without the need to redesign the internal layout of the carriage. Design 1.2 shows how plywood sheet could be used as a temporary arrangement, providing a platform for freight storage. Through evaluation, it was clear that this method would not provide the same freight capacity as other designs; however, this method could be used as a pilot scheme in the short term by Nexus, to identify whether such a scheme is viable for further investment.

The second design scenario presented a very versatile design in the form of a configurable carriage concept, allowing the internal design to be used for freight and passenger service with quick changes. Three designs were proposed for this scenario, in which Design 2.1 was selected as the highest performing design. This was due to the space that was made available by using the foldable, longitudinal seating. This design acts as the final recommendation due to its advantages in many areas with only minor changes to the carriage interior. Design 2.2 provided an alternative folding seat design, allowing transverse seating arrangements to be folded away to the side of the carriage. Both these designs created large amounts of space in the centre of the carriage. This addition of space would benefit the passenger service by allowing the seats to be folded away during extremely busy times such as after a concert or football game. Design 2.1 proved to be the best solution for this scenario as the longitudinal seating arrangement was more desirable for an interior layout due to the reduction in dwell time in peak hours and increase in standing passenger capacity.

The final scenario looked into the suggestion of using an automated freight delivery service, coinciding with the metro passenger service. This design scenario developed a concept for an automated guided vehicle (AGV) to pick up, transport and deliver the goods to the customer via the metro system. This solution shows promise for a future system, but many changes would need to be made to local infrastructure to accommodate the system. Large private and public investments would be needed to get such a project off the ground.

There is clear evidence to suggest that there is potential for a freight service on board the Tyne and Wear metro. The suggestions presented in this paper successfully provide a realistic and achievable solution to urban freight problems, which can be further investigated in the future.

10 Future Work

In order to provide a more detailed proposal, a cost–benefit analysis would need to be completed to show the funding needed to implement a metro freight service. Further analysis into the detail of the designs would need to be completed, such as a study into the components and materials. Engineering analysis into the effects of the increase in load on the rolling stock would need to be made in order to validate any technical uncertainties in the design. This paper briefly discussed how the freight would be stored at the end of the platforms, utilising the unused space; however, more research and design work would need to be completed with regard to the storage and distribution of the freight within stations. The lifts in the underground stations in the centre of Newcastle upon Tyne would need to be adapted to withstand the new demand of the freight network. An investigation into the potential for freight storage services at individual stations should be conducted in the near future.

Acknowledgements The authors acknowledge the contribution of Michael Bell, for providing in depth opinions with regard to an automated system throughout the duration of the study.

References

1. Independent Transport Commision: Improving the Efficiency of Freight Movements: the Contribution to UK Economic Growth. London (2014)
2. Department for Transport: Domestic Road Freight Statistics, United Kingdom 2015, UK (2016)
3. BBC News: UK Pollution, how bad is it? (2014) (online Blog). http://www.bbc.co.uk/news/uk-26851399. Accessed 13 Feb 2017
4. Crainic, T.G., Ricciardi, N., Storchi, G.: Models for Evaluating and Planning City Logistics Systems. CIRRELT (2009)
5. Airqualitynews.com: Pollution from HGVs 'Costs Europe Nearly £40 Billion (2013) (online). http://www.airqualitynews.com/2013/02/28/pollution-from-hgvs-costs-europe-nearly-40-billion. Accessed 14 Mar 2017
6. Cenex-Atkins: Low Carbon Truck and Refuelling Infrastructure Demonstration Trial Evaluation—First Annual Report to the DfT Executive Summary for publication, UK (2014)
7. European Commission: A Call to Action on Urban Logistics. Brussels (2013)
8. Taniguchi E, Thompson RG, Yamada T (2014) Recent trends and innovations in modelling city logistics. Proc. Soc. Behav. Sci. 125:4–14
9. Diziain D, Taniguchi E, Dablanc L (2014) Urban logistics by rail and waterways in France and Japan. Soc. Behav. Sci. 125:159–170
10. Dampier A, Marinov M (2015) A Study of the Feasibility and Potential Implementation of Metro-Based Freight Transportation in Newcastle upon Tyne. Springer, Newcastle Upon Tyne
11. Wallace School of Transport: What is the Difference Between LGV Training and HGV Training? (2016) (online). http://www.wallaceschool.co.uk/blog/HGV_or_LGV. Accessed 14 Feb 2017
12. HMRC: Department for Transport, Licensed Heavy Goods Vehicles, by Region: Great Britain and United Kingdom (2016) (online). https://www.gov.uk/government/statistical-data-sets/veh05-licensed-heavy-goods-vehicles. Accessed 12 Apr 2017
13. HMRC: Moving goods by road (2016) (online). https://www.gov.uk/guidance/moving-goods-by-road. Accessed 09 Apr 2017
14. Campaign for Better Transport: Air Pollution Matters (online). http://www.bettertransport.org.uk/air-pollution-matters. Accessed 10 Apr 2017
15. Crainic TG, Ricciardi N, Storchi G (2004) Advanced freight transportation systems for congested urban areas. Transp. Res. Part C 12:119–137
16. Transport for London: Electric Vehicles (2016) (online). https://tfl.gov.uk/modes/driving/electric-vehicles. Accessed 16 Apr 2017
17. Dutch Amsterdam: Amsterdam to Get Freight Trams (2007) (online). http://www.dutchamsterdam.nl/207-207. Accessed 16 Feb 2017
18. Shen, J., Qiu, F., Li, W., Feng, P.: A new urban logistics transport system based on a public transit service. In: CICTP, pp. 651–661 (2015)
19. Department for Transport: The Pathway to Driverless Cars: A Detailed Review of Regulations for Automated Vehicle Technologies, UK (2015)
20. Flickr: Tyne and Wear Metro (2013) (online). https://www.flickr.com/photos/emdjt42/10870475016. Accessed 16 Feb 2017
21. Monash University: An Innovative Approach to Metropolitan Train Carriage Interior Configuration—To Improve Boarding, Alighting, Dwell Time, Stability and Passenger Experience, Melbourne (2013)
22. 123 Royalty Free: London Underground (2017) (online). https://www.123rf.com/photo_18402815_interior-from-london-underground-train.html. Accessed 13 Feb 2017
23. Coxon, S., Burns, K., de Bono, A, Napper, R.: An Examination of Three Approaches to Metro Rolling Stock Design to Ameliorate Extended Dwell Times Due to Passenger Growth and Associated Crowding. Institute of Transport Studies, Monash University (2011)
24. Nexus: Metro Strategy 2030 (2014) (online). http://www.nexus.org.uk/sites/default/files/Metro%20Strategy%202030%20summary%20document.pdf. Accessed 19 Apr 2017
25. Wikiwand: Tyne and Wear Metro Rolling Stock (online Blog). http://www.wikiwand.com/en/Tyne_and_Wear_Metro_rolling_stock. Accessed 17 Feb 2017
26. Marinov, M., Toal, J.: A Study by Product Design into the Potential for a Check-In and Bag Drop. Newcastle upon Tyne (2016)
27. Railway-technology.com: Tyne and Wear Metro, UK (online). http://www.railway-technology.com/projects/tyne/. Accessed 19 Apr 2017
28. Marcle Leisure: Seat Bed Mechanism for Motor Homes and Caravans (online). http://www.marcle-leisure.co.uk/information/rock-n-roll-hinges.htm. Accessed 20 Mar 2017
29. Rockwell Collins: B/E Aerospace (2016) (online). http://beaerospace.com/. Accessed 02 Mar 2017
30. Telegraph, T.: Legroom Wars: The Seat to Change it All (2017) (online). http://www.telegraph.co.uk/travel/news/Legroom-wars-the-seat-to-change-it-all/. Accessed 02 Mar 2017
31. How Stuff Works: How RFID Works (2017) (online). http://electronics.howstuffworks.com/gadgets/high-tech-gadgets/rfid.htm. Accessed 08 Apr 2017
32. Motion Controls Robotics: Automatic Guided Carts (AGCs)—End-of-Line Automation—Integrated Material Handling Systems (online). http://motioncontrolsrobotics.com/robotic-applications/automated-material-handling/automatic-guided-carts-agc/. Accessed 08 Mar 2017
33. Technovelgy.com: What is RFID? (online). http://www.technovelgy.com/ct/technology-article.asp. Accessed 06 Apr 2017
34. Browning R, Parker E, Herron J, Kram R (2006) Effects of obesity and sex on the energetic cost and preferred speed of walking. J. Appl. Physiol. 100:390–398
35. The Journal: Bakery giant Greggs opens doors on £16.5 m site (2011) (online). http://www.thejournal.co.uk/business/business-news/bakery-giant-greggs-opens-doors-4417628. Accessed 25 Feb 2017
36. Traffic Technology International: Autonomous Vehicle Safety Regulation World Congress 2017 (2017) (online). http://www.autonomousregulationscongress.com/. Accessed 14 Apr 2017
37. Frevue: Freight Electric Vehicles in Urban Europe (2016) (online). http://frevue.eu. Accessed 16 Feb 2017
38. Maes, J.; Vanelslander, T.: The Use of Urban Rail Transport as Part of the Supply Chain in an Urban Logistics Context, Lisbon (2010)
39. Brice D, Marinov M, Rüger B (2015) A newly designed baggage transfer system implemented using event-based simulations. Urban Rail Transit 1(4):194–214
40. Marinov M, Giubilei F, Gerhardt M, Özkan T, Stergiou E, Papadopol M, Cabecinha L (2013) Urban freight movement by rail. J Transp Lit 7(3):87–116

13

Width Design of Urban Rail Transit Station Walkway: A Novel Simulation-Based Optimization Approach

Afaq Khattak[1] · Jiang Yangsheng[1] · Hu Lu[2] · Zhu Juanxiu[1]

Abstract The optimal design of the walkway at an urban rail transit station is a vital issue. The Transit Capacity and Quality of Service Manual (TCQSM) TCRP-100 report for the design of urban rail transit station walkway and the existing design models neglect the important factors such as randomness in the passenger arrival rate, randomness and state-dependent service time of the walkway and blocking phenomenon when the passenger flow demand exceeds the walkway capacity. There obviously exists a need to develop a design approach that overcomes these shortcomings. For this purpose, this paper details a *simulation-based optimization* approach that provides width design through automatic reconfiguration of walkway width during the simulation–optimization process based on phase-type (PH) distribution. The integrated PH/PH(n)/C/C discrete-event simulation (DES) model and optimization method that uses the genetic algorithm (GA) work together concurrently to obtain optimized (design) widths for different passenger flow and level of service (LOS) The numerical experiments are conducted to compare the proposed model with the existing design methods. It reveals that: (1) The width obtained by our proposed model is higher than the existing width design models; (2) when squared coefficient of variation of passenger arrival interval increases, the walkway width increases more for our proposed model than the existing design models; (3) when the arrival rate increases, the walkway width of our proposed model increases faster than the existing design models; (4) the increase in the length of walkway has no significant effect on the walkway width.

Keywords Urban rail transit station · Walkway *simulation-based optimization* · Queueing system · PH distribution · Genetic Algorithm

✉ Afaq Khattak
af.transpo@gmail.com

Jiang Yangsheng
jiangyangsheng@swjtu.cn

Hu Lu
hulu361@126.com

Zhu Juanxiu
zhjuanxiu@163.com

1 School of Transportation and Logistics, Southwest Jiaotong University, Chengdu 610031, Sichuan, China

2 Department of Industrial and Systems Engineering, National University of Singapore, Singapore 119077, Singapore

Editor: Xuesong Zhou

1 Introduction

1.1 Motivation

Building a new urban rail transit station involves a considerable amount of resource investment in a project. The design phase of the urban rail transit station project is of critical importance. It is arduous as well as costly to redesign the urban rail transit stations if there is a need to correct the flaws or other issues. The improper design of the urban rail transit station service facilities causes high-level congestion, longer travel time of passengers between service facilities, inefficient space utilization, resource wastage and increase in the waiting time of passengers. However, at the same time the urban rail transit station service facilities are obliged to hold the level of service (LOS) specified by the codes and design manuals.

The Transit Capacity and Quality of Service Manual (TCQSM) TCRP-100 report [1] presents the analysis and width design procedure of urban rail transit station walkways. The procedure involves computation of the walkway width based on a desirable LOS. The LOS of the walkway is based on the mean passenger space. The assessment of walkways LOS uses the calculation of the area occupied per passenger (m^2/ped) as the basis for LOS classification as shown in Table 1. According to TCQSM, the walkway width is obtained by using passenger arrival rate divided by walkway service rate per unit width under a given LOS. However, it ignores the important factors and has several flaws:

- It does not take into account the randomness in passenger arrival rate [2].
- It does not take into account the variation and state-dependent service time of the walkway (in reality, the walking/travel time of passengers on the walkway depends on the walking speed of passengers as well as on the number of passengers on the walkway).
- It does not consider the blocking of passengers from the entrance to the walkway when the passenger traffic demand approximates to or exceeds the capacity of the walkway.

As a result, the walkways designed by the TCQSM always show poor performance and experience heavy congestion even during the off-peak periods because of the variation in passenger traffic demand.

Thus, there exists an urgent need to establish a new approach for the width design of urban rail transit station that overcomes these shortcomings.

1.2 Literature Review

Substantial work has been done in the past to put forward the new design approaches for the walkways in urban rail transit stations as a well as in other buildings. Due to the inherent characteristics of walkways such as relations between the passengers and walkway facility (customers and server) and the randomness in passenger flow, many researchers described the walkway facility as a queuing system, based on which both the analytical and simulation models are developed as shown in Fig. 1.

Table 1 Passengers level of service (LOS) on the walkway [1]

Level of service (LOS)	Passenger space (m^2/ped)
A	≥3.3
B	2.3–3.3
C	1.4–2.3
D	0.9–1.4
E	0.5–0.9
F	<0.5

Many researchers have conducted their studies to evaluate and design the pedestrian facilities at urban rail transit station and in other buildings by analytical queuing models. For example, the G/M/1 queuing model by Jiang et al. [3] and the state-dependent M/G(n)/C/C queuing models by Yuhaski et al. [4], Cheah and Smith [5], Mitchell et al. [6], Chen et al. [7], Weiss et al. [8], Xu et al. [9] and the PH/PH(n)/C/C state-dependent queuing model by Hu et al. [10]. These researches showed that the queuing theory can properly illustrate and assess the pedestrian facility at urban rain transit station and other buildings. But the major drawback of all these models is fitting the exponential distribution to the passenger arrival process which is appropriate only when there is a free flow. The squared coefficient of variation (c^2) of the exponential distribution is equal to 1 which means that randomness in the arrival process is ignored but in reality there exist randomness and therefore do not depict the actual conditions. The analytical PH/PH(n)/C/C model captures the randomness in passenger's arrival rate as well as service time (passenger travel time in the walkway facility) using the phase-type (PH) distribution. The PH distribution is a probability distribution that can fit any positive random variable with various squared coefficient of variation while keeping the Markov characteristics. Thus, it has replaced the exponential distribution in several domains such as manufacturing and communication systems (Alfa et al. [11]; Miyazawa et al. [12]; Krishnamoorthy et al. [13]; Jiang et al. [14]). The existing analytical PH/PH(n)/C/C model takes the state dependence into account. However, this model is difficult to solve because it is based on the Quasi-Birth-Death process (QBS) and Matrix Analytical Scheme (MAS) [10] that involves a large number of matrix operations and iterative in nature. The complexity becomes larger when the system state n increase as it requires huge computer storage capacity. In addition to that, the analytical PH/PH(n)/C/C model cannot control the blocking probability during the width design process.

To take the advantage of queuing system for facility design and at same time eliminate the need to solve the large matrices and equation systems, many researchers established the simulation models for the facility analysis and design, for example, the G/M/1 queuing network simulation model by Lovas [15], the M/G(n)/C/C state-dependent simulation model by Cruz et al. [16], and Khalid et al. [17], the G/G(n)/C/C simulation model by Jiang et al. [18] and so on. In these researches, the queuing systems are translated into the Discrete-Event Simulation (DES) models. Based on the DES models, the facilities are evaluated and performance measures are estimated. However, facility

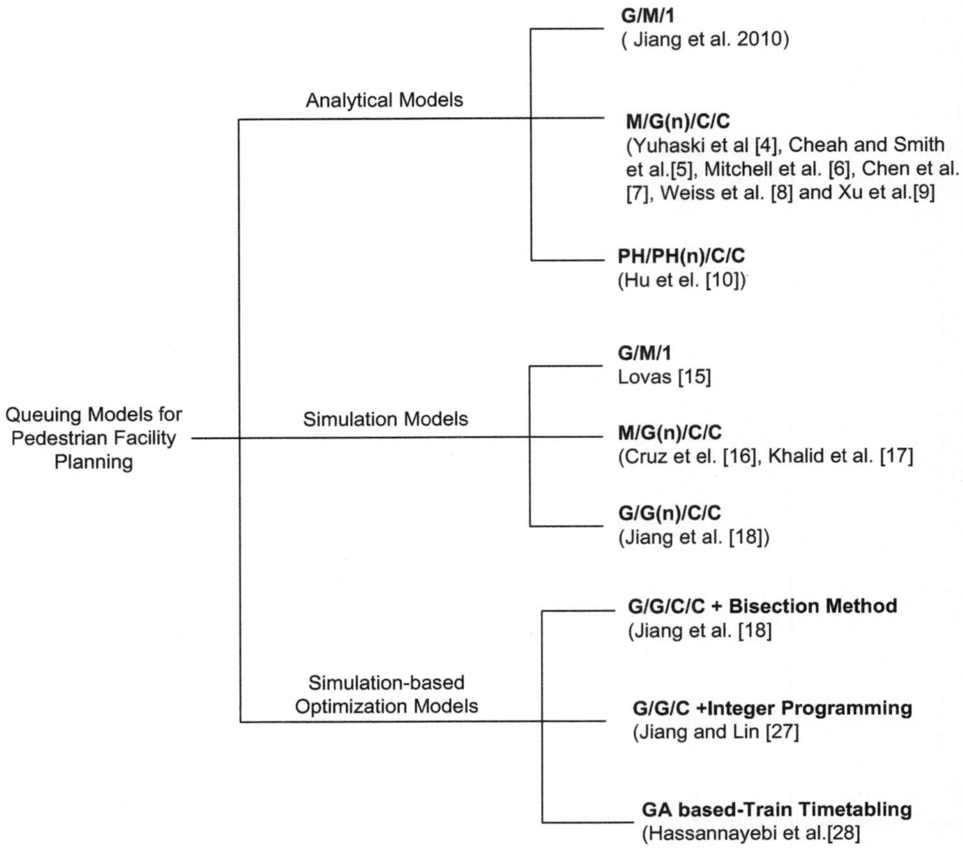

Fig. 1 Facilities analysis and design references

description still needs to be improved in the above researches. In addition, blocking probability is not controlled during the design process in all these previous researches.

Besides the DES, another well-known category of simulation is microscopic simulations. Microscopic simulation models are elaborate as they depict individual characteristics and behaviors of the pedestrians given by Teknomoet al [19] and Kaakai et al. [20]. However, they require extensive calibration work and larger computation time at the same time. On the contrary, DES does not require the specific physical environment and passenger entity, making it more efficient and easier to calibrate than microscopic simulation models. Therefore, DES is taken as an efficient and accurate simulation approach with a wide range of application. Another advantage of the DES is that *simulation-based optimization* can be carried out conveniently based on DES models due to its universality and efficiency.

Recently, *simulation-based optimization* has become a popular and efficient tool in many domains (Banks [21], Hagendorf et al. [22] and Figueira et al. [23]).It involves the optimization of model inputs by using simulation for the computation of parameters. Therefore, it is not necessary to provide an explicit analytical expression of the objective or constraint functions for optimization as in the case of analytical approach (Swisher et al. [24]; Fu [25]; Cassandras and Lafortune [26]). This is especially useful in some practical situations where the explicit analytical formulae are too complex to be deduced. In urban rail transit station domain, researchers used *simulation-based optimization* technique to evaluate and design different service facilities (Jiang et al. [18]; Jiang and Lin [27]; Hassannayebi et al. [28]).

To find the optimal width for the urban rail transit station walkway by a *simulation-based optimization* approach and overcome all the shortcomings, we need to develop a PH-based DES model of walkway in the *SimEvents*® Software. The genetic algorithm (GA) is chosen in this paper as an optimization approach and implemented in the MATLAB® Scientific Computing Environment. The *SimEvents*® and MATLAB® have been used in serval researches as DES modeling and optimization tool, respectively [29–33]. The MATLAB® offers a computational infrastructure for optimizing a hybrid, discrete-event and the time-based models, which allows for a great deal of flexibility in scripting and modifying the optimization objective function, while also making it easier to tie together parallel discrete simulation and optimization without the pain of the context switching into multiple software environments.

Moreover, there are several reasons for applying a GA rather than any other traditional optimization methods. One of the important reasons is its implicit parallelism (Swisher et al. [24]; Hubscher-Younger et al. [33]; Messac [34]. The GA searches parallel from a population of points. As GA has multiple offspring, it can explore the solution in different directions at a time giving it greater chance to find the optimal solution, while other traditional methods search from a single point and may trap in local optimal solution.

Based on the above analysis, we aim to propose a new *simulation-based optimization* approach for the width design of urban rail transit station walkway in this paper. The contribution of this paper falls into two aspects. First, we establish a PH/PH(n)/C/C DES model to describe the walkway as a queuing system in the urban rail transit station. The PH-based DES model captures the general randomness in passengers' arrival and walkway service time. It also takes the state-dependence of service time into consideration. Therefore, it can be used to accurately evaluate the performance of the walkways. Besides, it also serves as an important tool to validate the PH/PH(n)/C/C analytical model developed in Hu et al. [10]. Second, we design a *simulation-based optimization* approach by implementing the PH-based DES model and the GA to work concurrently. The *simulation-based optimization* approach determines the optimal widths for the walkways considering the requirements on both LOS and blocking probability. Therefore, the walkway designed by it enjoys higher service quality and less congestion.

1.3 Outline of the Paper

The rest of the paper is organized as follows. First, we describe the urban rail transit station walkway as a PH/PH(n)/C/C state-dependent queuing system in Sect. 2. Then the *simulation-based optimization* framework for the width design is established in the Sect. 3. The computational experiments between our proposed approach and other existing approaches are presented in Sect. 4. Finally Sect. 5 concludes the paper.

2 Modeling of Walkway as a Queuing System

In this section, the necessary notations and assumptions used in this paper are discussed first. Then we describe the PH/PH(n)/C/C state-dependent queuing system for the urban rail transit station walkways. Method for calibrating the PH arrival interval and PH service time is also presented.

2.1 Notations

This section presents almost all the notations we need for this paper.

Notation	Description
α	Initial probability vector
$\mathbf{D}$	Transient generator matrix
L	Length of the walkway (m)
W	Effective width of the walkway (m)
C	Capacity of the walkway
n	Number of passengers (system state)
λ	Passenger arrival rate (ped/h)
c_a^2	Squared coefficient of variation of arrival interval
q	Peak-hour volume
ε	Peak-hour factor
h	Mean headway between the trains (s)
c_h^2	Squared coefficient of variation of headway
μ_n	State-dependent service rate
T_n	State-dependent service time of walkway
V_n	State-dependent walkway speed of passengers
$c_{s,n}^2$	State-dependent squared coefficient of variation of walkway service time
P_c	Blocking probability
ES	Mean area occupied per passenger
f	Degree of Erlang distribution
U	Uniformly distributed random number

2.2 Assumptions

There are a few basic assumptions that should be presented before going further:

- The walkways are rectangular in shape with length (L) and width (W). The width W is the effective width of walkway facility. According to TCQSM, the total width is obtained by adding a buffer of 0.50 m on each side to the effective width.
- The passengers are uniformly distributed on the walkway. This phenomenon is quite rare from a practical point of view. But this assumption is used in many relevant studies such as [10, 18] which is important for queuing analysis.
- The queueing system to describe the walkway facility has a 'loss queue' which means that when the number of passengers equal to capacity of walkway then newly arrived passengers are lost or blocked. Therefore; we take into account blocking probability P_c and design the walkway by keeping the P_c below a certain value.

- The alighting passenger flow from the train is considered. Nevertheless, the proposed approach can also deal with the flow of passengers toward the walkway from entrance gate of the urban rail transit station.
- The assessment of LOS uses the mean area occupied per passenger (m^2/ped) as the basis for classification (see Exhibit 7-3 in Ref. [1]). It reflects proximity to other passengers and is therefore considered as an indicator of the passenger level of comfort and freedom to maneuver without conflict. It is generally desirable for the walkway of urban rail transit station to operate at LOS 'C' or above. In this paper, we design the walkways under the LOS 'B', 'C' and 'D'. The LOS values for walkways are given in Table 1.

2.3 Description of Walkway Queuing System

The urban rail transit station walkway and flow of passengers comprise a kind of unique finite capacity queuing system with the diversified arrival rate distribution (with no fixed squared coefficient of variation) and varying service time distribution (where the service time depends on the number of passengers). When the passengers enter into the walkway, they occupy the spaces (squares) (see Fig. 2). Each available space on the walkway acts like a server (service desk). The passengers spend some time (walking time) on the walkway and then exit. The passenger flow on the walkway can be viewed as a queuing system with passengers as customers, the spaces on the walkway facility as servers and the process of walking on the walkway as a service process.

The number of passengers 'n' changes on the walkway over the time. As the number of passengers on the walkway increases, the slower passengers block faster passengers. Thus, higher passenger densities reduce the individual passenger walking speed. The speed is reduced to 0 when the number of passengers 'n' reaches to the capacity of walkway '$C = 5LW$'. The passenger flow on the walkway can be viewed as stopped when the density of passengers is 5 ped/m^2. [35]. The walking speed variation phenomenon with an increase or decrease in the number of passenger 'n' on the walkway is known as state dependence. Hence, the walkway can be described as a state-dependent queuing system with passenger arrival rate represented by **A**, state-dependent service rate of the walkway by **B(*n*)** and the number of servers (available positions/spaces) **C** to accommodate the maximum **C** number of passengers, i.e., A/B(n)/C/C queuing system.

Since the value of **C** is generally very high in hundreds and even thousands. The queuing systems with a high value of **C** are difficult to simulate and cause serious problems in optimization such as low optimization efficiency. Therefore, it is necessary to simplify the A/B(n)/C/C queuing system. We use the idea of transformation which is also used in relevant researches ([10, 18]). The transformation

Fig. 2 Transformation of walkway to a single-server queuing system

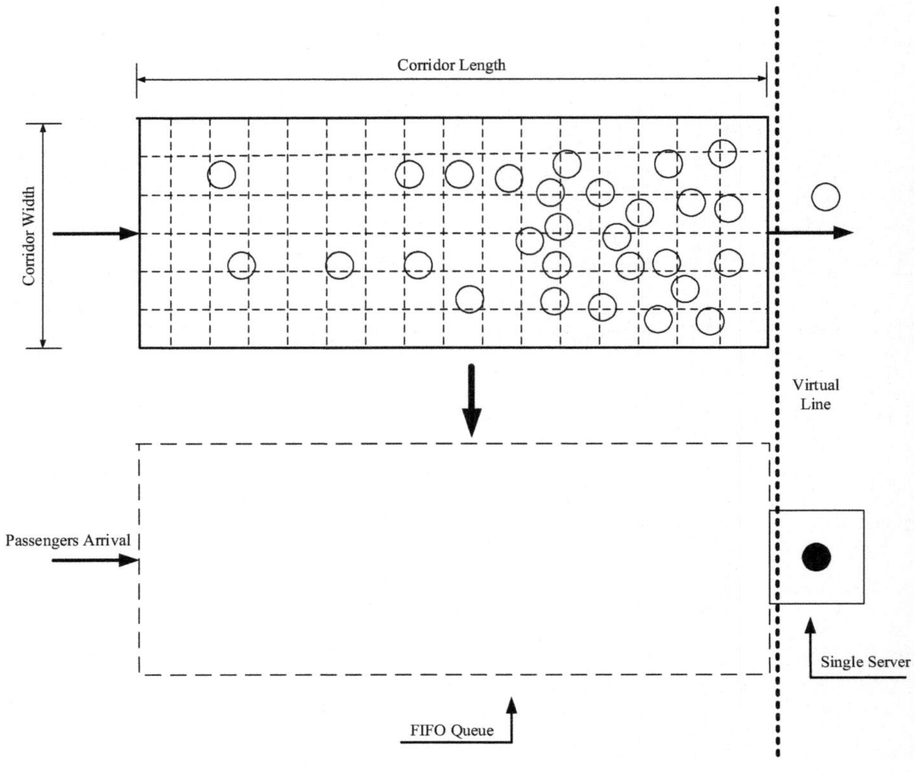

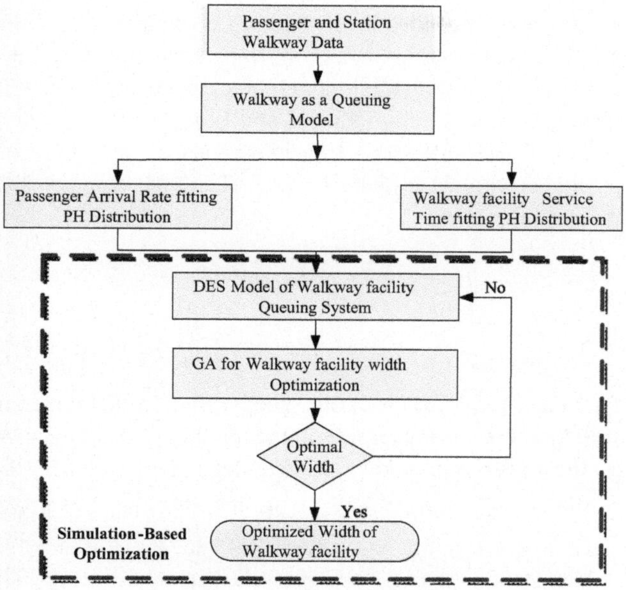

Fig. 3 *Simulation-based optimization* for the walkway width design

works as follows. A virtual line and a virtual server are set at the exit of the walkway, as shown in Fig. 3. When two sequential passengers pass through the virtual line, the time interval T_n is recorded. If the time at which the previous passenger leaves the circulation facility is viewed as the time the next passenger begins to be served, then the time interval between these two sequential passengers passing through the virtual line is equal to the service time of the virtual server. In this way, the A/B(*n*)/C/C queuing system with C parallel–serial servers can be transformed equivalently to an A/B$'$(*n*)/1/C queuing system with a single server. Note, the service time of the virtual server $B'(n) = B(n)/n$. Now how to calibrate the parameters for the PH arrival rate and service time will be discussed in the next subsection.

2.4 PH Arrival Rate and Walkway Service Time

The Phase-Type (PH) distribution was first introduced by Neuts [36] as the generalization of exponential distribution. It is constructed by the convolution of exponential distribution. A nonnegative random variable **X** has a PH distribution if its distribution function is given by;

$$f(x) = \alpha e^{\mathbf{D}x} d \tag{1}$$

$$\mathbf{F(x)} = 1 - \alpha e^{\mathbf{D}x}\mathbf{1} \tag{2}$$

Eqs. (1) and (2) are the probability density function (PDF) and cumulative distribution function (CDF) of PH distribution

Here; e: is the column vector with all the elements being 1. α: a sub-stochastic vector $\alpha = (\alpha_1, \ldots \alpha_n)$ of order n, i.e., α is a row vector, all elements of α are nonnegative and $\alpha e < 1$, where n is a positive integer. **D:** a sub-generator of order m, i.e., **D** is an $n \times n$ matrix such that (1) all the diagonal elements are negative; (2) all the off-diagonal elements are nonnegative; (3) all row sums are non-positive given as $\mathbf{D} = \begin{pmatrix} d_{11} & \cdots & d_{1n} \\ \vdots & \ddots & \vdots \\ d_{n1} & \cdots & d_{nn} \end{pmatrix}$ d: $d = \mathbf{D}.1$ and **1** is a column vector of one's of the appropriate size.

The 2-tuple (α, **D**) is called a phase-type representative (PH representative) of order n for the PH distribution. There are four conditions given for fitting PH distribution (Sadre and Haverkort [37], Sadre [38]) based on the squared coefficient of variation c^2

- If the c^2 for both the arrival and service processes is <1, a hypo-exponential distribution is used to fit the arrival and service processes with the number of phases given by $m = \frac{1}{c^2}$, the initial probability vector is $\alpha = (1, 0, \ldots, 0)$ and the matrix **D** is expressed by:

$$\mathbf{D} = \begin{pmatrix} -d_0 & d_0 & & & \\ & d_1 & -d_1 & & \\ & & \ddots & \ddots & \\ & & & -d_{m-2} & d_{m-2} \\ & & & & -d_{m-1} \end{pmatrix},$$

where

$d_j = \frac{m}{E[X]}$ for $0 \leq j < m-2$;

$d_{m-1} = \frac{2m\left[1+\sqrt{\frac{1}{2}m(mc^2-1)}\right]}{E[X](m+2-m^2c^2)}$;

$d_{m-2} = \frac{m\lambda_{m-1}}{2\lambda_{m-1}E[X]-m}$.

- If the c^2 is greater than 1 for both the arrival and service process, a hyper-exponential distribution is used for fitting with the number of phases $m = 2$, the initial probability vector is $\alpha = (g, 1-g)$ and the matrix **D** is given by:

$$\mathrm{D} = \begin{pmatrix} \frac{-2g}{E[X]} & 0 \\ 0 & \frac{-2(1-g)}{E[X]} \end{pmatrix} \text{ and } g = \frac{1}{2} + \frac{1}{2}\sqrt{\frac{c^2-1}{c^2+1}}$$

- If c^2 is equal to 1, then the approximation corresponds to an exponential distribution.
- If c^2 is very small, i.e., $c^2 \leq 1/30$, then the PH distribution with a large number of states is obtained and its approximation corresponds to an Erlang-30 distribution or deterministic distribution.

Jiang et al. [14] and Hu et al. [10] have achieved a good fitting effect for the passenger arrival interval from the train as well state-dependent service time of circulation facilities by using a PH distribution with any c^2. The four conditions show that we can determine the PH representation for the arrival rate and service time based on λ, c_a^2, μ_n and $c_{s,n}^2$. Note that $E[X]$ is the mean arrival interval and inverse of the arrival rate λ and the service rate μ_n. Therefore, the following content is about how to obtain these parameters.

The passenger arrival process to the urban rail transit station walkway is specified by the first and second moment of the inter-arrival interval. More specifically, the arrival rate λ and c_{a}^2 are the traffic descriptors that describe the passenger traffic flow. During the planning and design phase of urban rail transit stations, λ and c_a^2 can be deduced from peak-hour volume (q), peak-hour factor (ε), mean headway (h) between trains and squared coefficient of variation of the headway (c_{h}^2).

The passenger arrival rate can be obtained by using Eq. (3).

$$\lambda = \frac{q}{3600\varepsilon} \tag{3}$$

The value of c_a^2 of arrival rate can be calculated from Eq. (4) [10] and (5) [39]:

$$c_a^2 = \frac{e^{6.819\varepsilon}(\varepsilon - 1)^2}{4\varepsilon - 1} \tag{4}$$

$$c_a^2 = e^{0.503c_h^2}\left[\left(\frac{qh}{3600\varepsilon}\right) - 1\right] \tag{5}$$

We obtain this equation as follows: The peak-hour factor can be calculated by $\varepsilon = \frac{q}{4q_{15}}$. The q_{15} is the maximum flow rate in one quarter of the peak hour. The passengers' arrival intervals in the peak-15 min are divided into two parts.

(1) One is the time interval between passenger batches that are generated by the arriving trains that is known as headway 'h', and the data in this part account for $(900/h)/q_{15} = 3600\varepsilon/(qh)$ of the entire arrival interval data in peak-15 min.
(2) The other is the time interval between the passengers in batches from the same train. This interval is close to 0 as the alighting passengers arrive in a single batch from the same, and the data account for $1 - 3600\varepsilon/(qh)$ of the entire arrival interval in peak-15 min. If the arrival intervals between the trains on all the lines are the same, then the c_a^2 is equal to variance/(mean)^2, i.e.,

$$c_a^2 = \frac{(3600\varepsilon/qh)(h - 3600\varepsilon/q)^2 + (1 - 3600\varepsilon/qh)(0 - 3600\varepsilon/q)^2}{(3600\varepsilon/q)^2}$$

$$c_a^2 = (qh/3600\varepsilon) - 1$$

The correction factor ($e^{0.503c_h^2}$) is used in the above equation that is obtained by fitting the data by regression analysis that are obtained from field survey of stations in Beijing and Chengdu (North Railway Station on Line 1 and Tianfu Square Station which is transfer station at Line 1 and Line 2 in Chengdu and Chongwenmen Station of Beijing which is hub of Line 2 and Line 5). The reason to choose these urban rail transit stations is due to heavy passenger volume at these stations. The final equation that we get is:

$$c_a^2 = e^{0.503c_h^2}\left[\left(\frac{qh}{3600\varepsilon}\right) - 1\right]$$

According to TCQSM [1] and traffic flow theory, the passengers flow rate (μ) is given by Eq. (6):

$$\mu = kV \tag{6}$$

here k is the density of passengers and V is the passengers walking speed on the walkway. In case of urban rail transit station walkway, the passenger flow rate is the number of passengers passing through the walkway per unit time. The reciprocal of flow rate $1/\mu$ is referred as the time interval of the passengers leaving the walkway which is also the state-dependent service time of walkway T_n. Therefore, the state-dependent service time of walkway can also be expressed as Eq. (7):

$$T_n = 1/\mu = L/nV_n \tag{7}$$

The state-dependent service rate of walkway can be written as Eq. (8):

$$\mu_n = 1/T_n = nV_n/L \tag{8}$$

here L is the length and V_n is the state-dependent walking speed of passengers passing through the walkway. Yuhaski and Smith [4] developed an exponential model to describe the state-dependent walking speed in the corridors (walkways), shown by Eq. (9).

$$V_n = V_1 \exp\left[-\left(\frac{n-1}{\omega}\right)\right]^{\gamma} \tag{9}$$

where

$$\gamma = \left[\frac{\ln\left(\frac{v_a}{v_1}\right)}{\ln\left(\frac{v_b}{v_1}\right)} - \ln\left(\frac{a-1}{b-1}\right)\right] \quad \omega = (a-1)\left[\ln\left(\frac{v_1}{v_a}\right)\right]^{1/\gamma}.$$

Thus, Eq. (9) can now be written as:

$$\mu_n = nV_1 \exp\left[-\left(\frac{n-1}{\omega_i}\right)\right]^{\gamma}/L \tag{10}$$

In order to consider the randomness as well as state-dependent service rate of the walkway, the squared coefficient of variation ($c_{s,n}^2$) of service rate should be taken into account. The state-dependent $c_{s,n}^2$ of service rate of the walkway is given by Eq. (11) [10].

$$c_{s,n}^2=\left[\left(\frac{\delta_1}{v_1}\right)\exp\left(\left(\frac{n-1}{\omega}\right)^{\gamma}-\left(\frac{n-1}{\omega'}\right)^{\gamma'}\right)\right]^2 \tag{11}$$

where

$$\gamma'=\left[\frac{\ln\left(\frac{\delta_a}{\delta_1}\right)}{\ln\left(\frac{\delta_b}{\delta_1}\right)}-\ln\left(\frac{a-1}{b-1}\right)\right] \quad \omega'=(a-1)\left[\ln\left(\frac{\delta_1}{\delta_a}\right)\right]^{1/\gamma}$$

$\mathbf{v}_1$ mean walking speed when there is only one passenger on the walkway; δ_1 standard deviation of walking speed when there is only one passenger on the walkway; $\mathbf{v}_a$ mean walking speed when there are $a = 2LW$ passengers on the walkway; δ_a standard deviation of walking speed when there are $a = 2LW$ passengers on the walkway; $\mathbf{v}_b$ mean walking speed when there are $b = 4LW$ passengers on the walkway; δ_b standard deviation of walking speed when there are $b = 4LW$ passengers on the walkway.

After fitting the PH distribution, the passenger arrival process can be described by the initial probability vector α and the transient generator matrix $\mathbf{D}$ as:

$$A \sim \mathbf{PH}(\alpha, \mathbf{D}),$$

The state-dependent service process of the walkway can be described by the initial probability vector β_n and the transient generator matrix $\mathbf{H}_n$ as:

$$B(n) \sim \mathbf{PH}(\beta_n, \mathbf{H}_n), \quad n = 1, 2 \ldots C$$

The initial probability vectors and the transient generator matrix are used for the generation the PH random variates in PH/PH(n)/C/C DES model of urban rail transit station walkway.

3 Simulation-based Optimization Approach for the Width Design

The *simulation-based optimization* approach proposed in this paper performs the optimization by genetic algorithm (GA) using PH/PH(n)/C/C DES model as representative of the urban rail transit station walkway. A PH/PH(n)/C/C DES model of walkway is developed in the *SimEvents*® to estimate the necessary performance measures (blocking probabilities 'P_c' and mean area occupied per passenger 'ES') of walkway while the optimization method that uses the genetic algorithm (GA) is implemented in the *MATLAB*® programming environment. The proposed *simulation-based optimization* approach blends both the PH/PH(n)/C/C DES and GA to work together concurrently and find the best model parameters (optimized width of the walkway) by minimizing an objective function of the performance measure (mean area occupied per passenger 'ES') while keeping the LOS and blocking probability P_c in certain limits specified by TCQSM [1] and Hu et al. [10], respectively. The general flowchart of the *simulation-based optimization* for the urban rail transit station walkway width design is shown in Fig. 3. One of the main advantages of *simulation-based optimization* is that it not necessary to provide analytical expression of objective and constraints function as in the case of mathematical optimization. The first key ingredient of PH/PH(n)/C/C DES model is the efficient generation of PH random variates. Neuts [36] developed a 'Count Procedure' for the efficient generation of PH random variates.

3.1 Generation of Phase-Type (PH) Random Variates

PH distribution is proposed in this study to simulate the passenger arrival rate and state-dependent service rate of the walkway. The Neut's procedure relies on generating an Erlang-distributed sample with degree f and parameter λ given as:

$$\mathrm{Erl}(f, \lambda) = -\frac{1}{\lambda}\ln\left(\prod_{j=1}^{f} U_j\right)$$

where U is uniformly distributed random number [0, 1]. Let b_j represent the row vector with 1 at the position j. A pseudo-codes description of the PH random variates used in this work follows:

1) $x_{ph} := 0, f_j = 0$, for $j = 1, 2 \ldots\ldots n$, Draw an α-distributed discrete sample for the initial state.
2) The chain in the state j,
   i. $f_j += 1$
   ii. a $\mathbf{b}_j(\text{-}\mathbf{diag}\langle 1/\mathbf{d}_{jj}, 0\rangle \overline{\mathbf{D}}+\mathbf{I})$-distributed discrete sample is drawn for the next state,
   iii. in case the next state is an absorbing state then goes to 3 otherwise stay at 2 and repeat

   for j=1, 2....n;
3) do $x_{ph} += Erl(f_j, -d_{jj})$;

   done

Return x_{ph}.

3.2 PH/PH(n)/C/C Discrete-Event Simulation Model Architecture

In contrast to PH/PH(n)/C/C analytical queuing model [3], PH/PH(n)/C/C DES model eliminates the need to solve large matrices and equations systems to estimate the performance measures. A PH/PH (n)/C/C DES model of the urban rail transit station walkway is built in the *SimEvents*®. The walkway facility is described as a PH/PH(n)/C/C queueing system. An overview of the key components of the walkway facility is presented in Fig. 4.

The key components are *FIFO_Queue* blocks that represent the passenger queueing space, *Single-Server* block that stores the entities for certain period of time, (entities in our case are passengers), *Start and Read Timers* that report the time associated with the arriving passengers and the *Level-2 MATLAB*® *S-function* blocks that compute and update different parameters in the PH/PH (n)/C/C DES model. The *Event to Timed Signal* blocks and *Timed to Event Signal* blocks convert event-based signals to time-based signals and vice versa. *Constant* blocks are used to input different parameters values in the DES model while *Display* blocks show the performance measure (output). The *SimOut* blocks export the values of performance measures from *SimEvents*® simulation to *MATLAB*® programming environment.

3.2.1 Passengers Generation Phase

In our PH/PH/(n)/C/C DES model as shown in Fig. 4a, the passengers are first generated at the entrance of the walkway facilities. The PH random variates are programmed in *Level-2 MATLAB*® *S*-function blocks (designated as S4) at passenger generation phase using 'Count Procedure' given in Ref. [37]. The two input parameters for the computation of PH random variates are λ and c_a^2 which can be calculated by Eqs. (3), (4) and (5), respectively.

3.2.2 State-dependent Service Phase

After the generation of passengers in first phase, it must guarantee that the number of passengers that enter the walkway does not overcome its overall capacity $C = 5LW$. The passengers arriving at the entrance of the walkway facility form a queue and have to wait for free spaces (servers). At the same time, they reduce the free spaces on the walkway and affect the walking speed of passengers crossing the facility. To implement this condition, the generated passengers are stored in the *FIFO_Queue* block before being delayed by the *Single_Server* block and subsequently being sent to the successor facility.

Four *Level-2 MATLAB*® *S-function* blocks are used in this phase to calculate the state-dependent service rates based on PH random variates, mean area occupied per passengers '*ES*', blocking probability P_c and judging the number of passengers to prevent them from entry when maximum capacity $C = 5LW$ is reached as shown in Fig. 3b. The state-dependent service rate calculation depends on congestion of walkway area. The service rate calculation block takes capacity $C = 5LW$ and number of passengers (#n) from the *FIFO_Queue* block as input to compute the μ_n and $c_{s,n}^2$ using Eqs. (10) and (11), respectively.

The service rate calculation block dynamically updates the service rates as a function of number of passengers (#n)

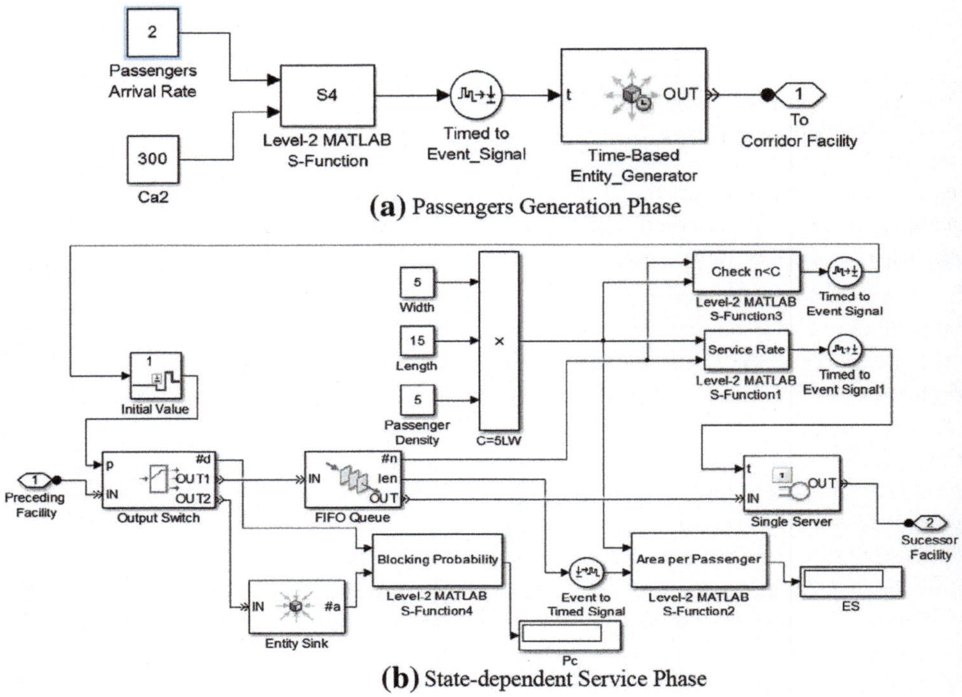

Fig. 4 PH/PH(n)/C/C discrete-event simulation model architecture

in the walkway. Similarly, mean area occupied per passenger '*ES*' is calculated by using area of the walkway $A = LW$ divided by mean queue length (len) obtained by *FIFO_Queue* block. When the capacity $C = 5LW$ has been reached in the walkway, i.e., $\#n = C$ the passengers are blocked to enter the *FIFO_Queue* block. At the same time, the second entity port (OUT2) of *Output Switch* block activates and registers the blocked passengers. The blocking probability P_c is calculated by using number of passengers departed via second entity port of *Output Switch* divided by total number of passengers departed via both first (OUT 1) and second ports (OUT 2).

3.2.3 PH/PH(n)/C/C DES Model Verification

Before developing the *simulation-based optimization* approach, we first verify the accuracy of our PH/PH(*n*)/C/C DES model. For this purpose, the simulation tests with different passenger arrival rate and squared coefficient of variation are performed and compared to the PH/PH(*n*)/C/C analytical model [10] The peak passengers arrival rate λ are taken as 0.21, 0.34, 0.4 0.67 1.25, 1.79 (ped/s) with squared coefficient of variation of arrival interval (c_a^2) 3.24, 4.29, 2.36 1.82 2.05, 2.91, respectively, that are according to actual conditions of urban rail transit station. A 30×2 m^2 horizontal urban rail transit station walkway is considered for all the experiments. The three representative points for the walking speed on the walkway are given as $(v_1 = 1.50,\ \delta_1 = 0.50)$, $(v_a = 0.64,\ \delta_a = 0.21)$ and $(v_b = 0.25,\ \delta_b = 0.08)$, respectively [3]. The three performance measures of the walkway such as mean number of passengers on walkway E[N], mean area per passenger '*ES*' and mean time spent by passenger on the walkway E[T] are calculated. The results of PH/PH(*n*)/C/C DES model are obtained after 10 replications with the simulation time of 20,000 time units. The PH/PH(*n*)/C/C DES model results are compared to the PH/PH(*n*)/C/C analytical queuing model. It is observed that the difference between PH/PH(*n*)/C/C DES model and PH/PH(*n*)/C/C analytical model is very small. The mean relative error of PH/PH(*n*)/C/C DES model is 2.1% for mean number of passenger on the walkway E [N], 2.9% for the mean area occupied per passenger ES and 3.2% for mean time spent by passenger on walkway E[T]. The small mean relative error indicates that PH/PH(*n*)/C/C DES model can be used with good accuracy in performance evaluation of urban rail transit station walkways.

The value of the mean number of passenger on walkway E[N] increases with the increase in arrival rate λ as shown in Fig. 5. Similarly; the mean area occupied per passenger ES decreases with increase in arrival rate. It is because of the fact that when more passengers occupy the spaces on the walkway, the relative distance between them reduces. The mean time spent by passenger on walkway E[T] increases with increase in arrival rate. It is quite obvious that increase in the number of passengers due to high arrival rate causes reduction in the walking speed of the passengers and thus passengers spend more time on the walkway.

3.3 Parallel Implementation of PH/PH(*n*)/C/C DES Model and GA

Based on the PH/PH(*n*)/C/C DES model, we develop a *simulation-based optimization* approach for the width design of urban rail transit station walkway, The GA is used as an optimization method in conjunction with PH/PH(*n*)/C/C DES model to determine the optimal width of the urban rail transit station walkway. The proposed *simulation-based optimization* based on PH/PH(*n*)/C/C DES model and GA is described as follow:

- The width W of the walkway is to be optimized under the LOS 'B', 'C' and 'D; for which the blocking probability P_c does not exceed the specified threshold p.
- The blocking probability.
- According to the TCQSM [1], the LOS of the walkway is reflected by the mean area occupied per passenger *ES*, which means that the *ES* on the walkway for a given LOS must fall within the range [$\mathrm{LOS_{UB}}$, $\mathrm{LOS_{LB}}$] where $\mathrm{LOS_{UB}}$ and $\mathrm{LOS_{LB}}$ are the upper and lower bounds of mean area occupied per passenger for the given LOS.
- The search space S for the optimal width W_{opt} is defined by the upper and lower bounds of the width.

$$S = [W_{UB}, W_{LB}]$$

$$W_{opt} \in S$$

- The performance measures (outputs) *ES* and P_c are estimated by running PH/PH (*n*)/C/C DES model of urban rail transit station walkway (see Fig. 4b).
- Mean area occupied per passenger *ES* on the walkway is $ES = LW/n$, from which we can see that mean area occupied per passenger will vary with the walkway width W. Therefore, the mean area occupied per passenger on the walkway can be expressed as a function of W, that is, $ES(W) = LW/n$.
- The wider the walkway width, the bigger will be the area occupied per passenger. So, the problem of finding the minimum (optimal) width W_{opt} is equivalently converted to find W_{opt} that makes the $ES(W) - LOS_{LB} = 0$. Therefore, the objective function (fitness

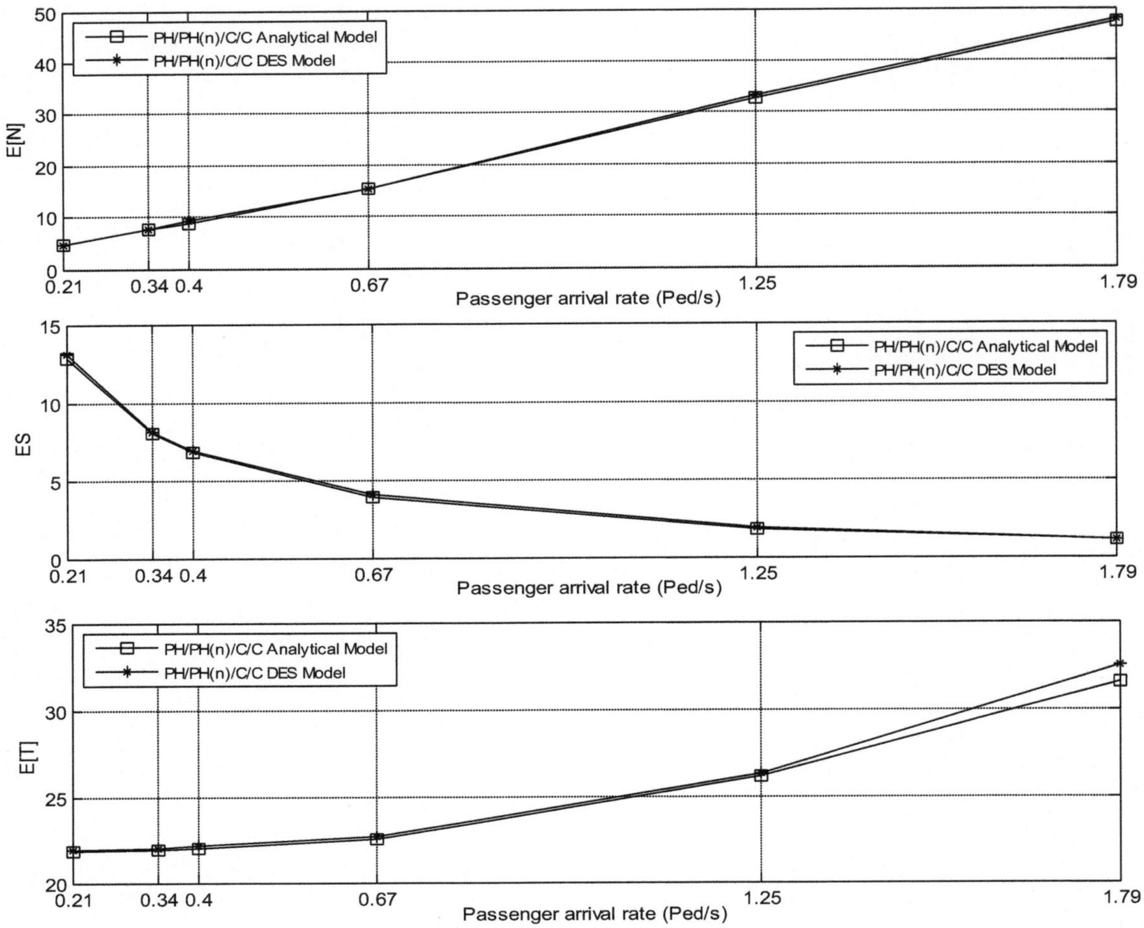

Fig. 5 Verification of the PH/PH(n)/C/C DES model

function) Z for the GA is established from output ES and the lower bound of the walkway LOS, $Z = (ES(W) - LOS_{\mathrm{LB}}) \rightarrow \Re+$.

- Similarly, the constraint function is established from blocking probability (output)P_c, $P_c(W) \leq p$.
- The GA search the search space S *to* find the parameter $W_{\mathrm{opt}} \in S$ with $Z(ES(W_{\mathrm{opt}}) - LOS_{\mathrm{LB}}) \leq Z(ES(W_i) - LOS_{\mathrm{LB}}) \quad \forall W_i \in S$

The *MATLAB*® programming environment is used for programming the constrained GA optimization script. The *MATLAB*® offers a computational infrastructure for optimizing the *SimEvents*® DES model as the *SimEvents*® is embedded in it. Moreover, *MATLAB*® offers parallel DES and optimization without the pain of context switching into multiple softwares. In Table 2, the parameters setting for the GA can be observed.

At first, the interval containing upper and lower bounds of walkway width $U = [W_{\mathrm{UB}}, W_{\mathrm{LB}}]$ is defined which is supposed to contain the optimal width W_{opt} of the walkway. We determine the upper and lower bounds of width by using the flow per unit width values for LOS 'B', 'C' and 'D" that are given in Exhibit 7-3 and Table 1 of the TCQSM [1]. It should be noted that this is the rough estimation of walkway width for setting the initial range. The PH/PH(n)/C/C DES model run initiates with an arbitrary value from the defined interval to simulate the performance of the walkway and obtain the necessary performance measure such as mean area occupied per passenger ES and blocking probability P_c when the simulation system reaches the steady state condition, i.e.,

Table 2 Options set for the genetic algorithm

GA options	Values
Population size	20
Maximum number of generations	10
Probability of crossover	0.8
Mutation probability	0.01
Elite count	10

(at 20,000 time units). Then the genetic algorithm (GA) that is programmed in *MATLAB*® adjusts the width W according to the value of fitness function $Z = ES(W)) - LOS_{\text{LB}}$ and the constraint function $P_c(W) \leq p$ until the optimal width W_{opt} is found. It is inefficient to find the exact value of W that makes the function $ES(W)) - LOS_{\text{LB}} = 0$; therefore, a function tolerance η is defined. If the relative change in the best fitness function value is less than or equal to the η, then the corresponding W can be approximately considered as the optimal width W_{opt}. If the difference is larger than η, the GA will replace W from the defined interval $[W_{\text{UB}}, W_{\text{LB}}]$ and set the new width value W in the PH/PH(n)/C/C DES model for the next iteration to obtain the ES and P_c by same means. The iterations continue until the relative change in the best fitness function value is less than or equal to η and the corresponding width W is the optimal width W_{opt}.

4 Computational Experiments

In this section, first we verify the accuracy of our PH/PH(n)/C/C *simulation-based optimization* approach for the width design of urban rail transit station walkway by comparing with the analytical model (Hu et al. [10]). Then, we use different LOS, arrival rates and peak-hour factors to compare our model with the existing design models.

4.1 Verification of PH/PH(n)/C/C Simulation-Based Optimization Approach

Tables 3, 4 and 5 show the comparison of widths obtained by our proposed PH/PH(n)/C/C *simulation-based optimization* approach and PH/PH(n)/C/C analytical model. Two different lengths of walkways, i.e., $L = 10$ and 30 m are considered for all the experiments under the LOS 'B', 'C' and 'D' requirements. The passenger arrival rate of 5,000 and 10,000 ped/h with the peak-hour factors of 0.9, 0.6 and 0.3 are considered. It should be noted that these parameters are only for illustration purposes and can be altered for different conditions. The results have shown clear consistency, and therefore our proposed PH/PH(n)/C/C *simulation-based optimization* approach can be used for the width design of urban rail transit stations.

4.2 Comparison with the Existing Design Approaches and Effect of Different Factors

The optimal (design) widths of the urban rail transit station walkway obtained by our proposed *simulation-based optimization* approach using PH/PH(n)/C/C DES model and GA are compared to the design widths obtained by existing M/G(n)/C/C ([5, 6, 17]) and D/D/1/C [1] analytical approaches. In TCQSM [1], the walkway width design is based on the fixed-length distribution. It means that the design procedure in TCQSM uses fixed arrival rate and a fixed service time which is similar to D/D/1/C analytical queuing model [10]. The squared coefficient t of variation c^2 is equal to 1/30 (0.03) as it neglect randomness and state dependence. We use the D/D/1/C queuing model to represent the width design procedure of TCQSM for the walkway facility. The required input parameters for the width design such as passenger arrival rate λ, peak-hour volume (q), headway between the trains (h), three representative points for the walking speed on the walkway [10] are predetermined. The c_a^2 is calculated by using Eq. (5) as we are dealing with the alighting passengers from the train. We consider mean headway (h) between the train as 90 s and c_h^2 as 0.1. The corresponding c_a^2 for the peak-hour factors 0.9, 0.6 and 0.3 are 140, 210 and 417, respectively. The width designed by the three approaches, PH/PH(n)/C/C *simulation-based optimization*, M/G(n)/C/C and D/D1/C, are presented in Fig. 6. The figures reveal some interesting and important findings:

- It has been observed that the width design by our PH/PH(n)/C/C *simulation-based optimization* is greater than the existing for all the LOS, arrival rates and squared coefficient of variation c_a^2.
- For all the peak-hour factors, when our proposed PH/PH(n)/C/C *simulation-based optimization* approach is

Table 3 Comparison of *simulation-based optimization* with analytical approach under LOS 'B'

Arrival rates q (ped/h)	Approaches	Lengths (m)					
		10			30		
		ε = 0.3	ε = 0.6	ε = 0.9	ε = 0.3	ε = 0.6	ε = 0.9
5000	PH/PH(n)/C/C analytical	9.78	4.87	3.45	9.77	4.86	3.44
	PH/PH(n)/C/C *simulation-based optimization*	9.74	4.83	3.39	9.72	4.85	3.38
10,000	PH/PH(n)/C/C analytical	18.34	9.06	6.24	18.34	9.04	6.24
	PH/PH(n)/C/C *simulation-based optimization*	18.26	8.94	6.17	18.25	8.94	6.19

Table 4 Comparison of *simulation-based optimization* with analytical approach under LOS 'C'

Arrival rates q (ped/h)	Approaches	Lengths (m)					
		10			30		
		ε = 0.3	ε = 0.6	ε = 0.9	ε = 0.3	ε = 0.6	ε = 0.9
5000	PH/PH(n)/C/C analytical	6.99	3.52	2.43	6.98	3.51	2.42
	PH/PH(n)/C/C *simulation-based optimization*	6.87	3.50	2.39	6.84	3.46	2.37
10,000	PH/PH(n)/C/C analytical	12.94	6.26	4.33	12.94	6.25	4.32
	PH/PH(n)/C/*C simulation-based optimization*	12.86	6.21	4.29	12.86	6.11	4.25

Table 5 Comparison of *simulation-based optimization* with analytical approach under LOS 'D'

Arrival rates q (ped/h)	Approaches	Lengths (m)					
		10			30		
		ε = 0.3	ε = 0.6	ε = 0.9	ε = 0.3	ε = 0.6	ε = 0.9
5000	PH/PH(n)/C/C analytical	5.59	2.71	1.77	5.57	2.70	1.75
	PH/PH(n)/C/C *simulation-based optimization*	5.51	2.65	1.74	5.50	2.67	1.74
10,000	PH/PH(n)/C/C analytical	10.48	4.93	3.48	10.47	4.92	3.47
	PH/PH(n)/C/*C simulation-based optimization*	10.41	4.88	3.43	10.41	4.85	3.44

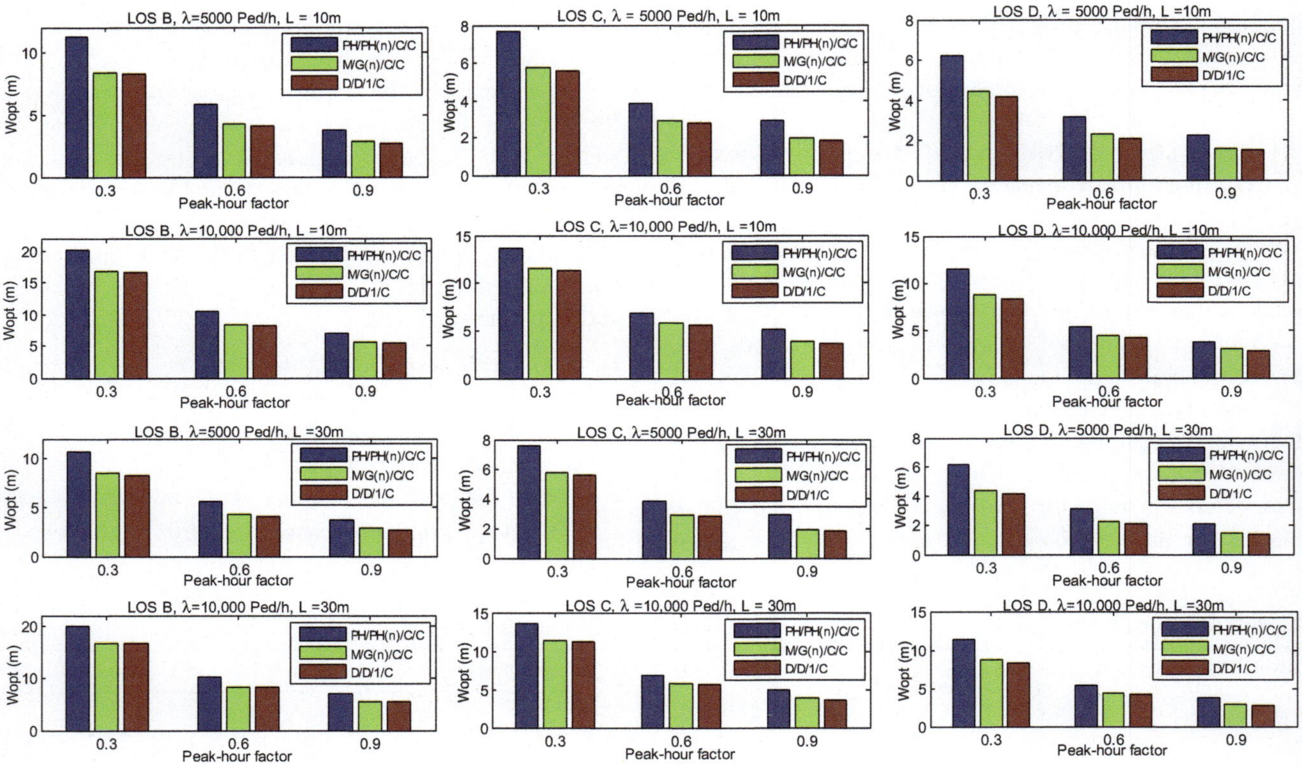

Fig. 6 Comparitive analysis of different design approaches

compared with the existing M/G(n)/C/C model, it has been observed that there is an average increase of 2.44 m in width for LOS B, 2.22 m for LOS C and 2.21 m for LOS D. Similarly, the average increase in width while comparing M/G(n)/C/C and D/D/1/C model is 0.12, 0.19 and 0.7 m for LOS B, LOS C and LOS D, respectively.

- For all the three design approaches, the width of the walkway increases with increase in passenger arrival rate even when squared coefficient of variation remains the same. It is expected because of the fact that these design approaches are sensitive to arrival rate. When the arrival rate increases from 5000 to 10,000 ped/h, for all the LOS, the average increase in our proposed PH/PH(n)/C/C *simulation-based optimization* model and M/G(n)/C/C is 2.45 m for $\varepsilon = 0.3$, 1.97 m for $\varepsilon = 0.6$ and 1.78 m for $\varepsilon = 0.9$. Similarly; for M/G(n)/C/C and D/D/1/C, the average increase is 0.45 m for $\varepsilon = 0.3$, 0.37 m for $\varepsilon = 0.6$ and 0.24 m for $\varepsilon = 0.9$. It reveals that the growth of width is higher for low peak-hour factors that eventually results in higher values of squared coefficient of variations (by using Eq. 5). The squared coefficient of variations accounts for the randomness in passenger arrival process as well as in service process of the walkway. The existing M/G(n)/C/C model and D/D/1/C model ignore this factor and result in underestimation of width. Therefore, the design methods based on M/G(n)/C/C and D/D/1/C models are not applicable in practical system where the squared coefficient of variation is very high. On the contrary, the width of our proposed model is sensitive to squared coefficient of variation.

- The increase in passenger arrival rate from 5000 to 10,000 ped/h results in almost doubling the design width for all the peak-hour factors and LOS. The increase in width value is quite obvious because the higher arrival rates cause high density of passengers on the walkway resulting in the slow movement of passengers and congestion. Therefore, in order to avoid the congestion and the blocking, the walkway width should be more for higher arrival rates to provide safe passage for the passengers to travel.
- The design width very slightly decreases, which is almost negligible when the length of the walkway is increased from $L = 10$ to 30 m. For all the LOS, the average width decrease of 0.03 m for $\varepsilon = 0.3$, 0.027 m for $\varepsilon = 0.6$ and 0.02 m for $\varepsilon = 0.9$ is observed when the length increases from $L = 10$ to 30 m and passenger's arrival rate from 5000 to 10,000 ped/h. It is observed that an increase in the length L has two opposing influences on the mean area occupied per passenger *ES*. The increase in the length means the increase in the capacity of the walkway. The *ES* increases with the increase in L since the area of the walkway increases; however, this advantage is adversely affected by the fact that increasing the length gives the passengers a longer distance to travel in the walkway from one end to another. Hu et al. [10] observed that due to these opposing influences, there is a threshold value of the L; when L is larger than that threshold value, the effect of the passengers having to travel a longer distance has a stronger overall effect on the congestion. If the value of L is smaller than the threshold value, then there is no significant effect.

5 Conclusion and Future Work

This paper proposes a *simulation-based optimization* approach using PH/PH (n)/C/C DES model and GA algorithm for the width design of urban rail transit station walkway. To reproduce the reality, the walkway is abstracted as a finite capacity queuing system. The corresponding PH/PH(n)/C/C model is developed in the *SimEvents*® software. Unlike the existing design codes (D/D/1/C) and M/G(n)/C/C model that ignore the randomness and state dependence, the proposed PH/PH(n)/C/C DES model considers the randomness and state dependence and overcomes these shortcomings. Then the GA algorithm, integrated with the PH/PH(n)/C/C DES model, determines the optimal width of the walkway with blocking probability control under the different LOS and passenger flow conditions.

The comparative experiments show that the walkway designed by our proposed model has larger widths compared with the designed by existing models. The width of the proposed model increases more with increase in c_a^2 of arrival interval compared to the widths of the design methods based on the M/G (n)/C/C and D/D/1/C models. The width of the proposed method increases faster than the other two methods when the arrival rate increases. The length of walkway has no significant effect on design width.

This new proposed *simulation-based optimization* approach, integrating DES and optimization, can help the planners and designers of urban rail transit station to make decisions regarding urban rail transit station design. The coefficient of variation (c_a^2) of arrival interval should be paid more attention during the design process of the walkway. To apply the proposed method, we only need the length of walkway and passenger forecast data, such as peak-hour factor (ε), passenger peak-hour volume (q), mean headway between the trains (h) and the c_h^2 of the arrival interval between the train.

Besides urban rail transit station walkway, the proposed method can also be applied to the stairs facilities in urban rail transit stations as well as corridors and stairs in other buildings such as shopping malls, stadiums and hospitals. The proposed method can be easily extended to a queuing network to design different service facilities such as a network of walkways and stairs. The alighting passenger flow from the train is considered. Nevertheless, the proposed approach can also deal with the flow of passengers toward the walkway from entrance gate of the urban rail transit station.

This paper considers rectangular walkway for evaluation and design purpose according to the TCQSM. More complicated facilities that are not rectangular can be divided into several rectangular facilities can be evaluated in similar manner. The principle procedure of transformation of walkway into a single-server finite capacity queuing system remains the same. In this research, we only consider the walkway facility. The escalator, platform and ticket facilities are not discussed in this research. These facilities can also be designed by abstracting into a queuing system. Moreover, the state-dependent arrival rate can also be considered in addition to state-dependent service rate in future research work.

Acknowledgements We would like to express our sincere acknowledgment to the National Natural Science Foundation of China (Serial No. 51578465 and 71402149), Basic Research Project of Sichuan Province, the Chinese government for funding of Ph.D. doctoral program at Southwest Jiaotong University and the colleagues of National United Engineering Laboratory of Integrated and Intelligent Transportation at Southwest Jiaotong University, Chengdu, for their support and valuable advices.

References

1. Kittelson, Associates, U. S. F. T. Administration, T. C. R. Program, T. D. Corporation, and N. R. C. T. R. Board (2003) Transit capacity and quality of service manual. Transportation Research Board of the National Academies
2. Hu L, Jiang Y, Zhu J, Chen Y (2013) Hybrid of the scatter search, improved adaptive genetic, and expectation maximization algorithms for phase-type distribution fitting. Appl Math Comput 219(10):5495–5515
3. Jiang Y, Hu L, G Lu (2010) Determined method of subway footway width based on queuing theory. J Traffic Transp Eng 10:61–67
4. Yuhaski SJ, Smith JM (1989) Modeling circulation systems in buildings using state dependent queueing models. Queueing Syst 4(4):319–338
5. Cheah J, Smith JM (1994) Generalized M/G/C/C state dependent queueing models and pedestrian traffic flows. Queueing Syst 15(1–4):365–386
6. Mitchell DH, MacGregor J (2001) Smith. Topological network design of pedestrian networks. Transp Res Part B: Methodol 35(2):107–135
7. Chen SK, Liu S, Xiaom X (2012) M/G/C/C-based Model of passenger evacuation capacity of stairs and corridors in urban rail transit station. J Chin Rail Soc 34:7–12
8. Weiss A, Williams L, MacGregor Smith J (2012) Performance and optimization of M/G/c/c building evacuation networks. J Math Model Algorithms 11(4):361–386
9. Xu X-Y, Liu J, Li H-Y, Hu J-Q (2014) Analysis of subway station capacity with the use of queueing theory. Transp Res Part C: Emerg Technol 38:28–43
10. Hu L, Jiang Y, Zhu J, Chen Y (2015) PH/PH(*n*)/C/C state-dependent queuing model for metro station corridor width design. Eur J Oper Res 240(1):109–126
11. Alfa AS, Zhao YQ (2000) Overload analysis of the PH/PH/1/K queue and the queue of M/G/1/K type with very large K. Asian Pac J Oper Res 17:122–136
12. Miyazawa M, Sakuma Y, Yamaguchi S (2007) Asymptotic behaviors of the loss probability for a finite buffer queue with QBD structure. Stoch Models 23(1):79–95
13. Krishnamoorthy A, Babu S, Narayanan VC (2008) MAP/(PH/PH)/c queue with self-generation of priorities and non-preemptive service. Stoch Anal Appl 26(6):1250–1266
14. Jiang Y, Hu L, Zhu J, Chen Y (2013) PH fitting of the arrival interval distribution of the passenger flow on urban rail transit stations. Appl Math Comput 225:158–170
15. Løvås GG (1994) Modeling and simulation of pedestrian traffic flow. Transp Res Part B: Methodol 28(6):429–443
16. Cruz FRB, MacGregor Smith J, Medeiros RO (2005) An M/G/C/C state-dependent network simulation model. Comput Oper Res 32(4):919–941
17. Khalid R, Nawawi MKM, Kawsar LA, Ghani NA, Kamil AA, Mustafa A (2013) A discrete event simulation model for evaluating the performances of an M/G/C/C state dependent queuing system. PLoS ONE 8(4):e58402
18. Jiang Y, Zhu J, Hu L, Lin X, Khattak A (2015) A G/G(*n*)/C/C state-dependent simulation model for metro station corridor width design. J Adv Transp 50(3):273–295
19. Teknomo K (2006) Application of microscopic pedestrian simulation model. Transp Res Part F: Traffic Psychol Behav 9(1):15–27
20. Kaakai F, Hayat S, El Moudni A (2007) A hybrid Petri nets-based simulation model for evaluating the design of railway transit stations. Simul Model Pract Theory 15(8):935–969
21. Banks J (2010) Discrete-event system simulation. Prentice Hall, Englewood Cliffs
22. Hagendorf O, Pawletta T, Larek R (2013) An approach for simulation-based parameter and structure optimization of MATLAB/Simulink models using evolutionary algorithms. *SIMULATION*
23. Figueira G, Almada-Lobo B (2014) Hybrid simulation–optimization methods: a *taxonomy and discussion[J]*. Simul Model Pract Theory 46:118–134
24. Swisher JR, Hyden PD, Jacobson SH, Schruben LW (2000) Simulation optimization: a survey of simulation optimization techniques and procedures. In: Presented at proceedings of the 32nd conference on Winter simulation, Orlando, Florida
25. Fu MC (2002) Feature Article: optimization for simulation: theory vs. practice. INFORMS J Comput 14(3):192–215
26. Cassandras CG, Lafortune S (2009) Introduction to discrete event systems. Springer, Berlin
27. Jiang Y, Lin X (2013) Simulation and optimization of the ticket vending machine configuration in metro stations based on anylogic software. In: Fourth international conference on transportation engineering, pp 754–760
28. Hassannayebi E, Sajedinejad A, Mardani S (2014) Urban rail transit planning using a two-stage simulation-based optimization approach. Simul Model Pract Theory 49:151–166
29. Hagendorf O, Pawletta T, Larek R (2013) An approach to simulation-based parameter and structure optimization of MATLAB/Simulink models using evolutionary algorithms. Simulation 89(9):1115–1127
30. Sebastiani MT, Lüders R, Fonseca KVO (2016) Evaluating electric bus operation for a real-world brt public transportation using simulation optimization. IEEE Trans Intell Transp Syst 17(10):2777–2786
31. Yan J, Wen J, Li L (2014) Genetic Algorithm Based Optimization for Energy-aware Hybrid Flow Shop Scheduling. In: Proceedings on the international conference on artificial intelligence (ICAI). 2014. The Steering committee of the world congress in computer science, computer engineering and applied computing (WorldComp)
32. Zelenka, J (2010) Discrete event dynamic systems framework for analysis and modeling of real manufacturing system in Intelligent

Engineering Systems (INES). In: 14th International conference on. 2010. IEEE
33. Hubscher-Younger T, Mosterman PJ, DeLand S, Orqueda O, Eastman D (2012) Integrating discrete-event and time-based models with optimization for resource allocation. In: Simulation Conference (WSC), Proceedings of the 2012 Winter, pp 1–15
34. Messac A (2015) Optimization in practice with MATLAB. Cambridge University Press, Cambridge
35. Tregenza P (1976) The design of interior circulation. Van Nostrand Reinhold, New York
36. Neuts MF (1981) Matrix-geometric solutions in stochastic models: an algorithmic approach. Dover Publications, New York
37. Sadre R, Haverkort B (2011) Decomposition-based queueing network analysis with FiFiQueues. In: Boucherie RJ, van Dijk NM (eds) Queueing networks, No. 154, Springer US, pp 643–699
38. Sadre R (2007) Decomposition-based analysis of queueing networks. University of Twente, Twente
39. Zhu J, Hu L, Jiang Y, Khattak A (2017) Circulation network design for urban rail transit station using a PH(n)/PH(n)/C/C queuing network model[J]. Eur J Oper Res 260(3):1043–1068

14

Research on Vertical Coupling Dynamics of Monorail Vehicle at Finger-Band

Zhouzhou Xu[1] · Zixue Du[2] · Zhen Yang[1] · Liang Chen[1]

Abstract In view of the fact that exhibits monorail transportation in recent years, the good situation and the lack of straddle-type monorail vehicle vertical dynamics research, the key problems of vertical vibration were studied in this paper: The vertical coupling is between walking tires and finger-band. A dynamic model based on the vertical equivalent stiffness of tire is proposed, the force mode of the tire is determined through the comparison calculation and the dynamics model between walking tires and finger band was established on ADAMS Dynamics Software. By dynamic calculation, the vibration of four running wheels in time and frequency domains is obtained, and the vibration of the front axle of the front bogie is studied in detail; the correctness of the model is confirmed by comparing with the real vehicle test. Finally, the limitations of this paper and the direction of follow-up work are discussed.

Keywords Monorail · Vertical coupling dynamics · Finger-band equivalent · Stiffness · Adams

1 Introduction

Since the opening of the Chongqing metro line 2 in 2004, the monorail has the advantages like small turning radius, small footprint, strong climbing ability, low noise , comfort, low cost (only about metro's 1/3–1/2). Monorail transporation got wide acclaim. In 2011, Chongqing line 3 with independently developed equipment and technology has been put into operation, which indicates that light rail technology in China is approaching maturity. The successful operation of large-capacity monorail transportation in Chongqing indicates that the technology of large-scale and large-span straddle-type monorail transportation has matured and formed a demonstration and promotion effect. Besides Chongqing, Guiyang, Liupanshui began planning for investment in the construction of straddle-type monorail transportation, Indonesia, India, Thailand, Malaysia, Pakistan and other countries are also in the planning of straddle-type monorail transportation [1].

Although the monorail was put in operation in 1970s, it has not been accepted much in the world for a long time; The maximum distance of the monorail operated in the whole world was less than 100 km until 2004, which is the cause that its research is not comprehensive, especially on straddle-type monorail track vertical coupling dynamics, so far, there were very few studies, and the works about the vertical dynamics of straddle-type monorail vehicle were almost zero.

As a standard of public transport, the vertical dynamic action of monorail determines many problems like the safety of train operation, the riding comfort and the vertical force of rail beam [2, 3]. The vertical dynamics is mainly caused by the interaction of the rail beam and the body (bogie), and the surface roughness of the track beam is the mechanism of its formation. The straddle-type monorail track beam is divided into two kinds of PC beam (prestressed reinforced concrete beam and steel structure beam), they are all prefabricated and finished, and so the surface smoothness is very good. (At least they are close to the US road A road surface, but because there is no relevant research, so we lack specific data.) The gap between

✉ Zhouzhou Xu
33243665@qq.com

1 Chongqing Jiaotong University, Chongqing, China

2 School of Mechanical and Vehicular Engineering, Chongqing Jiaotong University, Chongqing, China

Editor: Eryu Zhu

the beam and the beam, which called finger-band, is considered to be the largest vertical excitation source in the track beam, so if we want to study the vertical dynamics of straddle-type monorail, we must study its vertical dynamics at the finger-band. [4]

2 The Structure of Finger-Band

The finger-band is a bridge connecting different track beams, and its structure and mechanical model are shown in Fig. 1. Each set of finger-band consists of two plates of same structure. The finger-band is slightly wider than the running wheel, which fits exactly with the special structure of the two tires running side-by-side on each axle of the bogie. When the monorail vertical passes through the finger plate, each finger-band plate group carries the side of the traveling wheel. The finger-band is made up of steel plates, it is about 4 cm thick and fixed on the base of the concrete, it can be considered as a rigid body, and so the deformation of the finger-band is not considered in this paper [5, 6].

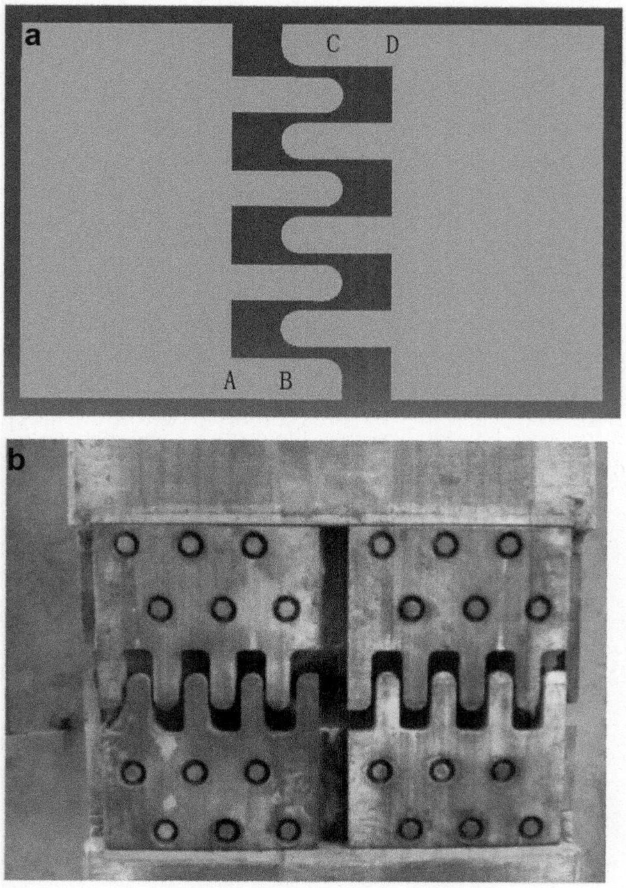

Fig. 1 **a** structural model diagram of finger-band, **b** the photograph of finger-band

3 The Dynamics Model of Running Wheel

As the running wheel uses rubber tires, the tire parts of the tire are subjected to uneven pressure during the crossing of the finger-band, resulting in a change in the amount of deformation and a vertical vibration of the tire [7, 8].

The force of the rubber tire is quite complex, and the tire is not uniform under the force of the finger-band. Therefore, the actual deformation of the tire at the finger plate cannot be explored by the force and deformation of each layer of the tire rubber. The force acting on a monorail vehicle passing through the finger-band is simulated by applying equal force on each axis of the bogie in this paper [9–12].

Compared to the steel wheel rails of ordinary trains, the wheel–rail coupling force of the monorail vehicle is much slower when rubber tracks are used on the vehicle [13, 14]. The material, physical and geometrical parameters of the wheel are shown in Tables 1 and 2. When the tire is running on a common beam, its ground mark average, but when it is running across the finger-band, as there is a gap in the finger-band, it must have uneven deformation at the ground, in the gap, it's no compressed, the original pressure of the tire will be released because there is no compression, but at the fingers, when the compression of the finger will be increased and the center of the tire will sag. The pressure in the vertical body (weight) unchanged, that is reflected in its vertical equivalent stiffness is smaller.

The original vertical stiffness of tire is: $k1 = 1{,}250{,}000$ N/M, and at the finger-band, it is assumed that the equivalent stiffness has changed to: $k2 = \gamma k1$.

ABAQUS was used to calculate the grounding mark of tire under static pressure, as shown in Fig. 2. When the tire is subjected to downward pressure of 5500 kg, the grounding mark is about 65,400 mm^2 and the mark is near rectangular. The tire is 240 mm wide, and the imprint is about 270 mm long. Known as the finger-band, the overall length of the finger is 146 mm, and it can be thought that when the tire passes through the finger-band region at regular intervals, the γ varies according to a rule; in order to simplify the calculation, it is set to do a uniform change, and its size can be between 1 and 0.75. (The actual situation is that, because the rubber tire tread rubber molecular interaction, deformation at the gap can not be fully released, so the gamma should be >0.75. Because this experiment cannot be done, the γ cannot be ascertained; here is the limit.)

According to the maximum value of the tire equivalent stiffness ($\gamma = 0.75$), If there is no vertical displacement of the bogie, the variation of the tire force is

$$\Delta F = \Delta\gamma \cdot k_1 \cdot x = 13750\,\text{N}$$

where x is the tire deformation under full load.

Table 1 The material parameter of running tire

Parameter	C_{10}/MPa	C_{01}/MPa	D_1/MPa	Density/(Kg/m^3)
Tread rubber	0.428	0.107	0.001	1113
Sidewall adhesive	0.326	0.082	0.001	1107
Cord rubber	0.581	0.145	0.001	1168
Triangular rubber	0.401	0.102	0.001	1108

Table 2 Section geometry parameter of running tyre

Mas of tire/kg	50
Diameter of tire /mm	1004
Width of tire /mm	340
Aspect ratio of tire section	0.85
Diameter of rim/mm	424
Tread pattern	Longitudinal groove pattern

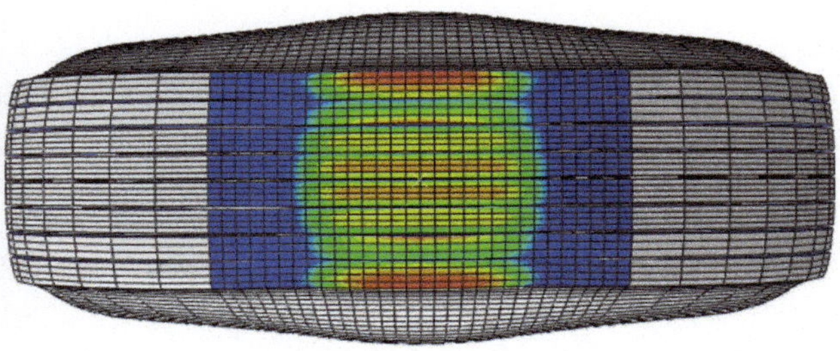

Fig. 2 Contact mark of walking wheel under the action of 5500 kg static pressure

According to the width of the finger-band and the speed of the vehicle, a single wheel passes through the finger-band at constant speed in 0.0122 s. It can be assumed that the γ is uniformly changed during this period of time. In the first 0.0061 s, γ varies from 1 to 0.75, and in the following 0.0061 s, from 0.75 to 1. Assuming that the bogie is not displaced vertically, the vertical load of the bogie changes from 0 to 13,750 N in the first 0.061 s and from 13,750 to 0 in the following 0.061 s.

According to simple dynamic calculation, under this action, the bogie moves downward vertically 1.24×10^{-4} M in 0.0122 s, and the effect of this displacement on the vertical load of the tire is 155 N, which is approximately 1%, compared with 13,750 N, and so it can be neglected. Therefore, it is considered that the force acting on the tire is mainly caused by the change in the vertical stiffness of the tire when the vertical center of the tire moves up and down.

So, by dynamics calculation, when the car body passes through the finger-shaped plate area, the tire force operation is as follows: the vertical acceleration of the tire (downward) is increased from 0 to 10 m/s^2 in the first 0.0061 s, and the vertical acceleration of the tire (downward) is reduced from 10 to 0 m/s^2 in the following 0.0061 s.

4 Walking Tires and Finger-band Dynamic Model Building

In this paper, ADAMS, one multi-body dynamics software, is used to calculate the wheel–rail coupling dynamics of straddle-type monorail vehicle.

For monorail vehicle, the bogie is the core of its structure, and also the key component of the wheel and rail contact force transmission. According to the mass distribution law of bogie and car body (Table 3), the transmission of bogie and vehicle, the transmission of tire and bogie, and the interaction between tire and rail surface, the model of bogie, vehicle and road model (Table 4) is built shown in Fig. 3, on the basis of fully considering the force of the bogie. In order to simulate the actual road surface, the road roughness is classified as class A according to the road spectrum in the USA.

The average speed of Chongqing monorail line 3 is about 43 km/h, so 43 km/h is used as the running speed of single track vehicle.

In order to simulate the force acting on the finger plate of a monorail train, an equivalent deformation force is added to all the axles of the vehicle to replace the equivalent stiffness change of the tire. The action, time, magnitude and direction of force are discussed in the second part of this paper. Since the length of the PC track beam is 22–24 m, the acting force is applied in accordance with the interval between each 2 s of the same set of tires passing through a finger plate. According to kinematics calculation, the force conditions of the two bogies, a total of four axles, are the same, and they will pass the same finger-band at the same time intervals. As the calculation is too large, only four finger-bands are selected for dynamic simulation modeling.

5 Simulation

Because the vehicle movement is applied on the tire, the vehicle is uniformly accelerated from 0, reaching a steady speed 43 km/s after 3 s. As this drive will affect the

Table 3 Parameter of vehicle

Length/m	15.5
Width/m	2.9
The height between the top of the air conditioner to the track surface /m	3.76
The mass of vehicle under full load /t	41.6
Average speed of line segments /m/s	11.94
The mass of bogie/t	5.5

Table 4 Parameter of the track

Width of running track/m	0.85
Side height/m	1.5
The Power density function of running track Gd	1.5e–6

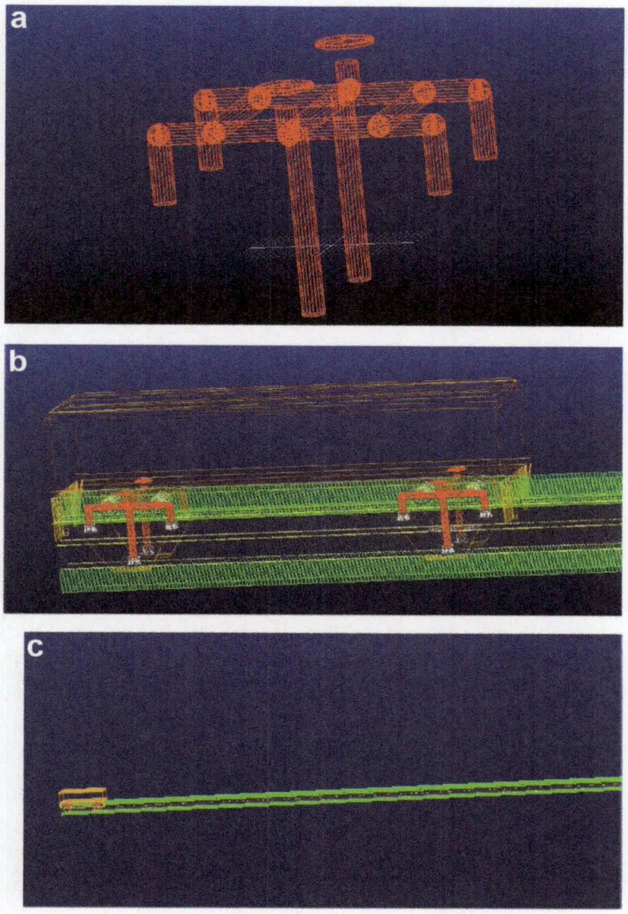

Fig. 3 Wheel–rail coupling dynamic model in Adams. **a** Bogie model, **b** vehicle model, **c** whole vehicle routing model

vertical motion of the vehicle in a short time, thus, the force on the front axle of the front bogie is applied from the 5 s to the end of the 11 s, and a total of four finger plate sections are experienced, the force acting on the other axles moves in time according to the kinematic equations. The simulation ends at 14 s.

As shown in Fig. 4, the simulation results show that the force of the front and rear bogies is symmetrical, that is, the vertical motion of the front axle of the front bogie and the front axle of the rear bogie is basically similar, and the rear axle of the front bogie is basically similar to the rear axle of the rear bogie. As the bogie structure is rigid, the the tire of the front axis through the finger-band, it causes the rear axle to produce an upward acceleration, as a result, the rear axle produces a downward force during the acceleration of the finger-band.

Because of the similarity of the vertical motion of the four axles, the vertical acceleration of the first axis of the front bogie (Fig. 4a) is only analyzed here. From its time-domain characteristic curve, its vertical acceleration is mainly within +5 m/s^2 and fully meets the large acceleration at a downward rate every 2 s. The acceleration of other time is due to the random vibration produced by the road roughness, and from the FFT it can be obtained by the frequency-domain characteristic curve shown in Fig. 5

As shown in Fig. 5, the vertical vibration of the bogie is mainly concentrated in the 2.5–13 Hz. If the frequency is too large or too small, the contribution is very low. In addition, its main frequency is about 4 Hz, and its contribution is about 0.43 m/s^2, it can be considered as the result of wheel–rail coupling in the finger-band region, and other parts should be mainly due to the random vibration produced by the road roughness.

Chongqing Orbital Corporation has entrusted the third party to do a real vertical running test of monorail train, and Fig. 6 shows the data of the vertical vibration acceleration of the center point of the front bogie front axle in the middle vertical (a vertical which is not the one placed in head or end of the monorail train). The data in Fig. 6a are the straight line to the time-domain vibration period, its ordinate unit is m/s2, and the horizontal axis is the sampling point. The sampling frequency is 2000 Hz. The data in Fig. 6b is from FFT transform to the data of Fig. 6a.

It can be seen from Fig. 6a, the front axle is still producing a larger impact every 2 s or so, about 10 m/s^2 in size and downward in direction, that should mean the impact of the finger-band on the tires. Figure 6b shows that the vibration frequency of the axle should be 3–13 Hz, in which the main frequency is about 4 Hz, and its contribution is about 0.8 m/s^2, and the rest should be mainly

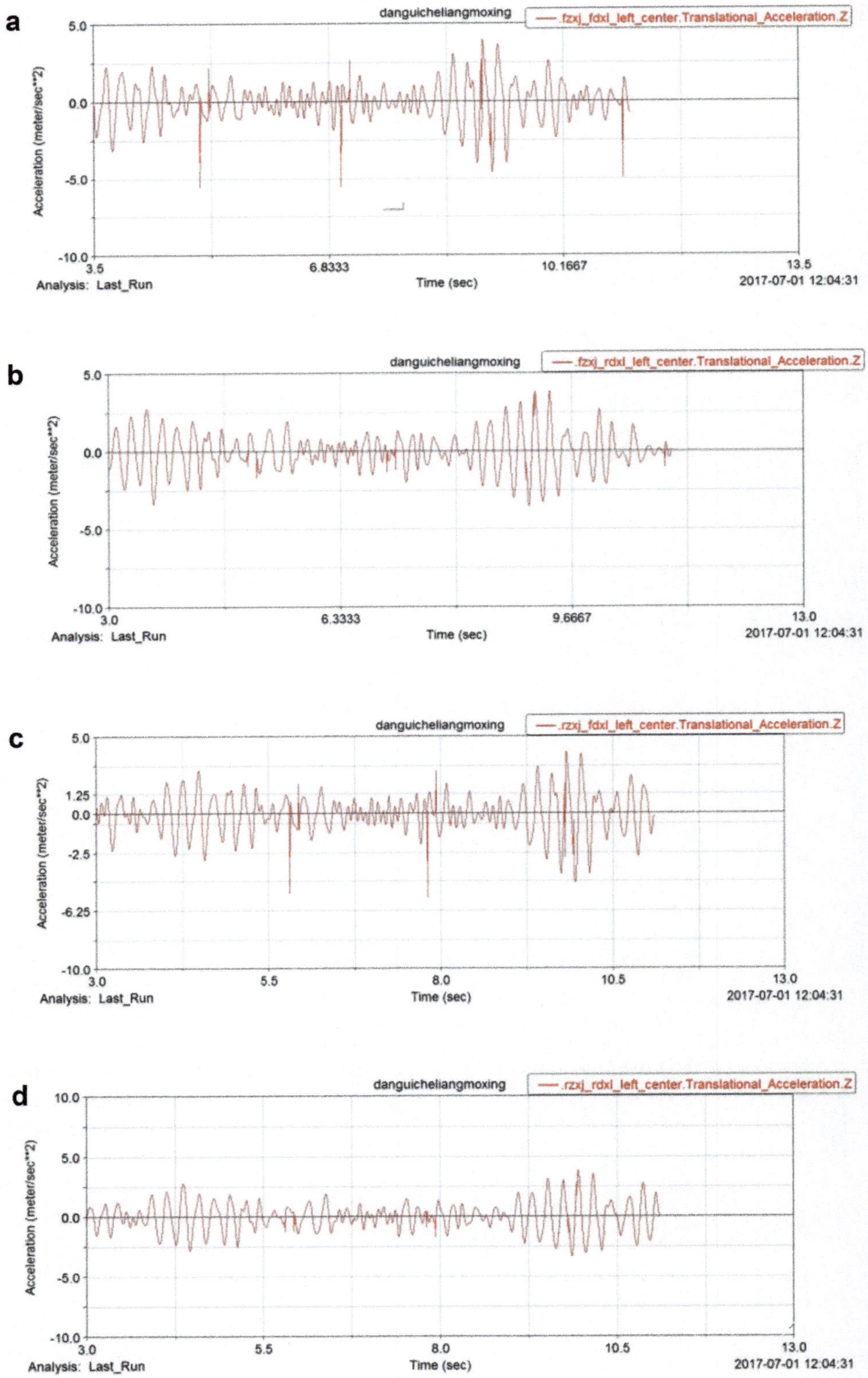

Fig. 4 Movement of the four bogies in time domain. **a** Vertical acceleration of front bogie front axle, **b** vertical acceleration of front bogie rear axle, **c** vertical acceleration of rear bogie front axle, **d** vertical acceleration of rear bogie rear axle

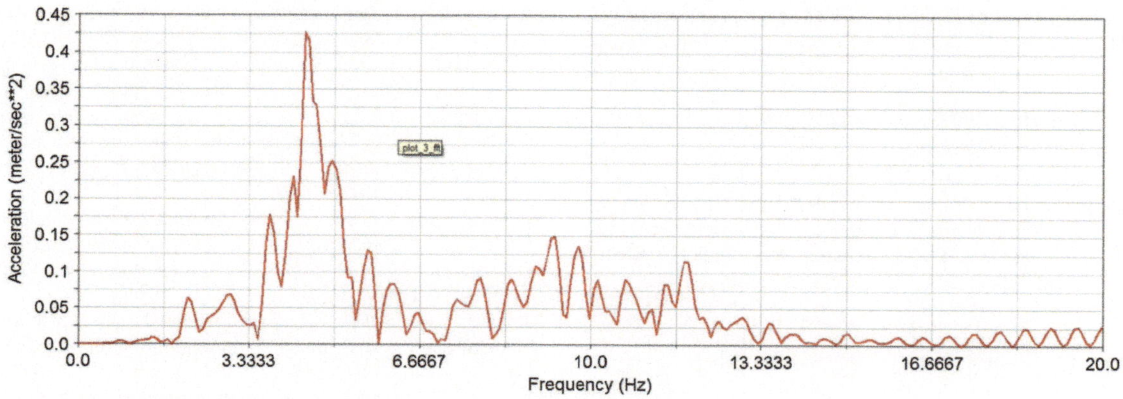

Fig. 5 FFT transformation results of vertical acceleration of front bogie front axle

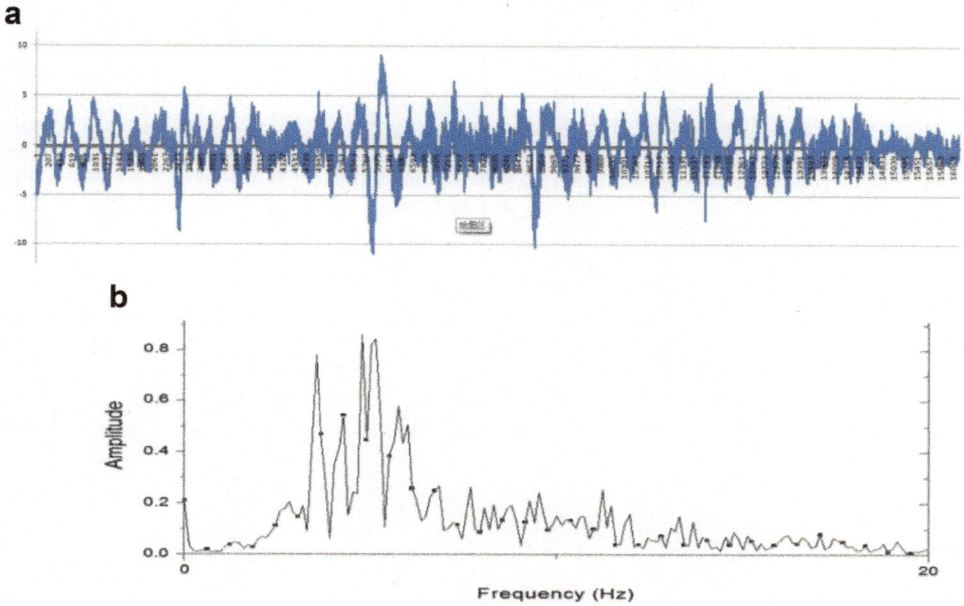

Fig. 6 Data of the vertical vibration acceleration of the center point of the front bogie front axle in the middle vertical. **a** Time-domain vibration, **b** data after FFT

caused by road roughness. This is very similar to the structure calculated by the kinetic model used in this paper.

6 Conclusion and Future Work

In this paper, the force and dynamic response of straddle-type monorail walking wheel passing through the finger plate are studied. An equivalent stiffness method for dealing with the dynamics of nonlinear mechanical materials such as rubber tires is presented, a dynamic model for the force and dynamic response of finger-shaped plate and walking tire is designed, their simulation models are constructed, the dynamic simulation calculation is carried out by using ADAMS, the results are in good agreement with the measured data, and the validity of the kinetic model is verified. According to the experimental data, this paper proved that finger-band is the maximum excitation source of vertical vibration of monorail vehicle by simulation, we can reduce the effective stiffness of the tire when crossing the band by change refers to the geometric structure and parameters of the finger-band, it has great effect to reduce the vertical vibration of vehicle.

There are still some shortcomings.

First, due to the lack of track beam road roughness test data, the simulation can only use the US road spectrum in the mean of grade A road roughness as ordinary excitation, it makes the measured values have certain discrepancy in the PC beam area compared with the simulation data, and it influences the superposition of acceleration at the finger-band area.

Second, for uneven deformation of the tire at the finger plate, due to the lack of necessary experimental data, shear stress between adjacent structures of the rubber body is neglected, only the geometry relation is processed simply, so that the simulation data and the measured value have certain discrepancy in the finger-band area, especially in frequency spectrum analysis, the contribution of each frequency vibration is different.

Acknowledgements Fund Project: National Natural Science Foundation of China. Project Name: Research on the Mechanism and Control Method of Monorail Trains Running Tires' Shoulder Wear. Project No. 51475062.

References

1. Zhong J, Du Z, He X (2013) Turnout structure and analysis of monorail vehicle. People Communications Press, London
2. Zhai W (2007) Vehicle track coupling dynamics. Science Press, Beijing
3. Yang S, Chen L, Li S (2012) Research on vehicle road coupling system dynamics. Science Press, Beijing
4. Penn GR (1968) An analog model of a powerplant coupled to a real car for structure dynamic development. Automotive engineering congress
5. Lyon D (1972) The calculation of track forces due to dipped rail joint, wheel flats and rail weids. The second ORE colloquium on technical computer programs
6. Jenkins H (1974) The effect of track and vehicle parameters on wheel/rail vertical dynamic forces. Railw Eng J 3(1):2–16
7. Markow J, Hedrick K (1988) Analyzing the interactions between dynamic vehicle loads and highway pavements. Transp Rec 1196:161–169
8. Myers L (1999) Measurement of contact stress for different truck tire types to evaluate their influence on near surface cracking and rutting. Transp Rec 1655:175–184
9. Li S, Yang S (2006) Primary resonance of a vehicle suspension with nonlinear stiffness and nonlinear damping. In: The 8th biennial ASME conference of engineering systems design and analysis, Torino, Italy
10. Li S, Yang S (2007) Dynamical analysis of a nonlinear vehicle suspension system under harmonic excitation. Dyn Contin Discrete Impuls Syst Ser B Appl Algorithms 14(Suppl 5):169–173
11. Li S, Yang S (2006) Primary resonance of nonlinear vehicle suspension using improved Bingham model. J Vib Shock 25(4):109–111, 128
12. Li S, Yang S, Li H (2007) Semi-active control of vehicle suspension based on ADAMS-MATLAB joint simulation. J Syst Simul 19(10):2304–2307
13. Zhang K, Zhu E-y (2008) Simulation analysis for the nonlinear seismic response of seismically isolated continuous bridge. In: 2008 international workshop on modelling, simulation and optimization. Computer Society Press, pp 364–367
14. Zhang K (2012) Vehicle bridge coupled vibration analysis of curved bridge with monorail. Beijing Jiaotong University, Beijing

15

Research on the Influence of Straddle-Type Monorail's Pantograph Head Parameter on Power Collection Quality

Zhen Yang[1] · Zixue Du[1] · Cheng Chen[1] · Xiaoxia Wen[1] · Zhouzhou Xu[1]

Abstract At present, the coupling relationship between the pantograph and catenary is not ideal for straddle-type monorail. The power collection quality is poor, and the wear of the contact strip of a pantograph is uneven. Therefore, the key problem of coupling relation between the pantograph and catenary is the dynamics of a pantograph–catenary system. This paper analyzed the modal of catenary through simulation with finite element model and the vibration modal of pantograph multi-rigid-body dynamic model by simulation in multi-body danamics software ADAMS. A rigid-flexible coupling dynamic model of pantograph–catenary system is established. Finally, extensive research has been made on the influence of pantograph head parameters on the power collection quality.

Keywords Straddle-type monorail · Power collection quality · Contact strip · Rigid-flexible coupling · Pantograph head parameters

✉ Zhen Yang
yangz0000@163.com

Zixue Du
aaadzx@163.com

Cheng Chen
705479475@qq.com

Xiaoxia Wen
963884811@qq.com

Zhouzhou Xu
33243665@qq.com

[1] Chongqing Jiaotong University, No. 66 Xuefu Road, Nan'an District, Chongqing, China

Editor: Eryu Zhu

1 Introduction

The track of straddle-type monorail consists of a PC beam which is made of concrete, and the positive and negative catenaries are rigidly installed along the two sides of the PC beam [1]. Because of the particular wheel–rail coupling relationship and the collection modes, the coupling relation between the pantograph and catenary in the straddle-type monorail is very different from other systems. Therefore, the aims to ameliorate coupling relation and improve the power collection quality between pantograph and catenary are of great significance and will serve to improve the safety of the operation of straddle-type monorail.

2 Power Collection Quality Evaluation Index Between Pantograph and Catenary

There is no power collection quality standard for straddle-type monorail. This paper proposed power collection quality standard for straddle-type monorail on the basis of railway standards TB/T 3271-2011 and the experience of straddle-type monorail operations in Chongqing.

2.1 The Average Value of Contact Force F_m

Contact force F is the most important technical index to characterize the dynamic parameters of the pantograph–catenary system. It is also the sum of the forces that are applied to the catenary by the pantograph. In the coupling process between the pantograph and catenary, the average value of the contact force between the pantograph and catenary is:

$$F_m = \frac{\sum F_i}{n} \tag{1}$$

where F_i is the sampling value of the contact force, $i = 1, 2, \ldots, n$;

n is the number of samples of contact force in the statistics section.

2.2 The Standard Deviation of Contact Forces

The standard deviation of contact force between pantograph and catenary can be calculated through statistical knowledge, and the standard deviation of the contact force is within a certain range:

$$s = \sqrt{\frac{\sum (F_i - F_m)^2}{n - 1}}. \tag{2}$$

2.3 Maximum and Minimum Contact Force

The maximum and minimum contact force is the actual maximum contact force and the minimum contact force exists between the pantograph and catenary when the vehicle passes the evaluation section.

2.4 Off-Line Rate

Off-line is defined as the condition that the contact force between the pantograph and catenary is zero. Off-line rate is calculated by evaluating the ratio of the sum of the off-line distance to the total distance length when the vehicle passes the evaluation section.

The contact force between the pantograph and catenary of straddle-type monorail should be in the range of 44–79 N in the whole working stroke, and in the range of 59^{+10}_{-10} N in normal working positions. Therefore, according to evaluation index system in this study, the maximum contact force between the pantograph and catenary should be less than 79 N, the minimum contact force between the pantograph and catenary should be greater than 44 N, and the optimum mean contact force should also be in the range of 59^{+10}_{-10} N [2]. The standard deviation of contact force directly reflects the fluctuation range of contact forces. The smaller the standard deviation is, the smaller the fluctuation range of the contact force is, and the better the power collection quality.

3 Dynamic Model of Pantograph

3.1 Two-Dimensional Model of Pantograph

Straddle-type monorail pantograph (Fig. 1) is mainly composed of a bottom plate, a lower frame, an upper frame, a connecting rod, a balance bar, a pantograph head seat, a pantograph spring, a pantograph head and cylinder parts (positive pantograph only) and so on.

Fig. 1 Pantograph

The pantograph is reduced to the model shown in Fig. 2. α is the angle between AC and X axis, β is the angle between BD and X axis, γ is the angle between CD and X axis, δ is the angle between AB and X axis, ε is the angle between CD and CG, θ is the angle between FG and X axis. Fc is the contact force between the pantograph head and catenary. The direction of the vehicle operation is defined as the positive direction of the X axis, the yaw direction of the vehicle is defined as the Y axis, and the vertical upward direction is the positive direction of the Z axis.

3.2 Dynamic Computational Model of Pantograph

The motion differential equations of pantograph head components are established on the basis of Newton's second law.

$$m_h \ddot{z}_h = F_h - F_c \tag{3}$$

where m_h is the weight of pantograph head component, Z_h is the vertical displacement of pantograph head component,

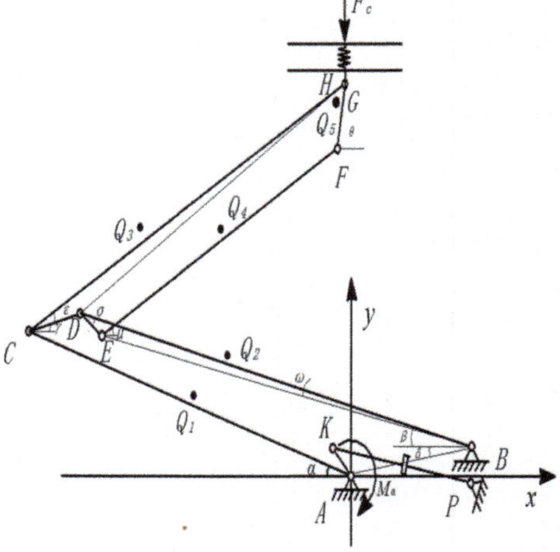

Fig. 2 Two-dimensional model of pantograph

F_c is the contact force between the head of the pantograph and the catenary

$$F_h = -k_h(y_h - y_G - \Delta l_0) - c_h(\dot{y}_h - \dot{y}_G) \quad (4)$$

where k_h, c_h are the stiffness and damping of elastic elements of the pantograph head, Δl_0 is the natural length of the elastic element.

The frame portion of the pantograph has only one degree of freedom, so the angle α can be regarded as an independent variable. According to Lagrange equation:

$$\frac{d}{dt}\frac{\partial L}{\partial \dot{\alpha}} - \frac{\partial L}{\partial \alpha} = Q_\alpha \quad (5)$$

where Q_α is the generalized force, $L = T - V$. T is the total kinetic energy of the frame system, V is the total potential energy of a frame system, $V = 0$.

$$T = T_{\text{Lower}} + T_{\text{LG}} + T_{\text{Upper}} + T_{\text{UG}} + T_{\text{H}} \quad (6)$$

where T_{Lower}, T_{LG}, T_{Upper}, T_{UG}, T_{H} are the kinetic energy of the lower frame, the connecting rod, the upper frame, the balance bar and the pantograph head support, respectively.

Kinetic energy of the lower frame:

$$\begin{aligned} T_{\text{Lower}} &= \frac{1}{2}J_1\dot{\alpha}^2 + \frac{1}{2}m_1\left(\dot{x}_{Q_1}^2 + \dot{y}_{Q_1}^2\right) \\ &= \frac{1}{2}J_1\dot{\alpha}^2 + \frac{1}{2}m_1\left[\dot{x}_A^2 + \dot{y}_A^2 + l_{AQ_1}^2\dot{\alpha}^2\right. \\ &\quad + 2l_{AQ_1}\sin(\alpha + \angle CAQ_1)\cdot\dot{x}_A\dot{\alpha} \\ &\quad \left.+ 2l_{AQ_1}\cos(\alpha + \angle CAQ_1)\cdot\dot{y}_A\dot{\alpha}\right] \end{aligned}$$

Kinetic energy of the connecting rod:

$$\begin{aligned} T_{\text{LG}} &= \frac{1}{2}J_2\dot{\beta}^2 + \frac{1}{2}m_2\left(\dot{x}_{Q_2}^2 + \dot{y}_{Q_2}^2\right) \\ &= \frac{1}{2}J_2\dot{\beta}^2 + \frac{1}{2}m_2\left[\dot{x}_A^2 + \dot{y}_A^2 + l_{BQ_2}^2\dot{\beta}^2\right. \\ &\quad + 2l_{BQ_2}\sin(\beta + \angle DBQ_2)\cdot\dot{x}_A\dot{\beta} \\ &\quad \left.+ 2l_{BQ_2}\cos(\beta + \angle DBQ_2)\cdot\dot{y}_A\dot{\beta}\right] \end{aligned}$$

Kinetic energy of upper frame:

$$\begin{aligned} T_{\text{Upper}} &= \frac{1}{2}J_3\dot{\gamma}^2 + \frac{1}{2}m_3\left(\dot{x}_{Q_3}^2 + \dot{y}_{Q_3}^2\right) \\ &= \frac{1}{2}J_3\dot{\gamma}^2 + \frac{1}{2}m_3\left[\dot{x}_A^2 + \dot{y}_A^2 + l_{AC}^2\dot{\alpha}^2 + l_{CQ_3}^2\dot{\gamma}^2\right. \\ &\quad + 2l_{AC}\sin\alpha\cdot\dot{x}_A\alpha - 2l_{CQ_3}\sin(\gamma + \varepsilon + \angle GCQ_3)\cdot\dot{x}_A\dot{\gamma} \\ &\quad + 2l_{AC}\cos\alpha\cdot\dot{y}_A\dot{\alpha} + 2l_{CQ_3}\cos(\gamma + \varepsilon + \angle GCQ_3)\cdot\dot{y}_A\dot{\gamma} \\ &\quad \left.+ 2l_{AC}l_{CQ_3}\cos(a + \gamma + \varepsilon + \angle GCQ_3)\cdot\dot{\alpha}\dot{\gamma}\right] \end{aligned}$$

Kinetic energy of the balance bar:

$$\begin{aligned} T_{\text{UG}} &= \frac{1}{2}J_4\dot{\mu}^2 + \frac{1}{2}m_4\left(\dot{x}_{Q_4}^2 + \dot{y}_{Q_4}^2\right) \\ &= \frac{1}{2}J_4\dot{\mu}^2 + \frac{1}{2}m_4\left[\dot{x}_A^2 + \dot{y}_A^2 + l_{BE}^2\dot{\beta}^2 + l_{EQ_4}^2\dot{\mu}^2\right. \\ &\quad + 2l_{BE}\sin(\beta - \omega)\cdot\dot{x}_A\dot{\beta} - 2l_{EQ_4}\sin\mu\cdot\dot{x}_A\dot{\mu} \\ &\quad + 2l_{BE}\cos(\beta - \omega)\cdot\dot{y}_A\dot{\beta} + 2l_{EQ_4}\cos\mu\cdot\dot{y}_A\dot{\mu} \\ &\quad \left.+ 2l_{BE}l_{EQ_4}\cos(\mu + \beta - \omega)\cdot\dot{\beta}\dot{\mu}\right] \end{aligned}$$

Kinetic energy of the pantograph head support:

$$\begin{aligned} T_{\text{H}} &= \frac{1}{2}J_5\dot{\theta}^2 + \frac{1}{2}m_5\left(\dot{x}_{Q_5}^2 + \dot{y}_{Q_5}^2\right) \\ &= \frac{1}{2}J_5\dot{\theta}^2 + \frac{1}{2}m_5\left[\dot{x}_A^2 + \dot{y}_A^2 + l_{BE}^2\dot{\beta}^2 + l_{EF}^2\dot{\mu}^2 + l_{FQ_5}^2\dot{\theta}^2\right. \\ &\quad + 2l_{BE}\sin(\beta - \omega)\cdot\dot{x}_A\dot{\beta} \\ &\quad - 2l_{EF}\sin\mu\cdot\dot{x}_A\dot{\mu} - 2l_{FQ_5}\sin(\theta + \angle GFQ_5)\cdot\dot{x}_A\dot{\theta} \\ &\quad + 2l_{BE}\cos(\beta - \omega)\cdot\dot{y}_A\dot{\beta} \\ &\quad + 2l_{EF}\cos\mu\cdot\dot{y}_A\dot{\mu} + 2l_{FQ_5}\cos(\theta + \angle GFQ_5)\cdot\dot{y}_A\dot{\theta} \\ &\quad + 2l_{BE}l_{EF}\cos(\mu + \beta - \omega)\cdot\dot{\beta}\dot{\mu} \\ &\quad \left.+ 2l_{BE}l_{FQ_5}\cos(\beta - \omega + \theta)\cdot\dot{\beta}\dot{\theta} + 2l_{EF}l_{FQ_5}\cos(\mu + \theta)\cdot\dot{\mu}\dot{\theta}\right] \end{aligned}$$

where m_1, m_2, m_3, m_4, m_5 are the weight of the lower frame, the connecting rod, the upper frame, the balance bar and the pantograph head support, respectively; J_1, J_2, J_3, J_4, J_5 are the inertia around the centroid of the lower frame, the connecting rod, the upper frame, the balance bar and the pantograph head support around the centroid, respectively.

4 Dynamic Model of Pantograph and Catenary System

4.1 Modal Analysis of Catenary

The paper analyzed the modal of catenary finite element model by Hyperworks [3, 4]. The characteristics of the vibration in structures are composed of linear sets of vibration corresponding to their respective natural modes. The dynamic characteristics of the structure are mainly determined by the low-frequency vibration modes. Therefore, the paper analyzed the modes of the first eight orders. As can be seen from Table 1, with the increase in modal order, the frequency of resonance increases nonlinearly. Figure 3 is the first six order vibration modes of catenary.

4.2 Dynamic Characteristics Analysis of Pantograph

The paper analyzed the vibration modal of a pantograph multi-rigid-body dynamic model by ADAMS. Figure 4 is the dynamic analysis of the characteristics of

Table 1 The first eight order natural frequencies of the catenary (unit: Hz)

Modal order	1	2	3	4	5	6	7	8
Spacing 2.5 m	47.73	56.05	129.38	153.23	196.81	251.23	301.71	373.53

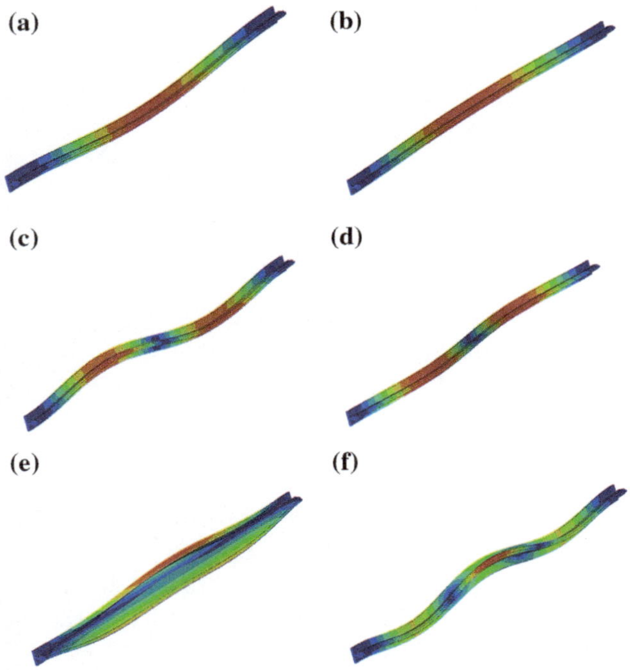

Fig. 3 The first six order vibration modes of the catenary. **a** First-order modal shapes of catenary. **b** Second-order modal shapes of catenary. **c** Third-order modal shapes of catenary. **d** Forth-order modal shapes of catenary. **e** Fifth-order modal shapes of catenary. **f** Sixth-order modal shapes of catenary

Table 2 The modal characteristics of the pantograph rigid body model

Modal	Vibration type	Frequency (Hz)
1	Transverse movement of pantograph head	1.51
2	Raise and fall pantograph	8.77

a pantograph. The modal characteristics of the pantograph rigid body model are shown in Table 2.

4.3 Dynamic Simulation Model of "Vehicle-Track Beam-Pantograph"

The bogie model equipped with a pantograph system is shown in Fig. 5; each bogie is fitted with two pantograph models at the pantograph mounting support [5, 6].

After each subsystem model is assembled together, the dynamic simulation model of single vehicle is then completed [7]. The dynamic simulation model of single vehicle is shown in Fig. 6.

5 Frequency Spectrum Analysis of Contact Force Response of Pantograph–Catenary

The road stochastic excitation is the input of pantograph–catenary coupling dynamics simulation in this paper, so the excitation of the pavement has an important influence on the response of the coupled system. In order to find the frequency range and dominant frequency of contact force vibration, this paper analyzed the frequency spectrum of contact forces under the condition of motion in curved track and straight track at different speeds [8–10].

Figure 7 shows the characteristics of the contact forces in response to the pantograph–catenary system's frequency when the vehicle passes through a straight track. The speeds of the vehicle are 20, 40, 60 and 80 km/h.

As shown in Fig. 7, when the vehicle runs at different speeds, the vibration of the contact force of pantograph–catenary subsystem mainly comes from the low-frequency contribution, and there is no high-frequency contribution. The dominant frequency of contact force is about 4.3 Hz.

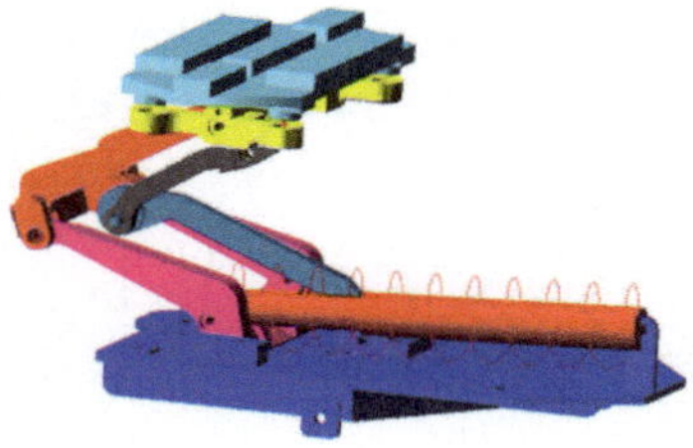

Fig. 4 Dynamic characteristics analysis of pantograph

Fig. 5 The bogie model equipped with pantograph system

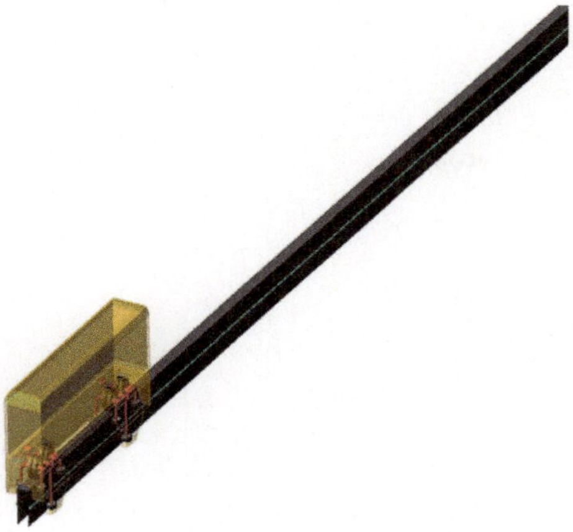

Fig. 6 Dynamic simulation model of single vehicle

Figure 8 shows the characteristics of contact forces in response to pantograph–catenary system's frequency when the vehicle passes through the curved track of R100. The speeds of the vehicle are 20, 30 and 40 km/h.

As it can be seen from Fig. 8, when the vehicle passes through the curved track of R100, with the increase in the running speed of a vehicle, the range of vibration frequency of the contact force between the pantograph and catenary is gradually enlarged.

As it can be seen from Figs. 7 and 8, when the vehicle is passing through the curved track, the vibration frequency of the contact force is larger than that which is passing through the straight track at the same speed.

6 The Influence of Pantograph Head Parameter on Power Collection Quality

This paper researched on the influence of pantograph head parameters on the power collection quality under the condition of motion in a curved track with a radius of 100 m. The pantograph head parameters are the weight of pantograph head parts, the stiffness of pantograph head elastic element and the damping of pantograph head elastic element.

6.1 The Influence of Weight of Pantograph Head Parts on Power Collection Quality

The contact force between the pantograph head and catenary is provided by the pantograph head spring system, so the weight of pantograph head will affect the following pantograph head.

In the actual operation of straddle-type monorail, the weight of the pantograph head is set to 1, 3, 5 and 7 kg, and then, the influence of the weight of the pantograph head on the power collection quality is analyzed.

As it can be seen from Figs. 9 and 10, with the increase in the weight of a pantograph head, the pantograph's standard deviation of contact force and the fluctuation of contact force become larger which occurs on the left and right of the bogie. Therefore, reducing the weight of pantograph head can improve the power collection quality.

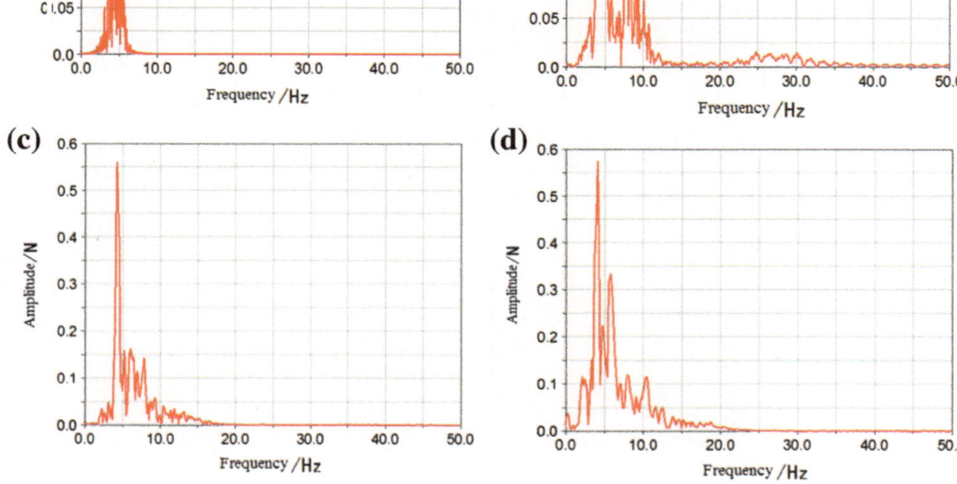

Fig. 7 Frequency spectrum of contact force in straight track. **a** Speed 20 km/h right pantograph. **b** Speed 40 km/h right pantograph. **c** Speed 60 km/h right pantograph. **d** Speed 80 km/h right pantograph

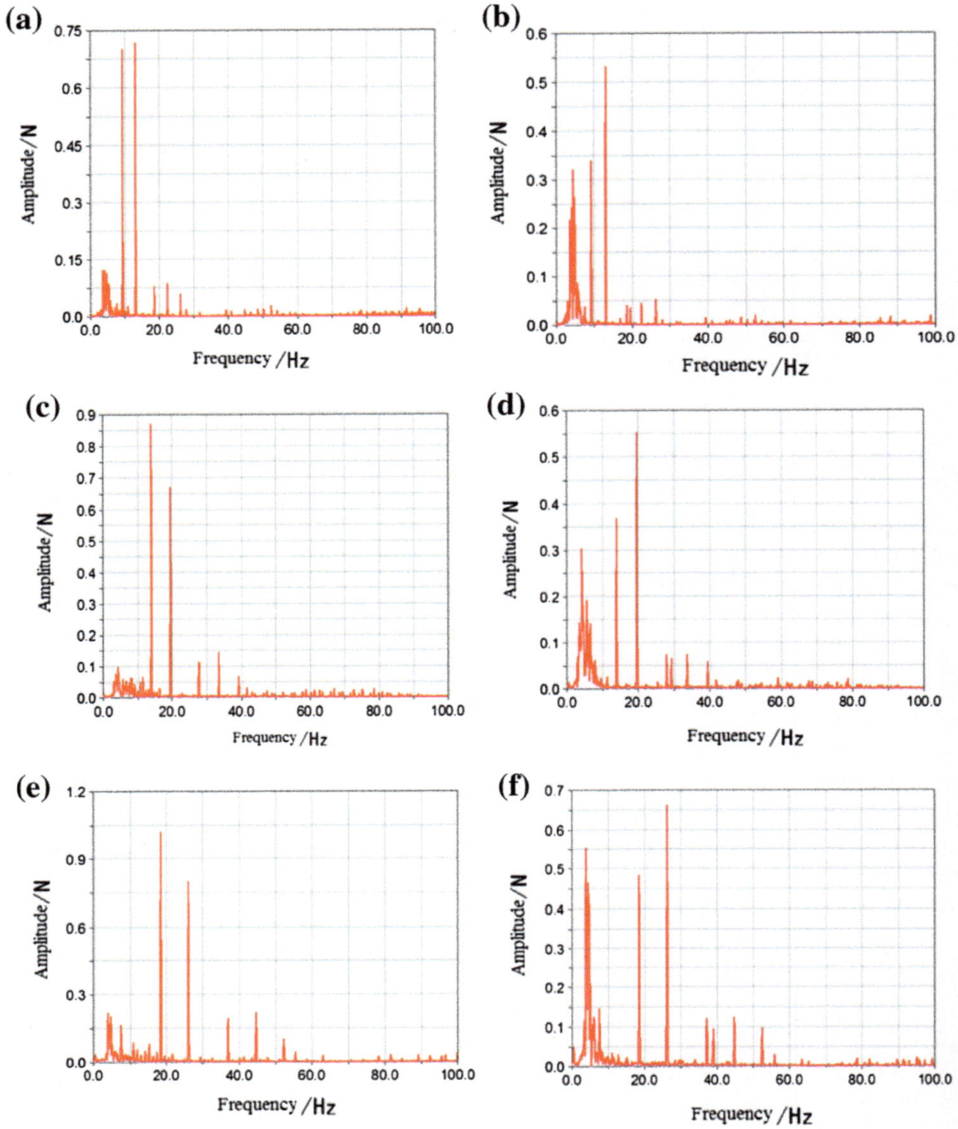

Fig. 8 Frequency spectrum of the contact force in curved track. **a** Speed 20 km/h left pantograph. **b** Speed 20 km/h right pantograph. **c** Speed 30 km/h left pantograph. **d** Speed 30 km/h right pantograph. **e** Speed 40 km/h left pantograph. **f** Speed 40 km/h right pantograph

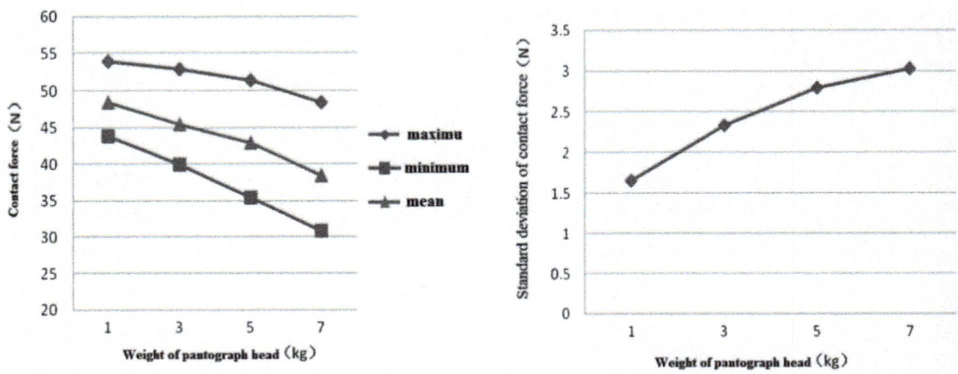

Fig. 9 The influence of the weight of left pantograph head parts on power collection quality in curve track

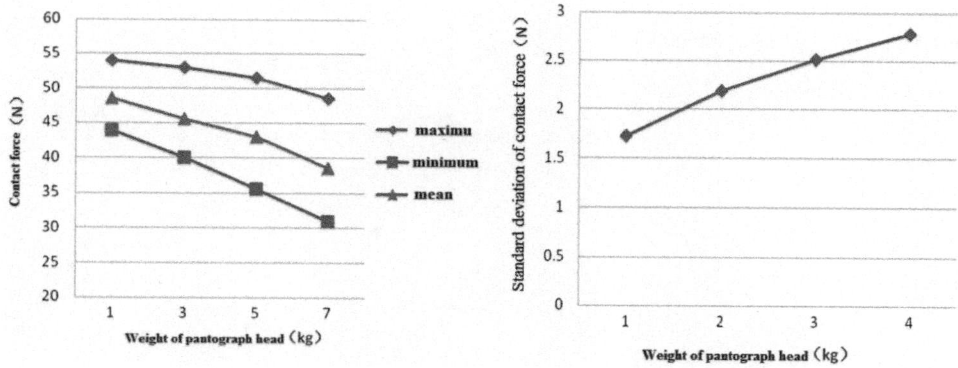

Fig. 10 The influence of weight of right pantograph head parts on power collection quality in curved track

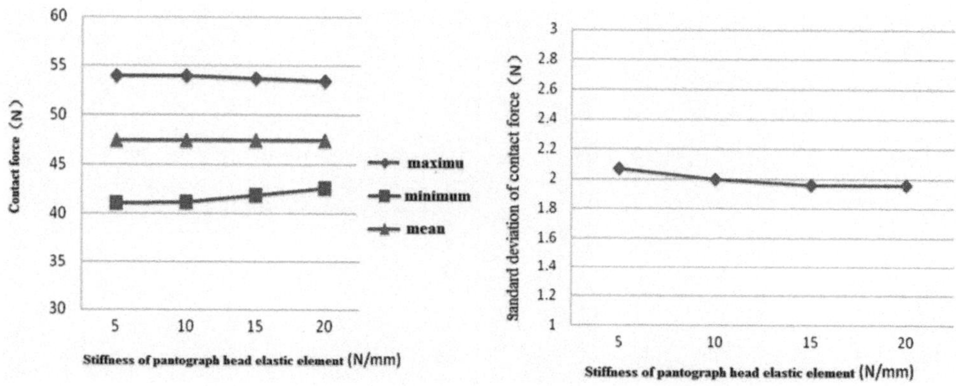

Fig. 11 The influence of stiffness of left pantograph head elastic element on power collection quality in curved track

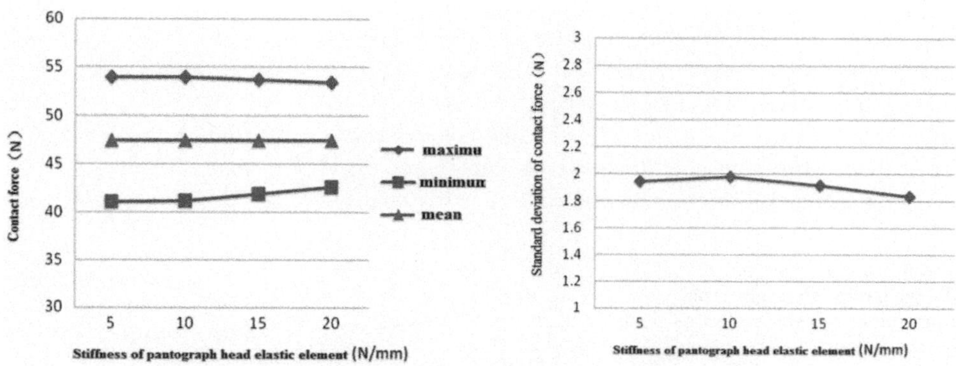

Fig. 12 The influence of stiffness of right pantograph head elastic element on power collection quality in curved track

6.2 The Influence of Stiffness of Pantograph Head Elastic Element on Power Collection Quality

The stiffness of the pantograph head elastic element is set to 5, 10, 15 and 20 N/mm, and the influence of stiffness of pantograph head elastic element on power collection quality is then analyzed.

As it can be seen from Figs. 11 and 12, with the increase in stiffness of pantograph head elastic element, the pantograph's standard deviation of contact force decreases and the fluctuation of contact force becomes smaller which occurs on the left and right of the bogie. Therefore, increasing the stiffness of the pantograph head elastic element can improve the power collection quality.

6.3 The Influence of Damping of Pantograph Head Elastic Element on Power Collection Quality

The damping of the pantograph head elastic element is set to 0, 50, 100, 150 and 200 Ns/m, and the influence of damping of pantograph head elastic element on power collection quality is then analyzed.

Fig. 13 The influence of damping of left pantograph head elastic element on power collection quality in curved track

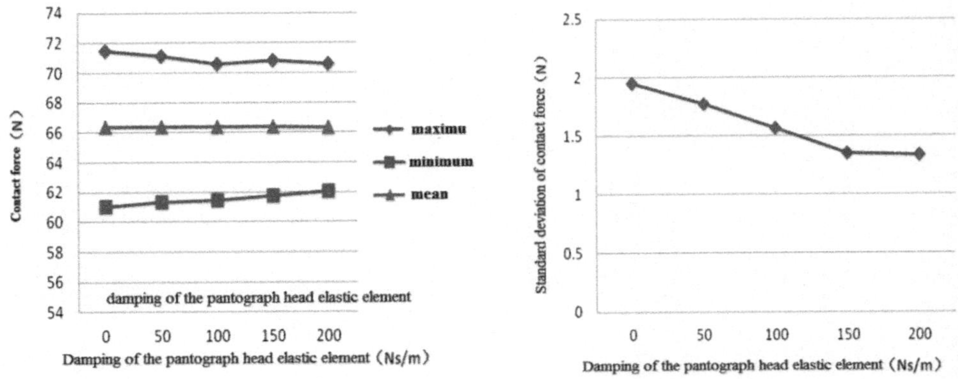

Fig. 14 The influence of damping of right pantograph head elastic element on power collection quality in curved track

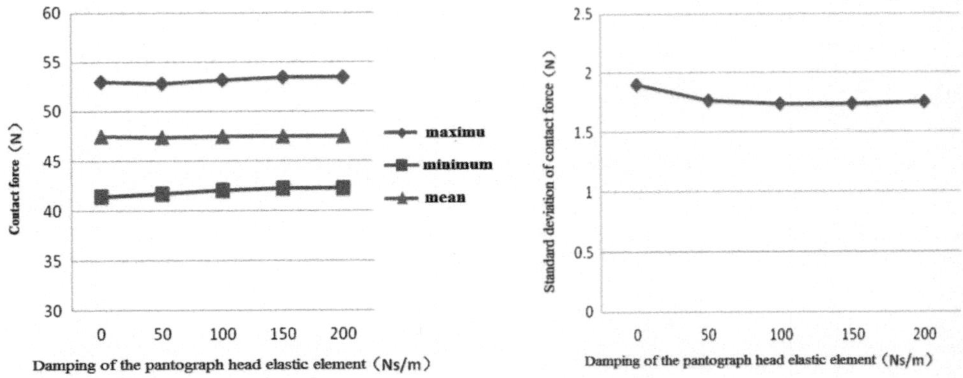

As it can be seen from Figs. 13 and 14: with the increase in damping of pantograph head elastic element, the pantograph's standard deviation of contact force decreases and the fluctuation of contact force becomes smaller which occurs on the left and right of the bogie. Therefore, increasing the damping of pantograph head elastic element can improve the power collection quality.

7 Conclusion

This paper established a dynamic computational model of pantograph and a rigid-flexible coupling dynamic model of pantograph–catenary system. Through an analysis of the computer simulation, when the vehicle runs at different speeds, the dominant frequency of the contact force of pantograph–catenary subsystem is about 4.3 Hz.

This paper researched on the influence of pantograph head parameters on the power collection quality. Reducing the weight of the pantograph head, increasing the stiffness of pantograph head elastic element or increasing the damping of pantograph head elastic element can improve the power collection quality to a certain extent.

But there are still some shortcomings:

Due to the limited conditions, when the rigid-flexible coupling dynamic model of pantograph–catenary system is being established, the stochastic excitation of the contact wire surface is not taken into account. Therefore, there is a certain deviation between the simulation results and the actual situation.

Acknowledgements The project titled Research on the Mechanism and Control Method of Monorail Trains Running Tires' Shoulder Wear, Project Number: 51475062 was funded by National Natural Science Foundation of China.

References

1. Zhong J, Du Z, He X (2013) Turnout structure and analysis of monorail vehicle. People Communications Press, Beijing
2. Duan Y (2009) Failure analysis of Chong Qing straddle type monorail vehicle's Pantograph. Chongqing University, Chongqing
3. Poetsch G, Evans J, Meisinger R et al (2007) Pantograph/catenary dynamics and control. Veh Syst Dyn 28(2):159–195
4. Rauter FG, Pombo J, Ambrósio J et al (2007) Multibody modeling of pantographs for pantograph–catenary interaction. IUTAM Bookseries 1:205–226

5. Zixue D, Li Y, Liang Z, Wen X (2016) Research on curve through safety of straddle-type monorail vehicle. Electric Drive Locomot 1:79–83
6. Zixue D, Chen C, Shen Z, Liu Y, Gao L, Zhong Y, Zhang Y (2015) Study of the negotiation stability for straddle-type monorail train under cross-wind condition. Electric Drive Locomot 5:60–64
7. Du Z, Cao D (2016) Influence of air spring failure on operation safety for straddle-type monorail vehicle. Electric Drive Locomot 5:76–80
8. Link DIM, von Berechnung Z (1980) Fahrleitungsschwingungen mit Hilfe frequenzabhängiger finiter Elemente. Arch Appl Mech 51(1):45–60
9. Brodkorb A, Semrau M (1993) Simulationsmodell des systems oberleitungskettenwerk und stormabnehmer. Elektrische Bahnen. 91(4):105–113
10. Reichmann T (2005) Simulation des systems oberleitungs-kettenwerk und stromabnehmer mit der Finite-Elemente-Methode. Elektrische Bahnen 103(1):69–75

16

A Straddle Monorail Running Gear with Single-Axle and Rotating Arm Axle Box Suspension

Rang Zhang[1] · Yuanjin Ji[1] · Lihui Ren[1] · Gang Shen[1]

Abstract In this study, a new type of single-axle bogie of straddle-type monorail vehicle is designed. The tumbler structures are adopted in the vertical suspension system. One end of the tumbler is fixed to the framework, while the other end supports the framework via air springs. In addition, the lateral suspension utilizes the suspender way, with longitudinal force transmitted via single pull rods. Guide wheels and steady wheels are connected to the end support arms of the framework through spindles around the horizontal direction. Rubber springs, the pre-compressed magnitudes of which could be adjusted, are set between the spindles and support arms. The multi-body dynamics software UM is adopted to construct a dynamic model and simulate related dynamic performance. The results of the simulation indicate that this new kind of bogie has a favorable performance when passing through curves and has a greater riding stability. The flexibility factor equation of this new monorail vehicle is deduced, with its flexibility factor obtained.

Keywords Monorail vehicle · Bogie · Structure · Dynamics

✉ Rang Zhang
1831446545@qq.com

Yuanjin Ji
jiyuanjin@tongji.edu.cn

Lihui Ren
renlihui@tongji.edu.cn

Gang Shen
elsg163@163.com

[1] Institute of Rail Transit of Tongji University, No. 4800. Cao an gong Road, Jia ding District, Shanghai, China

Editor: Eryu Zhu

1 Introduction

Due to advantages including strong adaptability to complex terrain, low land occupancy, moderate freight volume and low construction cost, straddle-type monorail transportation systems have already become the first choice of urban railway systems in small–medium cities, mountainous cities and cities with topographic and geological complexity [1]. The straddle-type monorail vehicle is mainly supported, guided and stabilized by a single-track beam. The vehicle straddles and runs on the track beam, with the center of gravity of its body located above the track. Comparing to traditional railway vehicles, the running gears of straddle-type monorail vehicles use rubber tires, with guide wheels and steady wheels set on both sides of the track beam. The track beam has a simple structure, playing the roles of bearing structure and track structure. Thus, the performance of the structure can be fully maximized while the maintenance workload of the track beam is of little requirement [2].

Currently, straddle-type monorail vehicles using Hitachi mode and Bombardier mode have both been introduced into China. Hitachi straddle-type monorail vehicles adopt double-axle bogies and thus possess a strong bearing capacity [3], track beam cross section of 850 × 1500 mm, and power supply of DC1500 V, as shown in Fig. 1. Meanwhile, single-axle bogies are adopted in Bombardier straddle-type monorail vehicles; as shown in Fig. 2, there are four guide wheels and two steady wheels in the bogie, and an hourglass rubber spring are used for secondary suspension paralleling vertical shock absorber and lateral shock absorber. Scomi running gear also uses single-axle structure and rotating arm structure; one end of the rotating arm is hinged with the frame, and the other end is supported by an air spring, and the carbody and bogie are

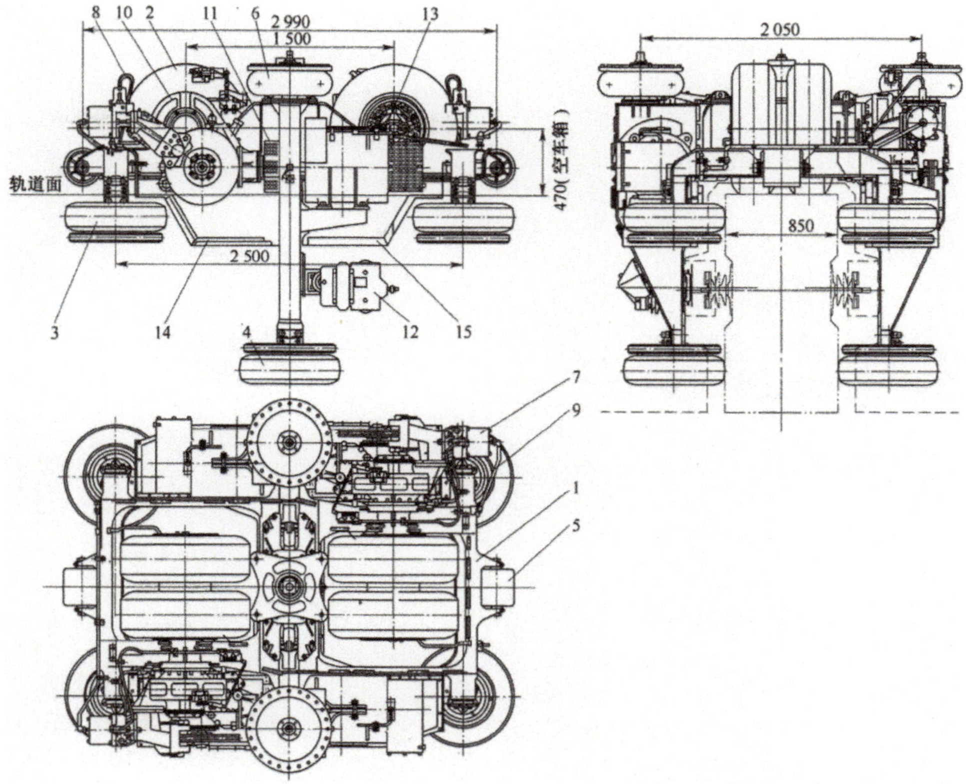

Fig. 1 Running gear of a Hitachi straddling monorail vehicle. 1 Framework, 2 walking wheels, 3 guiding wheels, 4 steady wheels, 5 auxiliary wheel, 6 second suspension, 7 and 9 break device, 8 parking break device, 10 driving device, 11 coupling, 12 collector shoe, 13 tire pressure detection device, 14 and 15 sound insulation plate

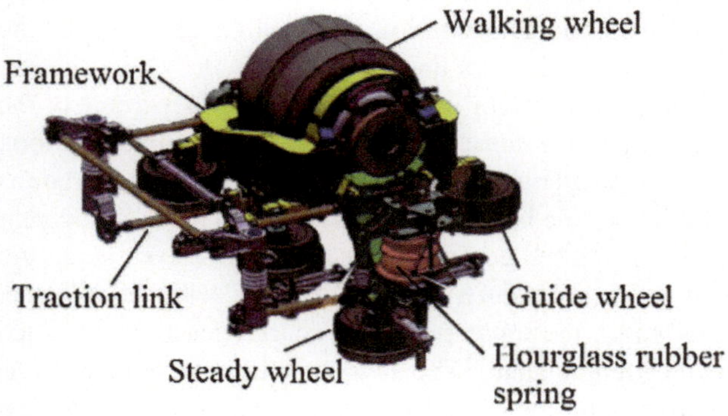

Fig. 2 Running gear of Bombardier

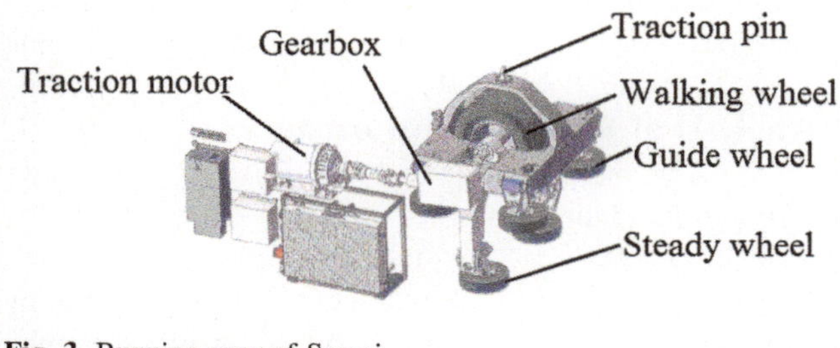

Fig. 3 Running gear of Scomi

connected by a traction pin, as shown in Fig. 3. Comparing to double-axle bogies, single-axle bogies have a stronger curve passing through capacity and can achieve the floor lowering of the passenger rooms. In addition, the vehicles also have a flexible marshaling capacity [4]. Different from wheel-track-type railway vehicles, the product types of the straddle-type monorail vehicles are less numerous, mainly due to the limited types of running parts. Therefore, this study designs a new type of single-axle bogie for straddle-type monorail vehicles and analyzes its dynamic performances, and aims to enrich the types of straddle-type monorail vehicles.

2 Structure of Running Gear

The bogie of new straddle-type monorail vehicles uses the single-axle structure type and mainly consists of one pair of walking wheels, four guide wheels, one pair of steady wheels, a suspension device and a framework, as shown in Fig. 4.

The walking wheels use the integral wheel pair structure, as well as the installation structure similar to brake discs. The hubs of two rubber pneumatic tires are installed onto the axle via push mounting. The tires rotate together with the axle. Axle boxes are designed outside of the walking wheel axle, to convert the rotation of walking wheel into translation. Tumbler structure is utilized in the axle box body. One end of the tumbler is fixed to the lower part of the framework via rotational joints, while the other end supports the framework via air springs. Holes are open

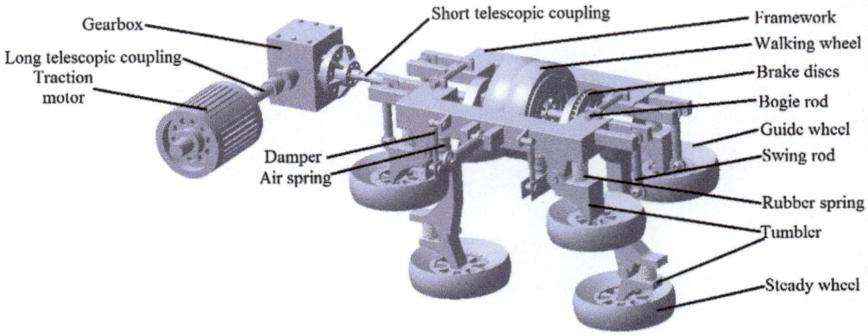

Fig. 4 3D schematic diagram of the bogie

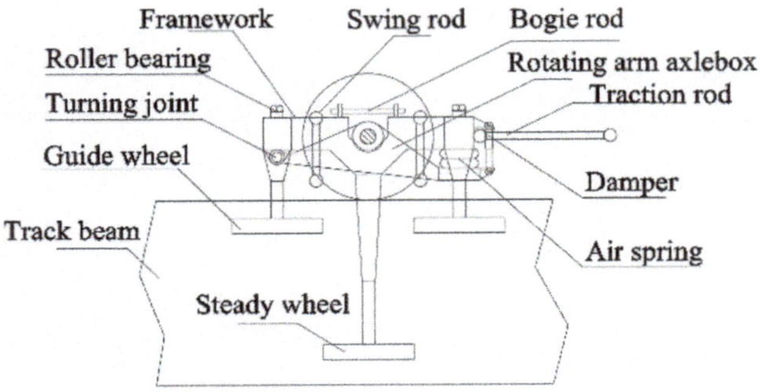

Fig. 5 Schematic diagram of the suspension structure

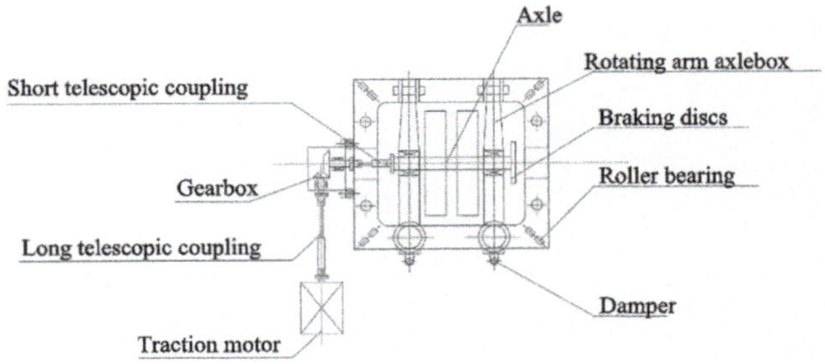

Fig. 6 Schematic diagram of the traction system

in the center of tumbler to install the axle box, as shown in Fig. 5. Via the tumbler axle box structure, the longitudinal location and vertical load bearing of walking wheels are achieved. Meanwhile, vertical cushion and vibration attenuation are realized by the air springs and vibration absorbers on one end of the tumbler. Lateral elasticity is realized by suspenders. The suspender top is connected to the framework, while its bottom is connected to the carbody. Through the lateral oscillation of suspenders, the lateral cushion is realized depending on the carbody gravity. At the same time, the deflection between the carbody and bogie during nonlinear motion when passing through curves can be adapted via the vertical oscillation of the suspenders. A single traction rod is assigned to the crossbeam in order to transmit the longitudinal tractive force and braking force.

Guide wheels and steady wheels are connected to the end support arms of the framework through spindles around the horizontal direction. Rubber springs are set between the spindles of the guide wheels, steady wheels and support arms. By adjusting the pre-compression amount of the springs, the pre-pressure between the guide wheels or steady wheels and the track beam can then be adjusted.

The main body of the framework presents a rectangular shape, mainly consisting of two side beams and two crossbeams. The support beams of the steady wheels are set below the center of the side beam in the framework, and the support beams of the guide wheels are set below the ends of the side beams. The center of the side beam in the framework concaves downward, to allow couplings to pass through. On both sides of the concaved part of the side beam within the framework, pull rod bases are set to install pull rods, aiming to improve the loading condition of the framework, as well as increase its strength.

The traction motor is installed on the under-frame of the carbody and is connected to the input shaft of the gearbox via a retractable universal spindle coupling. The gearbox is fixed to the side beam in the framework. Transmission gear groups are installed inside the gearbox. Bevel gears are used in the transmission gear groups, with small gears installed on the input shaft while large gears are installed on the output shaft. The output shaft of the gearbox is connected to the axle via another shaft coupling. The walking wheel is fixed on the axle via a hub and rotates together with the axle, as shown in Fig. 6. Thus, the bogie mass is reduced, and the bogie structure is simplified.

3 Analysis of Dynamic Performance

3.1 Dynamic Model

To verify the dynamic performance of the new monorail vehicle bogie, the UM software with multi-body dynamics is used to analyze the dynamic vehicle model. The dynamic model mainly consists of one carbody, two frameworks, four walking wheels, eight guide wheels and four steady wheels. The carbody and bogie frameworks are both treated as rigid bodies. Suspension devices, including air

Fig. 7 Topological structure of the vehicle dynamic model

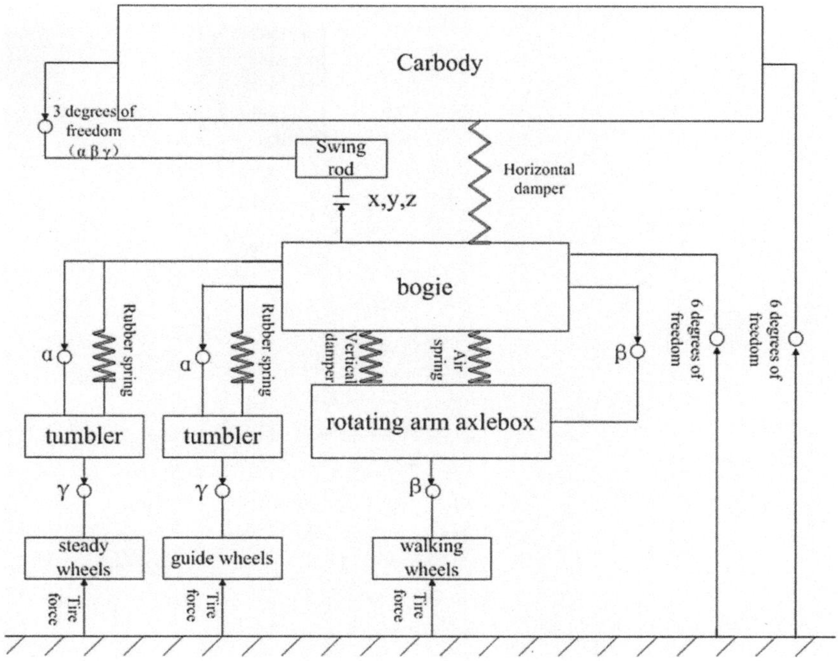

springs, suspenders, absorbers, are all simplified as spring-damper force units. Each bogie has four suspenders, distributed on both sides. Tires are modeled based on the Falia model, by taking the radial characteristics and cornering features of the tires into consideration. The walking wheel pairs are connected to the framework via hinge joints rotating around the y-direction. Guide wheels and steady wheels are connected to the corresponding tumblers via hinge joints rotating around the z-direction. The tumblers of the guide wheels and steady wheels are all connected to the framework via hinge joints rotating around the x-direction. The dynamic topological structure of the new monorail vehicle is shown in Fig. 7, with the vehicle dynamic model displayed in Fig. 8. Table 1 gives the main dynamic parameters of the vehicle.

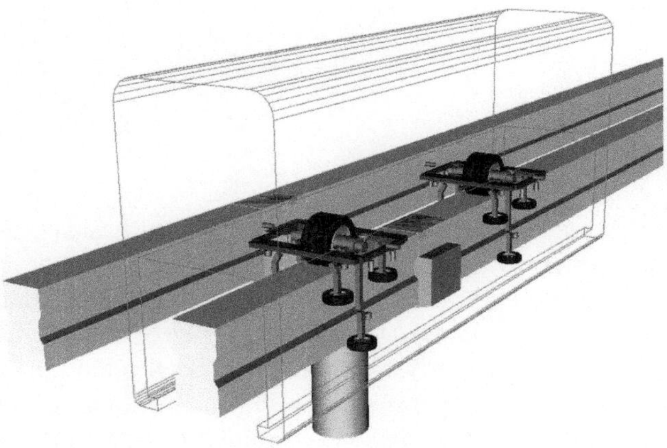

Fig. 8 UM simulation model of vehicle dynamics

3.2 Curve Passing-Through Performance

One of the advantages of the straddle-type monorail vehicles is their strong capacity of maneuvering around small corners. When a vehicle runs through a curve with a small radius, the orbital superelevation and running speed both have significant influences on its passing-through performance [5, 6]. Figure 9 shows the maximum value of carbody rolling angle, the maximum value of guide wheel radial force and the maximum value of steady wheel radial force, when the vehicle passes through curves with different radii. The superelevation is a total of 6%, and the vehicle speed is kept constant. In a curve with a radius of 60 m, the rolling angle of the carbody is at its greatest, reaching 0.012 rad. The rolling angle decreases as the curve radius increases. With the increase in the curve radius, the radial force of the guide wheels and steady wheels all reduces and the lateral force of the walking wheels also decreases.

Figure 10 shows the load variation regularity of the wheels in the front and back bogies, during the motion of the vehicle passing through a curve with a radius of 100 m. As shown in the figure, the left rear guide wheel of the front bogie, the right front guide wheel of the front bogie, the left rear guide wheel of the back bogie and the left front guide wheel of the back bogie all experience load increase. The right steady wheel of the front bogie and the left steady wheel of the back bogie both experience load increase.

Table 1 Vehicle dynamics parameters

Parameters	Values
Carbody mass	10,000 kg
Bogie mass	2100 kg
Radial stiffness of walking wheel	1.4 MN m
Cornering stiffness of walking wheel	0.35 MN m/rad
Radial stiffness of guide wheel	0.65 MN m
Radial stiffness of steady wheel	0.65 MN m
Stiffness of pre-compressed spring	12.5 MN m
Vertical stiffness of air spring	0.15 MN m
Suspender length	0.65 m
Damping coefficient of vertical absorber	50 kN/(m/s)
Damping coefficient of lateral absorber	50 kN/(m/s)
Length between vehicle centers	6.0 m
Distance between carbody gravity center and rail level	0.3 m
Distance between bogie gravity center and rail level	0.2 m
Distance between guide wheel center and rail level	0.15 m
Distance between steady wheel center and rail level	0.9 m
Half value of walking wheel transverse distance	0.15 m
Walking wheel radius	0.5 m
Guide wheel radius	0.27 m
Steady wheel radius	0.27 m
Pre-compression pressure of guide wheel and steady wheel	5000 N

3.3 Riding Stability

In this study, the Sperling index is used to evaluate the riding stability of monorail vehicles. Due to the lack of the track irregularity spectrum of the straddle-type monorail, the ISO8608 A-level road spectrum is used to simulate the track irregularity. After obtaining the acceleration of the vehicle during rectilinear motion, the Sperling index can be obtained based on treatment according to GB5599-85.

Figure 11 shows the lateral and vertical Sperling index values of the vehicle under different speed stage and under the excitation of the ISO8608 A-level road spectrum (the excitation spectrum coefficient is set as 0.5 and 1, respectively). As shown in Fig. 8, as the speed increases, the riding stability of the vehicle gradually decreases. When the track spectrum coefficient is set as 0.5, the lateral and vertical stability indexes of the vehicle within the speed range of 90 km/h are both lower than 2.5, corresponding to an excellent level of stability. When the track spectrum coefficient is set as 1, the speed is lower than 60 km/h, and the lateral and vertical stability indexes are lower than 2.5, corresponding to the excellent stability level. When the speed ranges between 70 and 90 km/h, the lateral and vertical stability levels are fine.

4 Flexibility Coefficient

The lateral span of the walking wheels in straddle-type monorail vehicles is very short. Hence, steady wheels must be set to improve the capsizing resistance of straddle-type monorail vehicles. The vehicle capsizing resistance is usually evaluated based on the buoyant center height and flexibility coefficient. In the UIC505-5 regulation, the flexibility coefficient is defined as follows: When a vehicle stays on a line with a superelevation of δ, by assuming that the angle between the cross-section centerline of the vehicle and centerline and the line cross-section centerline perpendicular to the track top plane is ϕ , the ratio of ϕ to δ is the flexibility coefficient S.

According to the definition of the flexibility coefficient, the calculation formula for the flexibility coefficient of this monorail vehicle can be deduced. Figure 8 shows the loading model of the monorail vehicle rolling under lateral force. The carbody and bogies are connected via suspenders. Hence, the carbody and bogies can be considered as a whole. The coordinate system YOZ, consisted by the track beam centerline and the horizontal line of track beam upper surface, is selected as the reference coordinate system. The degrees of freedom of the loading model contain overall lateral moving amount y and overall rolling angle ϕ. By converting the vertical

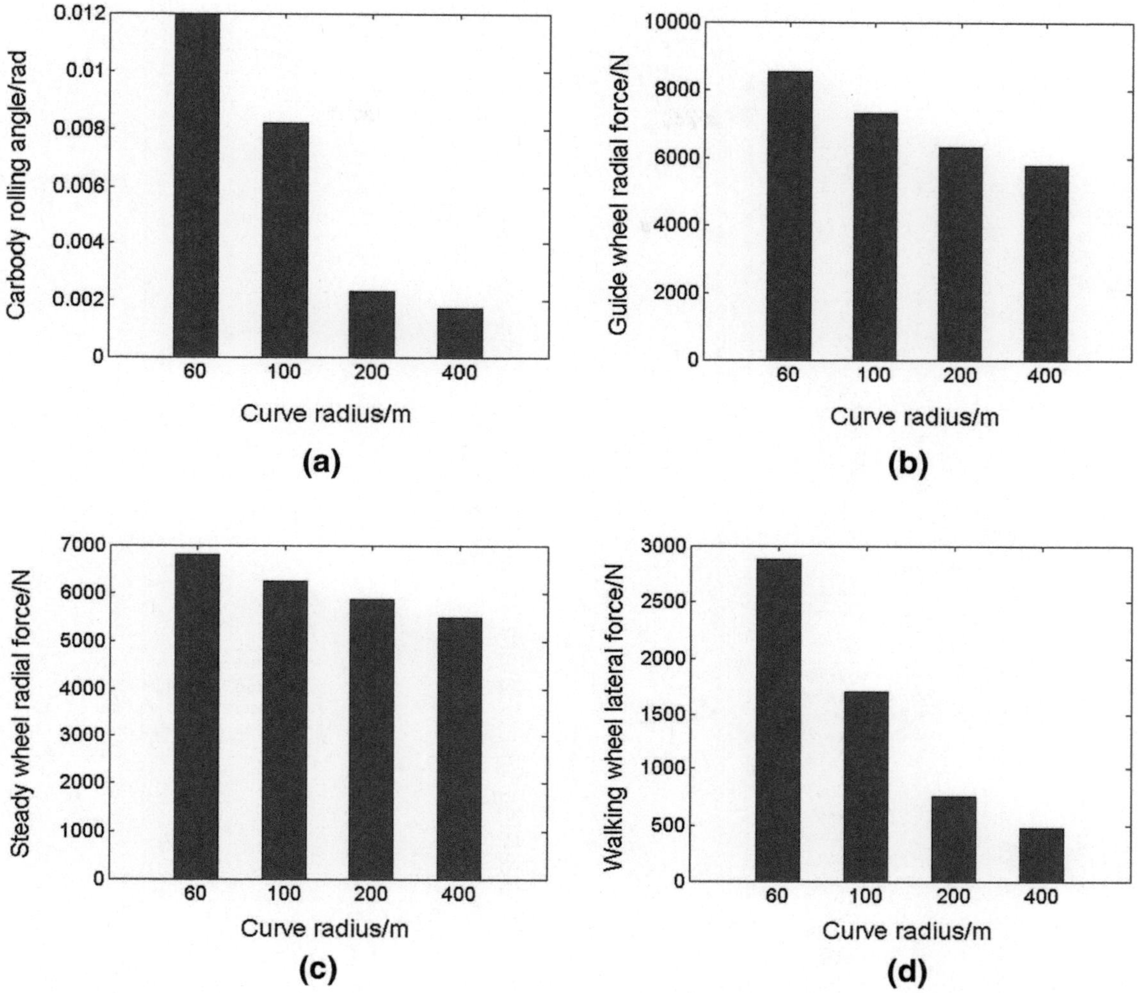

Fig. 9 Calculation results of curves with different radii, **a** carbody rolling angle, **b** guide wheel radial force, **c** steady wheel radial force, **d** walking wheel lateral force

stiffness of air springs on the tumbler axle box into the stiffness of walking wheel, the equivalent primary suspension stiffness K_{1z} can be obtained. By converting the vertical stiffness of spring on the tumbler of guide wheels into the radial stiffness of guide wheel, the equivalent radial stiffness of guide wheel K_g can be obtained. The equivalent radial stiffness of steady wheels K_{st} can also be obtained by converting the vertical stiffness of spring on the tumbler of steady wheels into the radial stiffness of steady wheel (Fig. 12).

According to the force analysis model in Fig. 8, the statics equilibrium equation during monorail vehicle rolling can be listed as follows:

1. Overall lateral movement equation

$$2K_g(y - h_3\phi) - 2(y - h_4\phi)K_{st} + (G_1 + G_2)\delta = 0 \tag{1}$$

2. Overall side rolling equation

$$\begin{aligned} &-2K_g(y - h_4\phi)h_3 - 2(y - h_4\phi)K_{st}h_4 - 2K_{1z}\phi b^2 \\ &- G_1[(h_1 + h_2)\delta + y] + G_2(h_2\delta + y) \\ &= 0 \end{aligned} \tag{2}$$

By combining the above two equations, the flexibility coefficient equation of this monorail vehicle can be obtained as:

$$S = \frac{\phi}{\delta} = \frac{A_1 - A_2}{B_1 - B_2} \tag{3}$$

where $A_1 = [2(K_g h_3 + K_{st}h_4) + G_1 + G_2]$

$A_2 = [2G_1(h_1 + h_2) - G_2 h_2](K_g + K_{st})$

$B_1 = 2[G_1 + G_2 - 2(K_g h_3 - K_{st}h_4)](K_g h_3 + K_{st}h_4)$

$B_2 = 4[G_1 + G_2 - (K_g h_3^2 - K_{st}h_4^2 - K_{1z}b^2)](K_g h_3 + K_{st}h_4)$

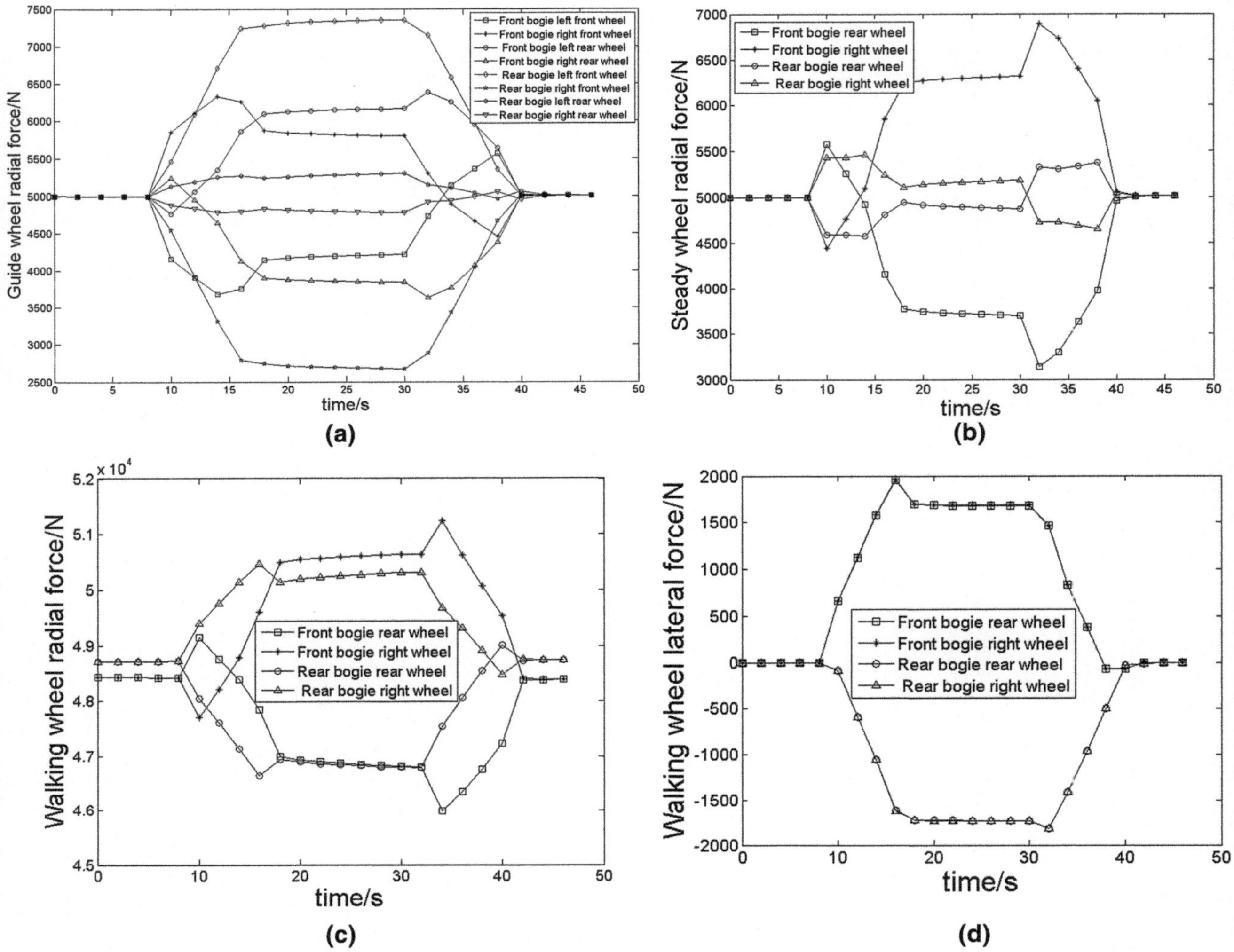

Fig. 10 Load variation regularity of each wheel in the bogie, **a** guide wheel radial force, **b** steady wheel radial force, **c** walking wheel radial force, **d** walking wheel cornering force

According to parameters listed in Table 1, the flexibility coefficient of this new straddle-type monorail vehicle can be calculated based on Eq. 3. When all the guide wheels are in contact with the steady wheels at the same time, the flexibility coefficient is 0.284. When steady wheels on one side and guide wheels on the opposite side leave the track surface, the flexibility coefficient is 0.651. According to the definition of the flexibility coefficient, the greater the flexibility coefficient, the easier it is for the vehicle to capsize on the curves. Hence, to ensure a favorable anti-overturning capacity of the straddling monorail vehicles, a certain degree of preload pressure should be applied on the guiding or stabilizing wheels to guarantee a consistent contact between the wheels and track beam. [7]

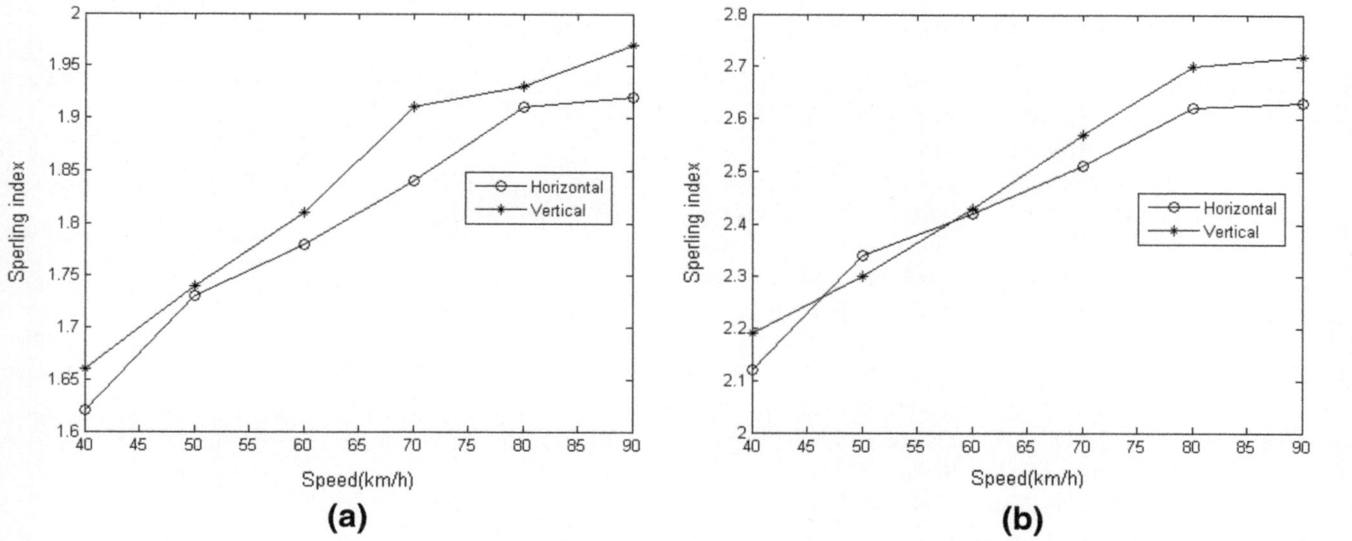

Fig. 11 Sperling indexes of vehicle, **a** track spectrum coefficient of 0.5, **b** track spectrum coefficient of 1

Fig. 12 Force analysis model of the vehicle during rolling

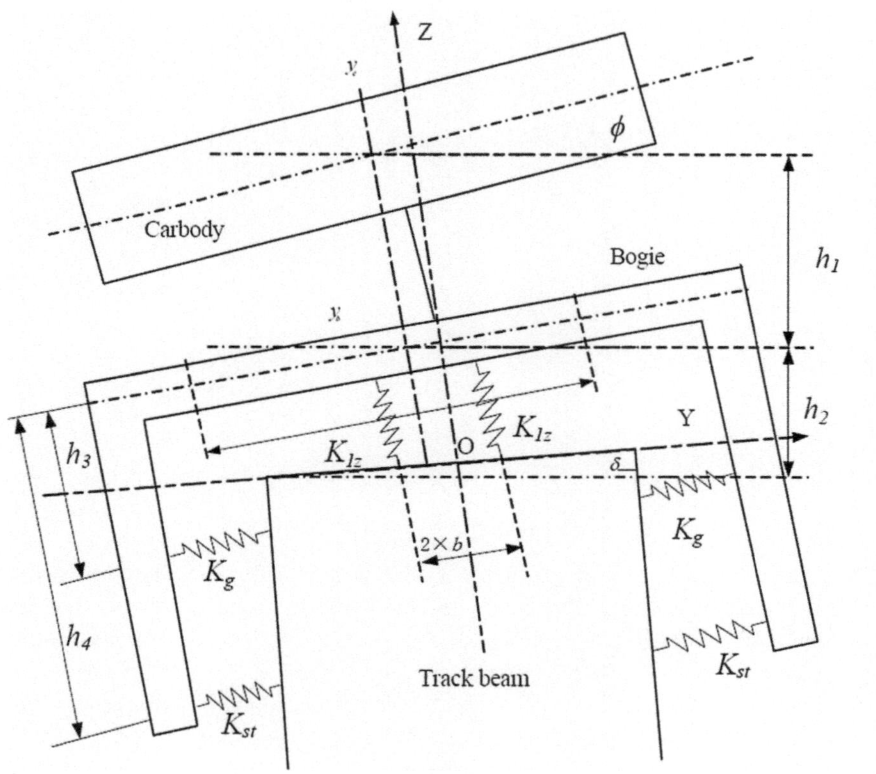

5 Conclusion

In this study, a new type of straddle-type monorail vehicle single-axle bogie is developed. The vertical suspension system uses tumbler structure. One end of the tumbler is fixed to the framework bottom via rotational joints, while the other end supports the framework via air springs. The lateral suspension system uses suspenders. Lateral elasticity was provided by the lateral oscillation of the suspenders. The deflection between the carbody and bogie during the motion when passing through curves is adapted by the vertical oscillation of the suspenders. The guide wheels and steady wheels rotates around the horizontal axle at the end of the framework. Rubber springs are set between the spindles and framework, and make it easier to adjust the pre-compressive pressure between the steady wheels and guide wheels.

The multi-body dynamics UM software is utilized to construct the single-vehicle dynamic model. Via simulation, its dynamic performance was analyzed. Dynamic analysis results indicated that the straddle-type monorail vehicle designed in this study has a favorable curve passing-through motion performance. When the A-level road spectrum coefficient is 0.5, the stability index of the monorail vehicle is excellent within the speed range of 90 km/h.

The calculation formula of the flexibility coefficient of this new monorail vehicle was deduced, with related flexibility coefficients obtained. When all the guide wheels are in contact with steady wheels at the same time, the flexibility coefficient is 0.284. When steady wheels on one side and guide wheels on the opposite side leave the track surface, the flexibility coefficient is 0.651.

References

1. Liu Y (2007) Development model of monorail transit system. Chin Railw 6:69–71
2. Jun X (2013) Analysis on the adaptability of monorail transit. Electr Locomot Mass Transit Veh 4:20–23
3. Liu S (2003) Development of the straddle type single-rail bogies in Chongqing. Roll Stock 9:18–21
4. Xia Z (2016) Improvement and innovation of bombardier monorail technology. Mod Urban Transit 2:97–100
5. Ren L, Zhou J, Shen G (2004) Dynamics model and simulation study of a straddle type monorail car. China Railw Sci 4:26–31
6. Du Z, Li N, Chen S (2012) Simulation of the curve negotiation performance of straddle-type monorail vehicle. Urban Mass Transit 7:22–25
7. Ren L, Ji Y Critical rolling angle of straddle-type monorail vehicle. J Tongji Univ (Natural Science Edition)

17

Methodology for Determining Dwell Times Consistent with Passenger Flows in the Case of Metro Services

Luca D'Acierno[1] · Marilisa Botte[1] · Antonio Placido[2] · Chiara Caropreso[1] · Bruno Montella[1]

Abstract The importance of a mobility system based on railway technology as the backbone of public transport is now widely acknowledged. Indeed, rail systems are green, high performing, smart and able to ensure a high degree of safety. Therefore, modal split should be steered towards rail transport by increasing the attractiveness of this transport mode. In this context, a key element is represented by the timetabling design phase, which must aim to guarantee an appropriate degree of robustness of rail operations in order to ensure a high degree of system reliability and increase service quality. A crucial factor in the task of timetabling entails evaluating dwell times at stations. The innovative feature of this paper is the analytical definition of dwell times as flow dependent. Our proposal is based on estimating dwell times according to the crowding level at platforms and related interaction between passengers and the rail service in terms of user behaviour when a train arrives. An application in the case of a real metro system is provided in order to show the feasibility of the proposed approach.

Keywords On-platform passenger behaviour · Dwell time estimation · Timetable design · Rail system simulation · Microscopic approach

✉ Luca D'Acierno
luca.dacierno@unina.it

[1] Department of Civil, Architectural and Environmental Engineering, Federico II University of Naples, via Claudio 21, 80125 Naples, Italy

[2] D'Appolonia S.p.A., via del Fiumicello 7, 80142 Naples, Italy

Editor: Xuesong Zhou

1 Introduction

The key role of the timetabling design phase in rail transport management lies in the fact that the results of this planning task have a direct impact on service quality, hence on passenger satisfaction [1]. The timetabling process is made ever more difficult by the constant increase in travel demand, and the consequent need to provide sufficient transport capacity, avoid train and platform congestion and guarantee a good level of service reliability is increasingly felt. It is therefore necessary to develop suitable methodologies to enable service providers to provide a timetable which is as robust and stable as possible, so as to increase the attractiveness of rail systems.

Timetable planning consists in establishing for each train line a feasible schedule of arrival and departure times at the stations served consecutively, consistent with the safety and signalling system, transfer connections and regulations. For this purpose, several factors have to be taken into account, such as train running times, blocking times and minimum headway between two successive convoys, dwell times at stations and possible buffer times. In particular, the paper focuses on the dwell time estimation problem. However, although mass transit agencies may adopt a statistical approach for determining dwell times as a function of confidence degrees (expressed, for instance, in terms of percentiles), the planning phase is implemented when the service is not yet in operation (for instance, in the case of a new line, new services or new timetables), and therefore, there are no available data for implementing such procedures. Hence, in these cases, it is necessary to adopt a modelling approach for simulating future and hypothetical conditions. Our proposal aims to provide an analytical tool for supporting planning phases. Moreover, since mass transit agencies may adopt some control strategies to increase the compliant between the real and the scheduled timetables, the use of a decision support system which allows to estimate actual dwell times may direct the use of proper intervention actions such as the adoption of additional rolling stock, the increase or reduction in stop duration of a convoy or the limitation of passengers flows on platform. In the literature, contributions on dwell times follow mainly two kinds of approaches, namely regression and micro-simulation methodologies. Methods based on the former approach aim to provide, on the basis of the data observed, a regression model which expresses dwell times as a sum of constant and variable predictors. In particular, fixed values are related to the unlocking, opening and closing times of doors together with any planned buffer times; indeed, they are invariant once rolling stock features and train dispatching times are set up. By contrast, variable parameters are a function of the users alighting and boarding times which, in turn, depend on passenger flows. Statistical techniques were the first kind of approach proposed in the literature [2, 3] and were borrowed from the bus service sector [4–6]. More recent contributions in this field are provided by [7–10]. However, broadly speaking, these methods are not generic enough to be applied in contexts other than those in which they were developed, and moreover, they provide no details about passenger behavioural rules when a train arrives.

On the other hand, methods based on a micro-simulation approach analyse pedestrian behaviour on platforms explicitly, especially in crowded conditions [11, 12], and relate it to delays and to other aspects of rail service performance.

In this context, [13] extended the simulation methodology proposed by [14], introducing the evaluation of the mutual interaction between dwell times and train delays. By contrast, [15] simulated the cooperation and negotiation process between boarding and alighting passengers by means of a cellular automata-based model and [16] proposed a multi-agent simulation method which is able to take into account passenger congestion both on the platform and inside trains. Moreover, in order to overcome the shortages of discrete approaches, [17] proposed an extension of a traditional floor field cellular automata pedestrian model, which is specifically developed for simulating high-density contexts. Furthermore, [18] addressed the problem of defining passenger service time and other related factors (i.e. user density on trains and platforms, pedestrian level of service and passenger dissatisfaction) by combining micro-simulation tasks with laboratory experiments.

Albeit based on statistical techniques, regression methods are mostly deterministic, in the sense that their results can be viewed as the expected values of dwell times which are required for completing the boarding and alighting process. By contrast, micro-simulation approaches are better able to take into account the stochasticity of the analysed phenomenon which lies in several factors: the temporal and spatial distribution of passengers, passenger and train driver behaviours, and train delays. Stochastic variations in dwell time are taken into account in [19], while [20] addressed the problem by determining dwell times by separating them into both deterministic and stochastic sub-processes.

Other methodologies proposed in the literature for estimating dwell times are based on the use of artificial neural networks: [21] modelled human behaviour and interaction between different groups of passengers by applying artificial intelligence-based techniques and especially by means of a fuzzy logic approach, while [22] addressed the problem by using the so-called *Extreme Learning Machine* (*ELM*), a very fast training speed algorithm described by [23].

From the above, we may identify some of the elements which influence dwell times such as rolling stock and passenger flows with their features. Rolling stock plays a role in the problem of dwell time estimation due to many factors: the control system, number and width of doors [24], kind of service performed [25], horizontal and vertical gaps between the train and platform [26, 27] and interior layout of the convoy seen as number and position of seats [28] or as passenger distance to exit doors and potential free space that the passenger is inclined to occupy [29]. As shown by [30], also fare collection methods may influence dwell times at stations.

Moreover, the importance of considering the number and characteristics of passengers (gender, age, mobility, luggage) for determining alighting and boarding times is self-evident [31]. This element is the hardest to evaluate, since it depends on uncertain conditions such as the interaction between different groups of passengers on the same platform and between passengers on the platform and those on the train [32].

An important factor to be taken into account is also the forecast of queue length of passengers who fail to take the train due to a lack of residual capacity on the convoy: these users will influence the next alighting and boarding process. In this regard, [33] proposed a methodology for supporting pedestrian flow management which is able to provide a theoretically quantitative prediction for the queue length of stranded passengers by means of probabilistic theory and discrete time Markov chain theory.

In addition, also station layout and rail operations affect dwell times. The former includes the position of access/egress facilities on the platform [34] and train stop type (e.g. short stop or large stations [35]). The latter (i.e. rail operation) presents a relationship of reciprocal influence with dwell times at stations, which are key factors in order to optimise the reliability of rail systems [36–39]. Dwell times play an important role especially in propagating delays [40–43]. Indeed, when a first delay occurs, if the timetable is neither stable nor flexible enough, the perturbation propagates and generates secondary delays, reducing service quality and hence passenger satisfaction. In this context, a careful estimate of dwell times, as a function of passenger flows, is the most critical task in order to design a robust timetable and make it able to absorb delays by avoiding disturbance propagation.

Clearly, this requirement becomes increasingly felt with the worsening of crowded conditions. Indeed, as shown by [44], the dynamic interaction between passengers and the rail service produces the so-called *snowball effect*: the number of passengers on the platform influences the dwell times of trains at stations which, in turn, cause increasing delays; this implies an increase in headways which could generate more passenger flows on the platform (generally proportional to the headway increase), producing a further extension of dwell times. However, the snowball effect does not evolve indefinitely, but converges towards an equilibrium state according to proper theoretical conditions, as shown in Sect. 2.

Given the close relationship between dwell times, the timetable and reliability of rail service [45, 46], together with the need to evaluate boarding and alighting times as a function of passenger flows, the majority of contributions in the literature have concentrated on dwell time estimation in the planning phase: [47] focused on the definition of run times and station dwell times in order to minimise transfer waiting times, while [48] analysed the possibility of adjusting dwell times so as to increase station capacity. Nevertheless, some have also proposed models for managing disruptions [49] and real-time rescheduling tasks [50], or even making more effective energy-saving measures [51].

In this context, this paper proposes a methodology whose aim is to support the timetabling planning process by providing an accurate estimation of train dwell times at stations, so as to increase timetabling robustness and hence rail service reliability. Dwell times are computed as a function of travel demand flows evaluated by simulating explicitly user behaviour on the platform when a train arrives. Moreover, the proposed methodology also takes account of station layout, assuming, as proposed by [34], that passengers during the boarding process choose the door on the basis of their exit position (for example, stairs or elevator) in the alighting stop so as to minimise the walking distance at their own destination station. Obviously, the choice of the preferred door cannot be always satisfied, especially in high crowding conditions. Indeed, in these cases, passengers tend to modify the selected door (choosing adjacent doors and/or adjacent coaches) as a function of the crowding degree in order to board as soon as possible. Hence, in these contexts, the assumption on the starting position of passengers on the platform (according to the approach proposed by [34] or other equivalent approaches) represents only the initialisation of the loading algorithm.

The paper is structured as follows. Section 2 describes the simulation framework for modelling the interaction between rail system components, focusing on the estimation of dwell times as flow-dependent variables and their influence on the achievable service performance. Section 3 shows the feasibility of the suggested approach by recourse to an application in the case of a real metro line. Finally, conclusions and future research are summarised in Sect. 4.

2 Simulation of Interaction Between Rail System Components

Rail service is characterised by a high degree of complexity because of the large amount of interactions existing between its components: infrastructure, signalling system, rolling stock and timetable. In addition, the reciprocal influence between rail service and travel demand has to be taken into account: passengers on the platform influence dwell times at stations and this represents one of the main disturbances of high-frequency service lines. As a result, a realistic analysis of rail operations should consider the effects of delays resulting from travel demand both in the phase of network design and in real-time management.

The proposed modelling framework is based on the adoption of three different models (Fig. 1):

- a Supply Model;
- a Travel Demand Model;
- a Service Simulation Model.

The first two models provide, respectively, supply characteristics and travel demand related to the analysed context. By means of their interaction, it is possible to evaluate passenger flows on the network and performance of elements of the transportation systems analysed.

In particular, the *Supply Model* (*SuM*) provides the features of all public transport systems within the study area, including the rail system. In this way, split demand among transport modes can be taken into account. Indeed, rail and metro lines, particularly within cities, are part of the public transportation system and cannot be considered individually. Hence, knowing the characteristics of the other transport modes can also provide better estimation of the arrival rate at each station.

The *Travel Demand Model* is the most innovative part of the procedure. It is split into two further sub-models: a *Pre-Platform Model* (*PPM*) and an *On-Platform Model* (*OPM*).

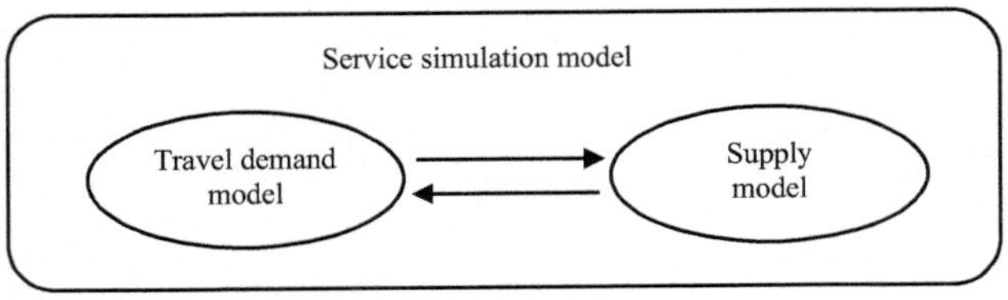

Fig. 1 Modelling structure

The former estimates the number of passengers arriving at stations as a result of the interaction with the *Supply Model* (*SuM*) by reproducing the decision-making process made by passengers who choose, among all possible alternatives (i.e. different transport modes), the one which maximises their utility. The results of this model allow arrival rates of passengers at platforms to be determined.

The latter explicitly simulates passenger behaviour on the platform by considering the maximum capacity of each train and estimating the dwell time required to complete the boarding/alighting process. This model is based on a *First-In First-Out* (*FIFO*) approach where it must be considered that passengers on the same platform may have different destinations and hence different alighting stations. In this way, travel demand is simulated dynamically according to rail service performance which, in turn, is influenced by passenger flows.

This reciprocal interaction between travel demand and rail service generates, as we will see, a fixed-point problem and takes place in a simulation environment provided by the *Service Simulation Model* (*SeSM*). Indeed, given their complexity, railway systems cannot be described in terms of a closed-form analytical solution, but it is necessary to rely on specific simulation techniques. The *SeSM* is a microscopic synchronous simulation model and is able to provide rail network performance as a function of infrastructure, rolling stock, signalling system, planned timetable and travel demand.

In our case, the simulation of the rail system can be performed by using commercial software (such as OpenTrack® software [52]), appropriately integrated with ad hoc external tools in order to include travel demand as input data and, therefore, to take into account variations in dwell times at stations just due to variations in travel demand. Moreover, both deterministic and stochastic simulations can be performed.

More detail on the formulation of these models can be found in [53, 54]. However, it is worth pointing out that by means of this new conception of the Travel Demand Model and especially through the *On-Platform* sub-model, we may capture the interaction between passenger flows and the rail service taking place at stations. This phenomenon is the key factor to be simulated in order to perform a reliable estimation of dwell times. This will, in turn, generate a robust timetable which is able to absorb delays, avoiding their propagation.

Therefore, evaluation of dwell time is a fundamental scheduling task of the wider process of timetabling which directly influences service reliability and passenger satisfaction. Herein lies the motivation behind our proposal.

In particular, the proposed methodology estimates dwell times at station by considering, as input, passenger flows, station configurations and a function expressing the dependence of dwell times on the number of passengers at the most loaded door.

Qualitatively, this function presents constant values of dwell time for low passenger flows, and then, the dwell time increases as the number of passengers rises. However, since the factors which influence this relationship vary for different railway networks, the function has to be suitably calibrated on the specific system analysed.

Moreover, although interchange with bus services may seem to transfer irregularity to a metro system, the low values of transferred passengers (the capacity of a bus generally has a lower order of magnitude than a metro context) do not yield significant effects. Likewise, if in the interchange node numerous bus lines converge, the overlap of irregularities tends to compensate, and therefore, also in this case transfer irregularities can be neglected. Finally, in the case of interchange with rail systems (regional trains, metros or funiculars), where the flows involved may be substantial, the regular structure of arrivals tends to stabilise travel demand variations. However, in order to prevent any kind of irregularity, we propose to perform analyses by considering different levels of travel demand in order to take even more severe conditions into account.

In planning dwell times, the greatest challenge is to be able to model the interaction between passengers and trains during the boarding/alighting process by capturing the above-mentioned snowball effect and the complex phenomenon of negotiation and cooperation between passengers. In particular, the proposed approach considers three levels of interaction.

First of all, on the basis of the station configuration, passengers move towards the door they prefer. In particular, the assumption is that passengers prefer to board the train with their final destination in mind so as to maximise their own utility, according to the theory of the rational decision maker (see, for instance, [55]). Indeed, especially commuters, who have experience of the system, choose the door of the train which is as close as possible to the stairs or the elevators so as to minimise their walking distance at the alighting station. When in front of a door there is a larger group of passengers than a set value (which, obviously, varies from case to case), they start moving to the adjacent doors in the same coach. In the case of crowded situations, the customer's target is to get on the train as rapidly as possible, trying to remain close to the first favourite door. Once the capacity constraints of the coach are reached, passengers move towards the other doors which attract flow according to their available capacity. This means that the emptier the coach, the more passengers will board it.

The assumption to be made, for the procedure in question, is that the door chosen to board the train will be the same as that to alight from it. This is supported by the lack of freedom of movement for passengers within the coach, especially in crowded conditions.

However, it is worth noting that, although in the literature there are few studies on the microscopic behaviour of passengers on the platform, we adopted the proposal of [34] for determining the initial position of passengers when a train arrives. Another equivalent approach would be based on distributing passengers uniformly on the platform. Obviously, in the cases of low crowding conditions, each passenger is able to board quickly through the preferred door without affecting the dwell time duration, and therefore, the two approaches would provide similar results. Likewise, in the case of high crowding conditions, the congestion of doors and coaches affects user choices aimed at reducing boarding times. Hence, also in these conditions, the initial position of passengers on platform being calculated with both approaches would provide similar results in terms of dwell time duration, since the loading algorithm converges to a solution depending on the train congestion level.

Therefore, by modelling the boarding/alighting process as a threefold interaction phenomenon, we may compute the number of passengers at the most loaded door and, by means of the calibrated function (expressing the relation between the number of passengers and dwell times for the specific context analysed), provide dwell times for each simulated train and at each station. In addition, by estimating the door chosen to board the train, the proposed model provides information about the crowding level within each coach (Fig. 2). This result is extremely useful for calculating the on-board disutility perceived by passengers and could enable a fleet composition to be planned which is closer to customers' needs. Furthermore, *Intelligent Transportation Systems* (*ITS*) could also be designed to inform customers about the crowding levels of the approaching train and to suggest positioning along the platform or which coach of the train should be preferred, so as to optimise the boarding/alighting process and reduce dwell times.

It is worth noting that, should the maximum capacity of the train be reached, the proposed methodology is able to estimate the passenger surplus (i.e. the number of passengers who are forced to remain on the platform due to

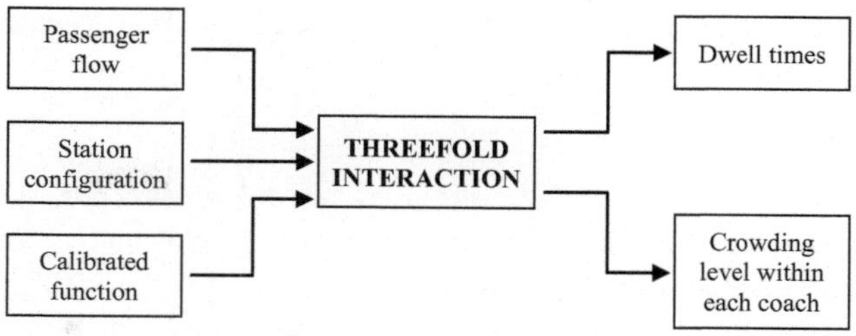

Fig. 2 Simulation framework

capacity constraints of the whole convoy being reached) who has to board following trains.

This task makes the simulation more realistic, especially in the case of breakdowns, because it allows for propagation of the disturbance. Hence, besides the planning phase addressed herein, the proposed model can be easily adjusted to deal with rescheduling measures in an off-line dimension (see, for instance, [56]).

From a mathematical standpoint, the snowball effect, and therefore the reciprocal dependence between headways and dwell times, results in a fixed-point problem formulation. For this purpose, let

$$\boldsymbol{dwt} = \vartheta(\boldsymbol{td}) \tag{1}$$

be a function to be calibrated which expresses the dependence of dwell times on the number of boarding/alighting passengers, where ***dwt*** and ***td*** represent, respectively, dwell time and travel demand vectors. Obviously, function $\vartheta(\cdot)$ has to consider that there is a threshold value of boarding/alighting passengers below which the dwell time is constant.

Likewise, let

$$\boldsymbol{hd} = \psi(\boldsymbol{dwt}) \tag{2}$$

be the relation providing headways (i.e. vector ***hd***) as a function of dwell times (i.e. vector ***dwt***) by means of the simulation performed by the *SeSM*.

Since the frequency of a metro rail service strongly affects the congestion level on the platform, assuming within a short time interval the arrival rate of passengers at station s as constant, the travel demand (i.e. the number of passengers waiting on the platform) at each station s may be calculated as:

$$td_{s,d,r} = upf_{s,d,r} \cdot hd_{r,s} \tag{3}$$

where $td_{s,d,r}$ is the number of passengers arriving at the platform of station s for travelling towards destination d during the time interval between run $r-1$ and run r; $upf_{s,d,r}$ is the arrival rate of passengers at platform of station s for travelling towards destination d during time interval between run $r-1$ and run r; $hd_{r,s}$ is the headway between run $r-1$ and run r at station s. Obviously, $td_{s,d,r}$ and $hd_{r,s}$ express, respectively, the component of vector ***td*** and vector $\boldsymbol{hd}$. Hence, Eq. (3) may be expressed in vector notation as:

$$\boldsymbol{td} = \gamma(\boldsymbol{hd}) \tag{4}$$

Obviously, the *PPM* provides arrival rates ($upf_{s,d,r}$) associated with each origin–destination station and each time interval between two successive runs.

Therefore, combining the above-mentioned functions, the following is obtained:

$$\begin{cases} \boldsymbol{dwt} = \vartheta(\boldsymbol{td}) \\ \boldsymbol{td} = \gamma(\boldsymbol{hd}) \\ \boldsymbol{hd} = \psi(\boldsymbol{dwt}) \end{cases} \tag{5}$$

or equivalently:

$$\boldsymbol{dwt} = \vartheta(\gamma(\psi(\boldsymbol{dwt}))) \tag{6}$$

The explicit formulation of relations in Eq. (5), or equivalently in Eq. (6), can be expressed in closed form in the case of function $\vartheta(\cdot)$ (whose details are provided in Sect. 3) and function $\gamma(\cdot)$ (whose formulation is provided in Eq. 3). Differently, function $\psi(\cdot)$ may not be expressed in closed form since it represents a system of differential equations whose solution requires the implementation of proper simulation software. In particular, in Sect. 3, we propose the use of the commercial software OpenTrack® [52] for solving this task numerically.

A first remark concerns the relationship between the travel demand (in terms of arrival rate) in a station and the related dwell time. Indeed, although Eq. (1) shows the direct dependence of dwell times on travel demand (i.e. the higher the travel demand, the higher the dwell time), Eq. (3) (or, equivalently, Eq. 4) shows the direct dependence of travel demand on headway which, by means of Eq. (2), in turn depends on dwell times in previous stations. Hence, the dwell time in a station is a function of the arrival rate in that station, the arrival rates in the previous stations and the framework of the travel demand (in terms of alighting flows). For this reason, in general, it is not possible to state that higher flows (in terms of arrival rates) necessarily imply higher dwell times.

However, the problem formulated by Eq. (6) highlights the need to find a dwell time vector which produces a headway vector which generates travel demand on platform which provides the initial dwell time vector. According to the theory of the fixed-point problem, this particular case is called 'compound fixed-point problem'. The mathematical conditions which ensure the existence of the solution of (6) are expressed by Brouwer's theorem [57]. In our case, since all conditions are satisfied, we may state that the snowball effect always evolves towards an equilibrium condition. Likewise, Banach's theorem [58] (or equivalently its version in the case of weaker conditions proposed by Cantarella [59]) allows us to state the uniqueness of the fixed-point solution. However, in our case, the uniqueness conditions are not satisfied and this has an impact on the selected resolution procedure. More specifically, two different methods were considered, namely an '*iterative algorithm*' and an '*MSA (Method of Successive Average) framework algorithm*' [59, 60]; for both of them it is worth analysing convergence properties.

With respect to the iterative algorithm, the convergence cannot be stated analytically, and hence, it cannot be excluded that the algorithm diverges. On the other hand, the issue of the convergence of the MSA algorithm requires a more detailed assessment. Indeed if, in general, the convergence of this algorithm for the resolution of the compound fixed-point problem is guaranteed by Blum's theorem [61]; in our case, the convergence of the algorithm cannot be demonstrated since, as explained above, the uniqueness of the solution cannot be assured mathematically. However, this does not mean that the solution is not unique, but that there are no mathematical bases to be sure of this assumption according to traditional proof proposed in the literature.

Hence, it is necessary to look for numerical evidence that the solution of this fixed-point problem could be unique. In particular, it is provided by [62] which applied both algorithms in the case of a real metro line and verified as well their convergence to the same configuration of dwell time.

Moreover, in contrast to the iterative algorithm, the MSA generates decimal values at each iteration. Obviously, such values need to be rounded up/down to the nearest integer before being set up within the simulation model. However, this process can cause theoretical problems for the achievement of the solution by causing a slowdown in reaching convergence.

Therefore, the iterative algorithm appears easier and faster than the MSA procedure, which was why it was selected as the resolution approach to be adopted. However, obviously, in order to prevent the algorithm from performing an infinite number of iterations, a termination test has to be established, based on the maximum number of solutions analysed.

3 Application of the Methodology in the Case of a Real Metro Line

In order to show its feasibility, the proposed approach was applied to Line 1 of the Naples metro system in its 17-station configuration. The line is about 18 km long and has a somewhat directional travel demand since it connects the suburbs (Piscinola) with the city centre (Garibaldi). Hence, morning flows are directed towards the city centre and afternoon/evening flows towards the suburbs. Figure 3

Fig. 3 Line 1 of the Naples metro system (Italy)

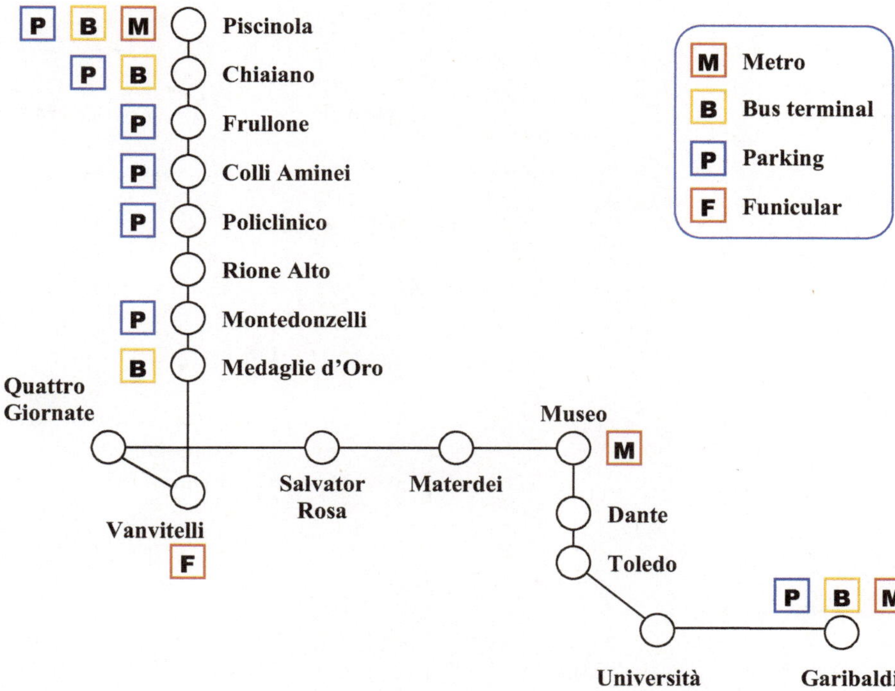

shows the framework of Line 1, highlighting interchange points with other transportation systems.

The choice of a metro network was no accident: due to the close distance between two successive stations, dwell times are frequently comparable to travel times. Thus, the need to provide a reliable estimation of dwell times becomes even more important.

Currently, for most of the days, there is a headway of 8 min between two successive convoys on the line. Together with the demand flows involved, this makes the snowball effects not very apparent. However, since our aim is to verify the capacity of the methodology to capture this phenomenon, in the application we stressed the system by simulating a denser timetable, up to a headway of 3 min.

However, the application of the proposed methodology has required two kinds of surveys: a survey of passenger flows for estimating travel demand and a survey of boarding/alighting flows and train stop durations for determining the passenger flow dwell time function.

The first approach has been performed periodically in all stations and in all time periods since 2012 by surveying both the count of passengers waiting on platforms and turnstile data. The second campaign was performed in a shorter period (May–June 2013) by collecting 172 data surveyed in 14 stations during the morning and afternoon peak hours.

Travel demand was estimated according to [63], and the simulation model was calibrated in terms of infrastructure, signalling system, rolling stock and timetable as shown by [64]. Moreover, in order to investigate the sensitivity of the methodology to different crowding conditions, the proposed approach was applied in relation to three different levels of travel demand obtained by means of the first type of surveys, whose results, expressed in terms of a distribution function, are shown in Fig. 4.

The capacity constraints set for simulating the threefold interaction between passengers and convoys in this specific network are as follows:

- 20 passengers per door (according to the experimental evidence during surveys);
- 216 passengers per carriage (according to the capacity of the rolling stock adopted on the line);
- wherever the capacity of the carriage was reached, the surplus is distributed in proportion to the residual capacity of the remaining wagons.

The function expressing the relation between passenger flows and dwell times in this specific network was calibrated by adopting the second kind of surveys (calibration details may be found in [56]), whose results are shown in Fig. 5. In particular, the analytical formulation is:

$$dwt_s = \begin{cases} 5 & \text{if } 0 \leq td_s^{max} \leq 3.5279 \\ 0.8602 \cdot td_s^{max} + 1.9653 & \text{if } 3.5279 < td_s^{max} \leq 2 \cdot Cap_{rc} \end{cases} \quad (7)$$

where td_s^{max} represents the sum of passengers boarding and alighting at the most loaded door (obviously expressed in terms of passenger number instead of passenger flow); Cap_{rc} is the capacity of the rail coach which represents the

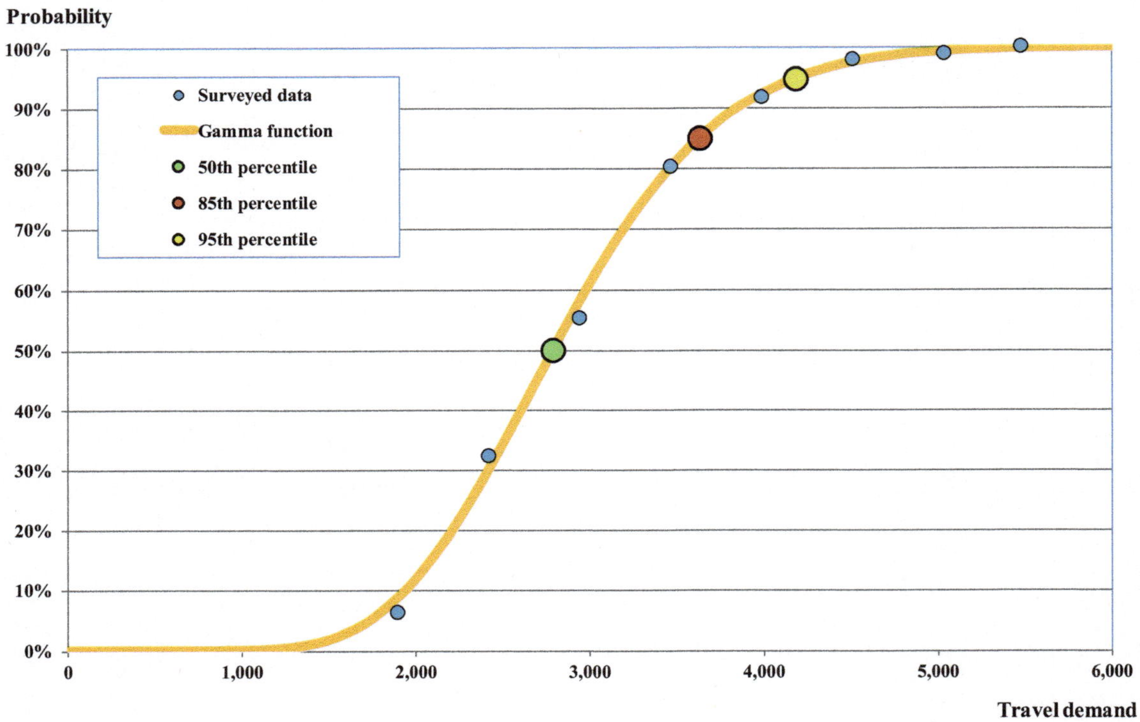

Fig. 4 Different travel demand levels

Fig. 5 Dwell time calibration function

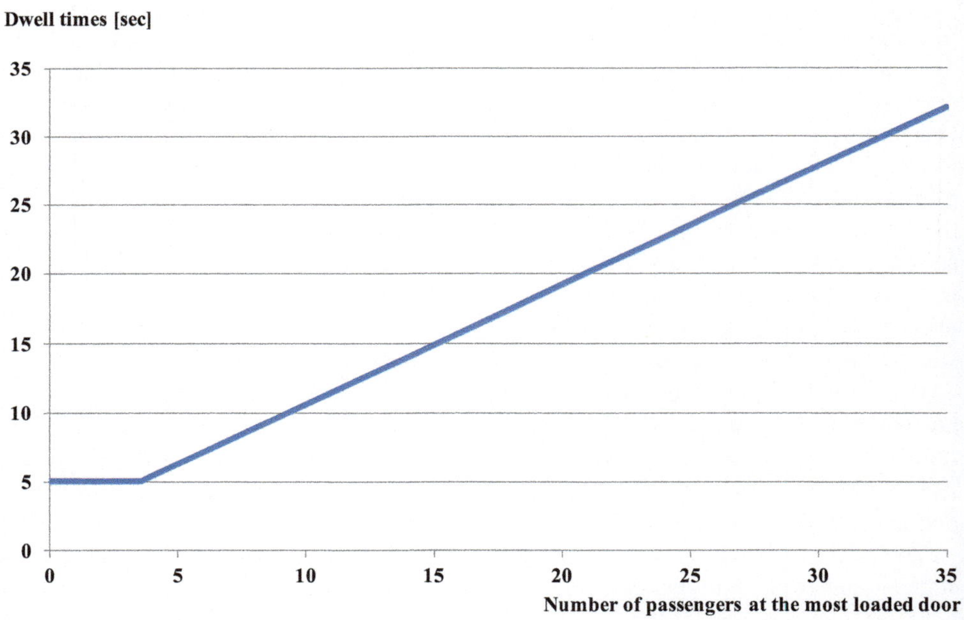

maximum number of boarding passengers or, equivalently, the maximum number of alighting passengers. Hence, the worst case consists in a completely full coach, which first unloads all passengers and then loads them again (in this case, the number of transiting passengers is equal to $2 \cdot Cap_{rc}$). This implies that, since td_s^{max} has a maximum value, the dwell time is upper bounded.

However, the R^2 test in the case of function (7) has provided a result equal to 0.81.

The basic idea of the following application is to stress the system in order to point out the presence of the snowball effect and the fact that dwell times need to be planned according to travel demand flows, rather than as fixed values. For this purpose, we analysed different timetables with a decreasing value of headways between two successive convoys and a fixed planned dwell time for each station and for each run, without any differences between peak hours and off-peak hours.

Hence, for each timetable, by means of the above-mentioned modelling framework and the implementation of the iterative algorithm, we simulated the threefold interaction between passengers and trains and solved the fixed-point problem (6), estimating dwell time as well for each station and for each run as a flow-dependent value.

Our simulation focused on the morning peak hour (6.00–9.00). As we will see, the resulting dwell times, which are estimated by simulating rail service performance dynamically according to passenger flows, represent much more reliable values to be set into the timetable, in order to improve its robustness and resilience to delays.

The steps of the procedure are as follows. First of all, a starting random value of the dwell time vector $\boldsymbol{dwt}^0$ is set up, and then, the *SeSM* evaluates the new headways according to the dwell time vector $\boldsymbol{dwt}^i$ provided at iteration i. The next step consists in simulating the threefold interaction between passengers and the train so as to estimate a new dwell time vector, that is $\boldsymbol{dwt}^{(i+1)}$, based on the headways obtained by *SeSM*.

At this point, the termination test has to be verified, that is:

$$\max_j \left(\frac{dwt_j^{i+1} - dwt_j^i}{dwt_j^i} \right) < 0.01 \quad \text{or} \quad i > M \qquad (8)$$

where M is a set number which expresses the maximum number of iterations (for instance, $M = 1000$). Therefore, if the test is satisfied the algorithm stops; otherwise, it is necessary to calculate the new headways. However, to be more accurate, it is worth noting that the proposed procedure has the following limitation: since the dwell time is a random variable, the value obtained should be considered as the expected value of the dwell time required for the boarding/alighting process, and thus, information about its statistical distribution is not available.

The iterative algorithm, whose flow chart is shown in Fig. 6, was implemented for each planned headway analysed (i.e. from 8 to 3 min) and for each level of travel demand (i.e. 50th, 85th and 95th percentile), amounting to a total of 18 processed scenarios (i.e. algorithm implementations).

The number of runs for each planned headway is shown in Table 1 (detailed for the outward and return trip), together with the number of iterations which need to be implemented to solve the fixed-point problem in the case of each planned headway and each travel demand level. Interestingly, the maximum number of iterations, which is set as a termination test requirement, is never reached, and

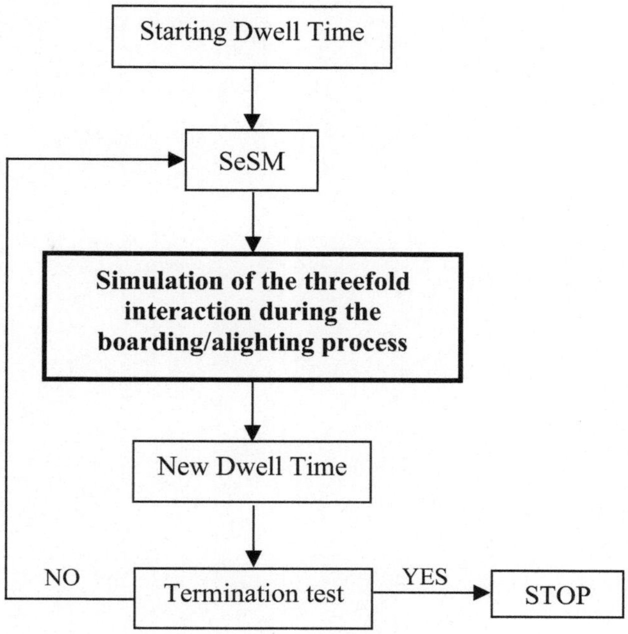

Fig. 6 Graphic representation of the iterative algorithm

therefore, the algorithm reaches the convergence in each scenario analysed. As can be seen, the number of iterations varies from a minimum of 6 to a maximum of 18, presenting a significant increase with the reduction in the planned headway due to growing system instability. Obviously, the lower is the planned headway to be analysed, the higher the number of runs and hence the computational time. However, since our proposal addresses a planning task, time is not the main variable to be considered.

In terms of analysis of results, it is worth noting the asymmetry between outward and return trips due to the asymmetry in the network. In particular, the depot (i.e. the place where each rail convoy starts the service every day and is housed every night) is located at one end of the line (i.e. at the beginning of the outward trip). Hence, achievement of the frequency regime conditions requires a transitory phase which implies a different number of runs.

The outcome of the addressed fixed-point problem is represented by an estimation of dwell times for each station and for each run. By way of illustration, Tables 2, 3 and 4 show the converging dwell times for a planned headway of 8 min, detailed for the three levels of travel demand analysed. Obviously, this type of result was also produced for the other planned headways tested: the structure and the rationale are the same, but the numerical values provided are different and the number of runs rises as the headway decreases.

Moreover, since we also considered rolling stock constraints and hence the real use of trains, it is important to

Table 1 Number of runs for each planned headway and number of iterations for each analysed scenario

Planned headway [min]	Number of runs		Travel demand level	Number of iterations
	Outward trips	Return trips		
8	23	19	50th	6
			85th	7
			95th	6
7	26	21	50th	6
			85th	8
			95th	6
6	30	25	50th	7
			85th	13
			95th	10
5	36	29	50th	17
			85th	10
			95th	15
4	45	36	50th	8
			85th	10
			95th	18
3	60	48	50th	8
			85th	15
			95th	12

provide also dwell times in the case of the first station in order to show that the timetable was planned to ensure the preservation of headways (i.e. the train is able to perform the outward trip, the return trip and the outward trip again without being delayed).

The dwell times can be observed to vary both along the columns (i.e. the stations) and rows (i.e. the runs with a certain planned departure time associated). This implies that dwell times experience both spatial and temporal variabilities. Moreover, the same run at the same station could have a different dwell time on different days, according to the travel demand level at that time and hence the flows of boarding/alighting passengers involved.

As shown in the tables, dwell times in some cases can reach also values above 90 sec. The reason lies in the fact that these values represent the converging dwell times, i.e. the dwell times reached at the end of the evolution of the snowball effect, which, as mentioned before, magnifies the involved quantities.

Moreover, these values are amplified by the congestion which affects the alighting/boarding time and makes this process very chaotic. Indeed, as shown by [24, 65], this is a nonlinear phenomenon: a certain number of users, which constitutes a mixed flow (i.e. some users must board and others must alight), needs a greater time, with respect to the case that they represent an unidirectional flow, for going through the same door.

In addition, also the congestion inside the coach (i.e. standing passengers on-board close to doors) and on the platform extends dwell times. In particular, in the analysed context, the former is accentuated by the fact that doors do not open on the same side in all stations, hence could happen that passengers who boarded on the right side must to move inside the coach in order to alight on the left side or vice versa, while the latter is due to the fact that passengers are not aware of the exact position of each door while the train is approaching, and therefore, they walk on the platform in a chaotic way.

Furthermore, it is worth nothing that, even if generally dwell times increase as the travel demand increases, as stated in Sect. 2, the dwell time in a station depends on arrival rates in that station, the arrival rates in the previous stations and the framework of the travel demand (in terms of alighting flows).

Hence, it is not possible to identify a simple evolution in dwell time values because these values represent the converging values at the equilibrium state. Therefore, since the equilibrium points of investigated flows are different (because boundary conditions are different), the comparison between results related to different percentiles is nonsense. This makes each solved fixed-point problem as stand-alone and prevents a comparison between results related to different conditions. Therefore, results show the uniqueness of each condition and confirm the necessity to make use of suitable simulation techniques which are able to reproduce

Table 2 Converging dwell times (in sec) for a planned headway of 8 min—50th percentile

Station sequence	1	2	3	4	5	6	7	8	9	10	11	12	13	14	15	16	17
Outward run																	
1	44	37	13	12	29	19	39	43	59	45	11	28	29	76	36	32	50
2	44	37	13	12	29	19	39	43	59	45	11	28	29	76	36	32	50
3	44	37	13	12	29	19	39	43	59	45	11	28	29	76	36	32	50
4	44	37	13	12	29	19	39	43	59	45	11	28	29	76	36	32	50
5	44	37	13	12	29	19	39	43	59	45	11	28	29	76	36	32	50
6	44	37	13	12	29	19	39	43	59	45	11	28	29	76	36	32	50
7	44	37	13	12	29	19	39	43	59	45	11	28	29	76	36	32	50
8	44	37	13	12	29	19	39	43	59	45	11	28	29	76	36	32	50
9	44	37	13	12	29	19	39	43	59	45	11	28	29	76	36	32	50
10	44	37	13	12	29	19	39	43	59	45	11	28	29	76	36	32	50
11	44	37	13	12	29	19	39	43	59	45	11	28	29	76	36	32	50
12	44	37	13	12	29	19	39	43	59	45	11	28	29	76	36	32	50
13	44	37	13	12	29	19	39	43	59	45	11	28	29	76	36	32	50
14	44	37	13	12	29	19	39	43	59	45	11	28	29	76	36	32	50
15	44	37	13	12	29	19	39	43	59	45	11	28	29	76	36	32	50
16	44	31	8	8	26	17	39	42	57	44	11	27	29	74	36	32	49
17	44	38	9	13	36	24	30	73	96	43	11	32	29	77	36	32	50
18	38	35	8	12	35	22	29	69	88	38	16	30	17	73	37	29	49
19	38	36	8	12	35	23	30	71	90	38	22	31	19	76	58	59	63
20	38	36	8	12	35	23	30	70	90	38	22	31	18	76	39	40	40
21	38	36	8	12	35	23	30	70	90	38	22	31	18	76	39	40	40
22	38	36	8	12	35	23	30	48	65	27	20	13	9	26	32	35	34
23	38	36	8	6	22	6	17	23	31	15	15	7	7	17	24	25	16
Return run																	
1	65	27	36	45	22	23	28	19	48	50	25	26	43	24	22	52	80
2	65	27	36	45	22	23	28	19	48	50	25	26	43	24	22	52	80
3	52	22	26	33	59	19	24	17	62	39	16	14	53	16	15	61	65
4	52	22	27	33	60	19	24	17	62	39	16	14	53	16	15	61	66
5	52	22	27	33	59	19	24	17	62	39	16	14	53	16	15	61	65
6	52	22	27	33	59	19	24	17	62	39	16	14	53	16	15	61	65
7	52	22	27	33	60	19	24	17	62	39	16	14	53	16	15	61	66
8	52	22	27	33	59	19	24	17	62	39	16	14	53	16	15	61	65
9	52	22	27	33	59	19	24	17	62	39	16	14	53	16	15	61	65
10	52	22	27	33	60	19	24	17	62	39	16	14	53	16	15	61	66
11	52	22	27	33	59	19	24	17	62	39	16	14	53	16	15	61	65
12	52	22	27	33	60	19	24	17	62	39	16	14	53	16	15	61	66
13	52	22	27	33	59	19	24	17	62	39	16	14	53	16	15	61	65
14	52	22	27	33	60	19	24	17	62	39	16	14	53	16	15	61	66
15	52	22	27	33	59	19	24	17	62	39	16	14	53	16	15	61	65
16	52	22	27	33	60	19	24	17	62	39	16	14	53	16	15	61	63
17	52	22	27	33	59	19	24	17	62	39	16	14	52	15	14	58	58
18	52	22	27	33	60	19	24	17	50	33	9	9	41	10	10	39	34
19	52	22	24	25	16	10	10	6	25	13	6	6	16	6	6	16	20

the complexity of phenomenon and catch its unpredictable evolution.

The possible variation in dwell times between two successive runs may generate variations in headway in each station. However, the timetable was planned so as to ensure that headway was constant on average. In order to highlight this phenomenon, Figs. 7, 8 and 9 show for each planned headway and for each travel demand level, the maximum

Table 3 Converging dwell times (in sec) for a planned headway of 8 min—85th percentile

Station sequence	1	2	3	4	5	6	7	8	9	10	11	12	13	14	15	16	17
Outward run																	
1	44	35	14	18	43	17	22	52	66	55	11	35	19	65	36	34	59
2	44	35	14	18	43	17	22	52	66	55	11	35	19	65	36	34	59
3	44	35	14	18	43	17	22	52	66	55	11	35	19	65	36	34	59
4	44	35	14	18	43	17	22	52	66	55	11	35	19	65	36	34	59
5	44	35	14	18	43	17	22	52	66	55	11	35	19	65	36	34	59
6	44	35	14	18	43	17	22	52	66	55	11	35	19	65	36	34	59
7	44	35	14	18	43	17	22	52	66	55	11	35	19	65	36	34	59
8	44	35	14	18	43	17	22	52	66	55	11	35	19	65	36	34	59
9	44	35	14	18	43	17	22	52	66	55	11	35	19	65	36	34	59
10	44	35	14	18	43	17	22	52	66	55	11	35	19	65	36	34	59
11	44	35	14	18	43	17	22	52	66	55	11	35	19	65	36	34	59
12	44	35	14	18	43	17	22	52	66	55	11	35	19	65	36	34	59
13	44	35	14	18	43	17	22	52	66	55	11	35	19	65	36	34	59
14	44	35	14	18	43	17	22	52	66	55	11	35	19	65	36	34	59
15	44	35	14	18	43	17	22	52	66	55	11	35	19	65	36	34	59
16	44	35	9	14	31	13	23	55	67	60	11	31	19	66	36	34	59
17	44	50	10	16	41	18	42	64	72	47	11	31	18	68	36	34	58
18	44	45	10	16	45	32	47	84	87	48	8	31	18	79	31	32	56
19	33	42	9	16	46	34	50	82	92	50	12	32	25	74	54	60	66
20	33	42	9	16	46	34	50	82	92	50	12	32	25	74	36	41	44
21	33	42	9	16	46	34	50	82	91	50	12	32	25	74	36	41	44
22	33	42	9	16	46	34	50	54	54	33	9	14	21	44	29	36	31
23	33	42	9	7	28	7	22	29	24	19	8	8	20	14	24	31	20
Return run																	
1	51	28	48	57	28	15	19	19	47	43	28	28	30	24	22	44	62
2	51	28	48	57	28	15	19	19	47	43	28	28	30	24	22	44	62
3	55	30	36	34	38	22	22	20	53	50	21	20	53	20	18	59	71
4	55	30	37	32	49	22	25	24	55	52	21	19	58	20	18	72	78
5	55	30	37	32	49	22	25	23	54	52	21	19	57	19	18	71	76
6	55	30	37	32	49	22	25	23	54	52	21	19	57	19	18	71	76
7	55	30	37	32	49	22	25	23	54	52	21	19	57	19	18	71	76
8	55	30	37	32	49	22	25	23	54	52	21	19	57	19	18	71	76
9	55	30	37	32	49	22	25	23	54	52	21	19	57	19	18	71	76
10	56	31	38	33	49	22	25	24	55	53	21	19	58	20	18	72	79
11	54	29	36	30	50	21	25	22	54	50	20	18	56	19	18	70	73
12	55	30	37	32	49	22	25	23	55	52	21	19	57	19	18	71	77
13	55	30	37	32	49	22	25	23	54	52	21	19	57	19	18	71	76
14	55	30	37	32	49	22	25	23	54	52	21	19	57	19	18	71	76
15	55	30	37	32	49	22	25	23	54	52	21	19	57	19	18	71	76
16	55	30	37	32	49	22	25	23	54	52	21	19	57	19	18	71	74
17	55	30	37	32	49	22	25	23	54	52	20	18	55	17	16	65	65
18	55	30	37	32	49	22	25	15	38	44	12	12	40	10	10	39	33
19	55	30	15	17	22	13	13	7	17	17	7	7	22	7	7	22	13

and the minimum actual headways obtained by implementing the converging dwell times in the simulation model and then selecting the maximum and minimum values among all the resulting headways (i.e. for each run and for each station). Moreover, it is shown that average values of actual headways coincide perfectly with planned ones.

This outcome points out the fact that an erroneous estimation of dwell times makes the system unstable and

Table 4 Converging dwell times (in sec) for a planned headway of 8 min—95th percentile

Station sequence	1	2	3	4	5	6	7	8	9	10	11	12	13	14	15	16	17
Outward run																	
1	51	50	16	15	35	16	24	44	59	35	10	39	19	63	37	34	58
2	51	50	16	15	35	16	24	44	59	35	10	39	19	63	37	34	58
3	51	50	16	15	35	16	24	44	59	35	10	39	19	63	37	34	58
4	51	50	16	15	35	16	24	44	59	35	10	39	19	63	37	34	58
5	51	50	16	15	35	16	24	44	59	35	10	39	19	63	37	34	58
6	51	50	16	15	35	16	24	44	59	35	10	39	19	63	37	34	58
7	51	50	16	15	35	16	24	44	59	35	10	39	19	63	37	34	58
8	51	50	16	15	35	16	24	44	59	35	10	39	19	63	37	34	58
9	51	50	16	15	35	16	24	44	59	35	10	39	19	63	37	34	58
10	51	50	16	15	35	16	24	44	59	35	10	39	19	63	37	34	58
11	51	50	16	15	35	16	24	44	59	35	10	39	19	63	37	34	58
12	51	50	16	15	35	16	24	44	59	35	10	39	19	63	37	34	58
13	51	50	16	15	35	16	24	44	59	35	10	39	19	63	37	34	58
14	51	50	16	15	35	16	24	44	59	35	10	39	19	63	37	34	58
15	51	50	16	15	35	16	24	44	59	35	10	39	19	63	37	34	58
16	51	51	11	10	26	11	24	44	59	28	10	37	19	61	37	34	58
17	51	41	15	12	46	13	37	50	55	42	10	32	18	58	37	34	57
18	51	38	36	16	42	19	40	69	55	44	8	33	18	52	25	32	55
19	39	33	29	44	50	23	43	70	70	47	12	41	26	81	32	41	58
20	39	34	11	50	43	22	43	67	63	49	12	39	26	61	19	27	36
21	39	34	11	40	39	22	46	75	70	52	12	36	26	79	19	27	36
22	39	34	11	18	51	17	51	46	31	28	9	16	22	36	12	22	23
23	39	34	10	8	31	8	24	32	12	21	9	9	22	16	12	22	23
Return run																	
1	51	36	51	55	28	17	21	19	49	40	30	30	30	24	23	41	70
2	51	36	51	55	28	17	21	19	49	40	30	30	30	24	23	41	70
3	65	35	38	59	37	23	26	21	40	33	24	24	52	21	20	59	65
4	65	35	27	39	39	22	23	19	53	50	20	20	55	21	20	62	77
5	65	35	27	39	49	23	26	28	57	62	25	23	56	22	20	80	94
6	65	35	27	39	49	23	26	27	56	61	24	22	55	21	19	77	90
7	65	35	27	39	49	23	26	27	56	61	24	22	55	21	20	78	90
8	65	35	27	39	49	23	26	27	56	61	24	22	55	21	20	78	90
9	65	35	27	39	49	23	26	27	56	61	24	22	55	21	20	78	90
10	65	35	27	39	49	23	26	27	56	61	24	22	55	21	20	78	90
11	65	35	27	39	49	23	26	27	56	61	24	22	55	21	20	78	90
12	65	35	27	39	49	23	26	27	56	61	24	22	55	21	20	78	90
13	65	35	27	39	49	23	26	27	56	61	24	22	55	21	20	78	90
14	65	35	27	39	49	23	26	27	56	61	24	22	55	21	20	78	90
15	65	35	27	39	49	23	26	27	56	61	24	22	55	21	20	78	90
16	65	35	27	39	49	23	26	27	56	61	24	22	55	21	20	78	88
17	65	35	27	39	49	23	26	27	56	61	24	22	52	19	17	70	76
18	65	35	27	39	49	23	26	15	36	51	13	13	35	11	11	40	39
19	65	35	18	20	26	15	15	8	20	20	8	8	26	8	8	26	16

hence unreliable in the minds of passengers, since the planned timetable is not being respected. The phenomenon which takes place is so-called *platooning*: trains no longer travel evenly spaced but in bunches. Indeed, the headway between two successive convoys fluctuates so strongly between low and high values that some trains are very close, while others are very far apart. This entails an increase in both the mean and variance of passenger

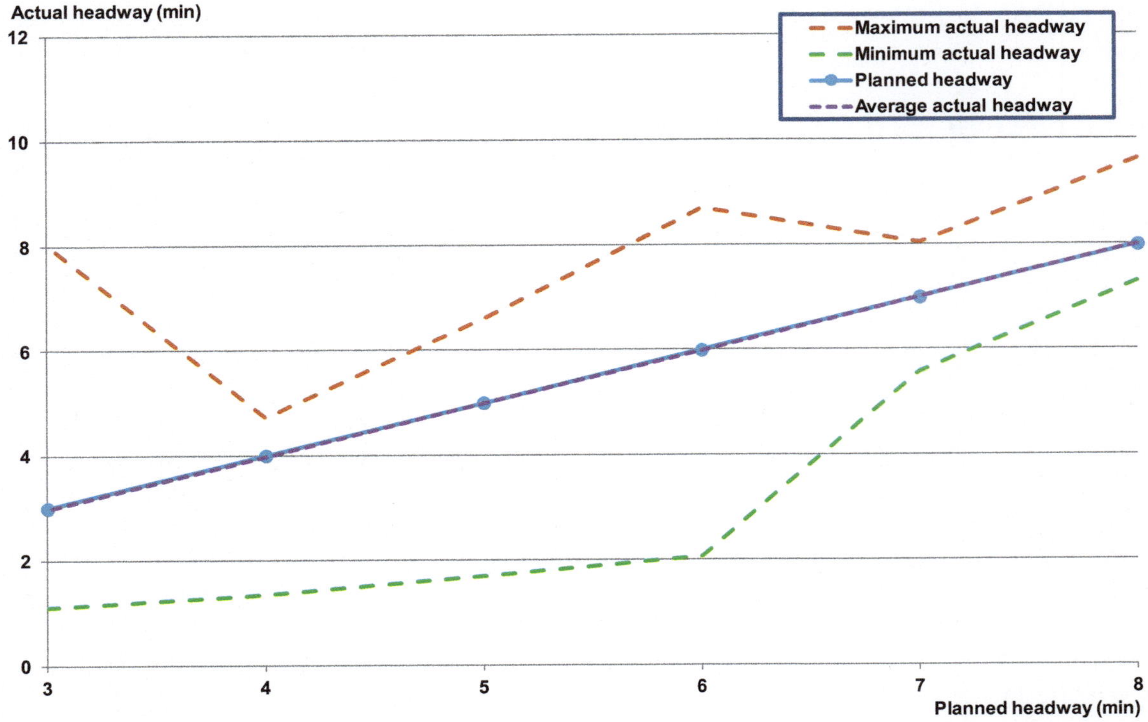

Fig. 7 Fluctuation band of actual headway—50th percentile

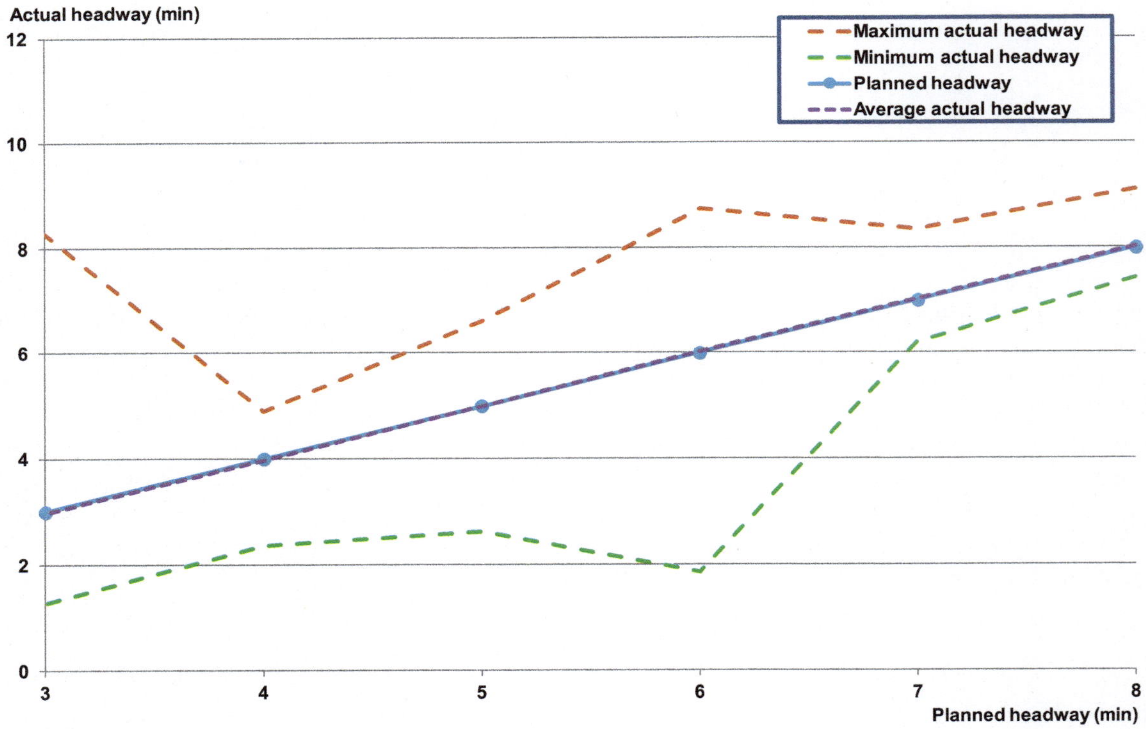

Fig. 8 Fluctuation band of actual headway—85th percentile

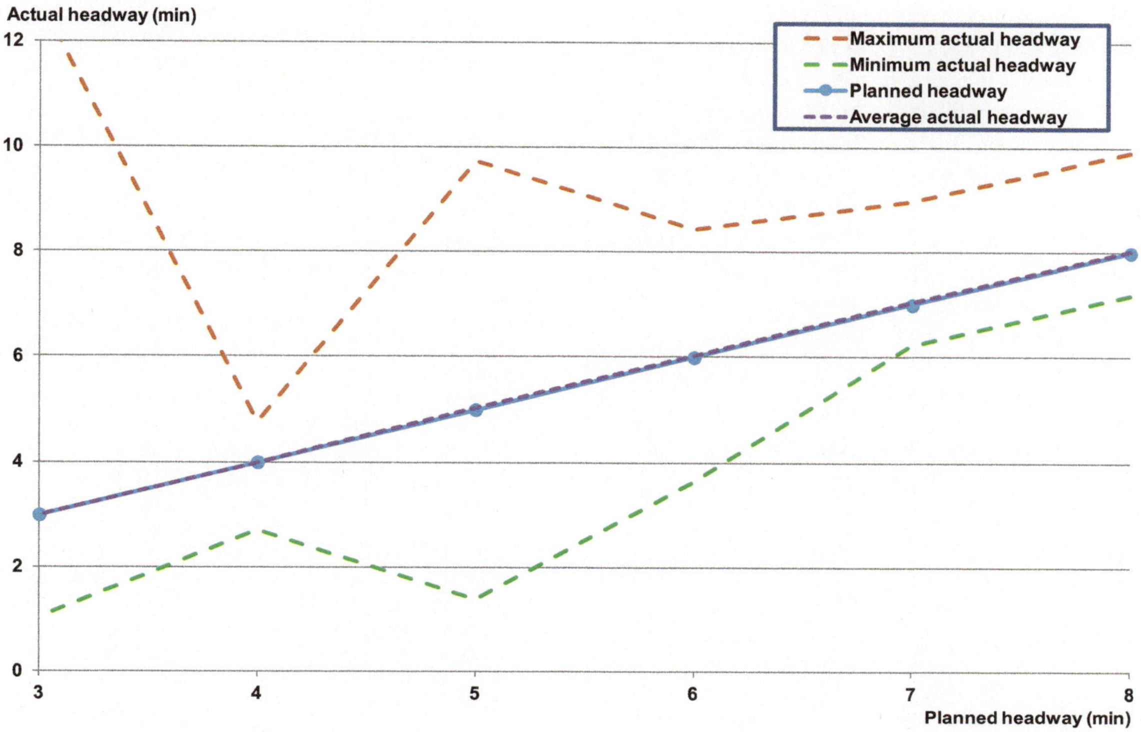

Fig. 9 Fluctuation band of actual headway—95th percentile

waiting times and the presence of overcrowded trains followed by empty ones, with a further decrease in the provided level of service.

This proves the importance of computing dwell times as flow-dependent factors so as to optimise the timetabling design phase and increase system reliability.

Finally, although it is not possible to state that there is a simple dependence between arrival rate in a station and related dwell time at the equilibrium conditions, it is possible to highlight that in some interchange points (i.e. Piscinola, Vanvitelli and Garibaldi which are, respectively, the 1st, the 9th and the 17th station in the outward trip) we obtained great values of dwell times. However, the generally low values in the other transfer stations (i.e. Chiaiano, Medaglie d'Oro and Museo which are, respectively, the 2nd, the 8th and the 13th station in the outward trip) may be related to low interchange flows due to the low quality of bus (Chiaiano and Medaglie d'Oro) and metro (Museo) services. Obviously, similar remarks can be carried out also in the case of the return trip.

4 Conclusions and Research Prospects

The paper addressed the importance of estimating dwell times at rail stations as flow-dependent factors, rather than as fixed values, providing a methodology which is able to capture the threefold interaction between passengers and trains and the associated snowball effect. The effectiveness of the proposed approach was confirmed by a test application on a real metro line.

Providing an accurate estimate of dwell times is a fundamental task to perform in order to design a robust timetable which presents a high degree of flexibility and resilience to propagation of disturbances. It is thus possible to ensure a high-quality service and increase the attractiveness of the system.

Given the significance of dwell times within the timetable planning phase, the suggested approach has considerable potential which is worth investigating in greater depth in forthcoming works. First of all, it would be possible to include the rail crowding theory in the proposed modelling architecture by providing further important information which would enrich the analysis, enabling an estimate of en route passenger discomfort, perhaps even differentiating between passengers standing or sitting. This could contribute to improve the traditional approach proposed in the literature, according to which waiting time is worth three times on-board time.

Moreover, the rail crowding model would also enable overcrowding conditions to be simulated, in which passengers could decide not to take the first arriving train, but wait for the following one (hoping it will be less crowded), although this would entail an increase in their waiting time. In this way, the simulation could become even more realistic. Another important research prospect consists in operationalising the proposed simulation tools for supporting the planning of energy-efficient driving measures so as to increase their effectiveness, or for assisting the design of dedicated ITS systems which could inform passengers about crowding conditions on the train. Furthermore, it would be interesting to strengthen the random component of the procedure by enabling it to estimate the probability distribution of dwell times and not only their average value. Finally, as future developments, we propose to investigate other resolution methods for the fixed-point problem and test the procedure on different network contexts.

Compliance with Ethical Standards

Conflict of interest The authors declare that they have no conflict of interest.

References

1. Niu H, Zhou X, Gao R (2015) Train scheduling for minimizing passenger waiting time with time-dependent demand and skip-stop patterns: nonlinear integer programming models with linear constraints. Transp Res Part B 76:117–135
2. Wirasinghe SC, Szplett D (1984) An investigation of passenger interchange and train scheduling time at LRT stations: (ii) estimation of standing time. J Adv Transp 18(1):13–24
3. Lam WHK, Cheung C-Y, Lam CF (1999) A study of crowding effects at the Hong Kong light rail transit stations. Transp Res Part A 33(5):401–415
4. Levinson HS (1983) Transit travel time performance. Transp Res Rec 915:1–6
5. Guenthner RP, Hamat K (1988) Transit dwell time under complex fare structure. J Transp Eng 114(3):367–379
6. Levine J, Torng G (1994) Dwell-time effects of low-floor bus design. J Transp Eng 120(6):914–929
7. Harris NG, Anderson RJ (2007) An international comparison of urban rail boarding and alighting rates. Proc Inst Mech Eng Part F J Rail Rapid Transit 221(4):521–526
8. Vuchic VR (2005) Urban transit operations, planning and economics. Wiley, Hoboken
9. Kecman P, Goverde RMP (2015) Predictive modelling of running and dwell times in railway traffic. Public Transp 7(3):295–319
10. Hansen IA, Goverde RMP, van der Meer DJ (2010) Online train delay recognition and running time prediction. In: Proceedings of 13th international IEEE annual conference on intelligent transportation systems (IEEE-ITSC 2010), Funchal, Portugal, September 2010, pp 1783–1788
11. Lam WHK, Cheung CY, Poon YF (1998) A study of train dwelling time at the Hong Kong mass transit railway system. J Adv Transp 32(3):285–295
12. Tirachini A, Hensher DA, Rose JM (2013) Crowding in public transport systems: effects on users, operation and implications for the estimation of demand. Transp Res Part A 53:36–52
13. Jiang Z, Xie C, Ji T, Zou X (2015) Dwell time modelling and optimized simulations for crowded rail transit lines based on train capacity. Promet 27(2):125–135
14. Jiang Z, Li F, Xu R, Gao P (2012) A simulation model for estimating train and passenger delays in large-scale rail transit networks. J Cent South Univ 19:3603–3613
15. Zhang Q, Han B, Li D (2008) Modeling and simulation of passenger alighting and boarding movement in Beijing metro stations. Transp Res Part C 16(5):635–649
16. Yamamura A, Koresawa M, Aadchi S, Inagi T, Tomii N (2013) Dwell time analysis in railway lines using multi agent simulation. In: Proceedings of 13th world conference on transportation research (WCTR 2013), Rio de Janeiro, Brazil, July 2013
17. Bandini S, Mondini M, Vizzari G (2014) Modelling negative interaction among pedestrians in high density situations. Transp Res Part C 40:251–270
18. Seriani S, Fernandez R (2015) Pedestrian traffic management of boarding and alighting in metro stations. Transp Res Part C 53:76–92
19. Larsen R, Pranzo M, D'Ariano A, Corman F, Pacciarelli D (2014) Susceptibility of optimal train schedules to stochastic disturbances of process times. Flex Serv Manuf J 26(4):466–489
20. Longo G, Medeossi G (2013) An approach for calibrating and validating the simulation of complex rail networks. In: Proceedings of transportation research board 92nd annual meeting, Washington DC, USA, January 2013, pp 1–19
21. Berbey A, Galan R, Sanz Bobi JD, Caballero R (2012) A fuzzy logic approach to modelling the passengers' flow and dwelling time. WIT Trans Built Environ 128:359–369
22. Chu W-J, Zhang X-C, Chen J-H, Xu B (2015) An ELM-based approach for estimating train dwell time in urban rail traffic. Math Probl Eng 2015:1–9
23. Huang G-B, Zhu Q-Y, Siew C-K (2006) Extreme learning machine: theory and applications. Neurocomputing 70:489–501
24. Weston JG (1989) Train service model—technical guide. London Underground operational research note 89/18
25. Jong J, Chang E (2011) Investigation and estimation of train dwell time for timetable planning. In: Proceedings of 9th world congress on railway research, Lille, France, May 2011
26. Buchmuller S, Weidmann U, Nash A (2008) Development of a dwell time calculation model for timetable planning. WIT Trans Built Environ 103:525–534
27. Wiggenraad PBL (2001) Alighting and boarding times of passengers at dutch railway stations analysis of data collected at 7 stations in October 2000. Dissertation, TRAIL Research School, Delft University of Technology, Delft, The Netherlands
28. Harris NG (2006) Train boarding and alighting rates at high passenger loads. J Adv Transp 40(3):249–263
29. Baee S, Eshghi F, Hashemi SM, Moienfar R (2012) Passenger boarding/alighting management in urban rail transportation. In: Proceedings of the joint rail conference (JRC'12), Philadelphia, USA, April 2012, pp 823–829
30. Fletcher G, El-Geneidy A (2013) The effects of fare payment types and crowding on dwell time: a fine-grained analysis. Transp Res Rec 2351:124–132
31. Daamen W, Lee Y, Wiggenraad P (2008) Boarding and alighting experiments: overview of setup and performance and some preliminary results. Transp Res Rec 2042(1):71–81
32. Puong A (2000) Dwell time model and analysis for the MBTA red line. Research Memo, Massachusetts Institute of Technology, Cambridge (MA), USA
33. Xu X, Liu J, Li H, Zhou Y (2013) Probabilistic model for remain passenger queues at subway station platform. J Cent South Univ 20:837–844
34. Kunimatsu T, Hirai C, Tomii N (2012) Train timetable evaluation from the viewpoint of passengers by microsimulation of train operation and passenger flow. Electr Eng Jpn 181(4):51–62
35. Li D, Goverde RMP, Daamen W, He H (2014) Train dwell time distributions at short stop stations. In: Proceedings of 17th international IEEE conference on intelligent transportation systems (IEEE-ITCS 2014), Qingdao, China, October 2014, pp 2410–2415

36. Hadas Y, Ceder A (2010) Optimal coordination of public-transit vehicles using operational tactics examined by simulation. Transp Res Part C 18:879–895
37. Carey M, Carville S (2000) Testing schedule performance and reliability for train stations. J Oper Res Soc 51:666–682
38. Heimburger DE, Herzenberg AY, Wilson NHM (1999) Using simple simulation models in operational analysis of rail transit lines: case study of Boston's red line. Transp Res Rec 1667:21–30
39. Dewilde T, Sels P, Cattrysse D, Vansteenwegen P (2014) Improving the robustness in railway station areas. Eur J Oper Res 235:276–286
40. Cui Y, Martin U, Zhao W (2016) Calibration of disturbance parameters in railway operational simulation based on reinforcement learning. J Rail Transp Plan Manag 6:1–12
41. Ceder A, Hassold S (2015) Applied analysis for improving rail-network operations. J Rail Transp Plan Manag 5:50–63
42. Burdett RL, Kozan E (2014) Performance profiling for predictive train schedules. J Rail Transp Plan Manag 4:98–114
43. Büker T, Seybold B (2012) Stochastic modelling of delay propagation in large networks. J Rail Transp Plan Manag 2:34–50
44. Kanai S, Shiina K, Harada S, Tomii N (2011) An optimal delay management algorithm from passengers' viewpoints considering the whole railway network. J Rail Transp Plan Manag 1:25–37
45. Pouryousef H, Lautala P (2015) Hybrid simulation approach for improving railway capacity and train schedules. J Rail Transp Plan Manag 5:211–224
46. Sato K, Tamura K, Tomii N (2013) A MIP-based timetable rescheduling formulation and algorithm minimizing further inconvenience to passengers. J Rail Transp Plan Manag 3:38–53
47. Wong RCW, Yuen TWY, Fubg KW, Leung JMY (2008) Optimizing timetable synchronization for rail mass transit. Transp Sci 42(1):57–69
48. Landex A, Jensen LW (2013) Measures for track complexity and robustness of operation at stations. J Rail Transp Plan Manag 3:22–35
49. Kepaptsoglou K, Karlaftis MG (2010) A model for analyzing metro station platform conditions following a service disruption. In: Proceedings of 13th international IEEE annual conference on intelligent transportation systems (IEEE-ITSC 2010), Funchal, Portugal, September 2010, pp 1789–1794
50. Li D, Daamen W, Goverde RMP (2016) Estimation of train dwell time at short stops based on track occupation event data: a study at a Dutch railway station. J Adv Transp 50:877–896
51. Xiaoming X, Keping L, Xiang L (2016) A multi-objective subway timetable optimization approach with minimum passenger time and energy consumption. J Adv Transp 50:69–95
52. Nash A, Huerlimann D (2004) Railroad simulation using OpenTrack. Comput Railw 9:45–54
53. D'Acierno L, Gallo M, Montella B, Placido A (2013) The definition of a model framework for managing rail systems in the case of breakdowns. In: Proceedings of 16th international IEEE annual conference on intelligent transportation systems (IEEE-ITSC 2013), The Hague, The Netherlands, October 2013, pp 1059–1064
54. D'Acierno L, Placido A, Botte M, Montella B (2016) A methodological approach for managing rail disruptions with different perspectives. Int J Math Models Methods Appl Sci 10:80–86
55. Domencich TA, McFadden D (1975) Urban travel demand: a behavioural analysis. American Elsevier, New York
56. Placido A, D'Acierno L, Botte M, Montella B (2015) Effects of stochasticity on recovery solutions in the case of high-density rail/metro networks. In: Proceedings of 6th international conference on railway operations modelling and analysis, Tokyo, Japan, March 2015
57. Brouwer LEJ (1912) Uber Abbildung der Mannigfaltigkeiten. Math Ann 71:97–115
58. Banach S (1922) Sur les opérations dans les ensembles abstraits et leur application aux équations intégrales. Fund Math 3:133–181
59. Cantarella GE (1997) A general fixed-point approach to multimodal multi-user equilibrium assignment with elastic demand. Transp Sci 31:107–128
60. Sheffi Y, Powell WB (1981) A comparison of stochastic and deterministic traffic assignment over congested networks. Transp Res Part B 15:53–64
61. Blum JR (1954) Multidimensional stochastic approximation methods. Ann Math Stat 25:737–744
62. Placido A (2015) The definition of a model framework for the planning and the management phases of the rail system in any kind of service condition. Ph.D. dissertation, Federico II University of Naples, Naples, Italy
63. Ercolani M, Placido A, D'Acierno L, Montella B (2014) The use of micro-simulation models for the planning and management of metro systems. WIT Trans Built Environ 135:509–521
64. D'Acierno L, Gallo M, Montella B, Placido A (2012) Analysis of the interaction between travel demand and rail capacity constraints. WIT Trans Built Environ 128:197–207
65. Douglas N (2012) Modelling train and passenger capacity. Report to transport for NSW by DOUGLAS economics, Sydney, Australia

18

Dynamic Responses of Low–Medium Speed Maglev Train–Simply Supported Beam Interaction System

Xiaozhen Li[1] · Jie Geng[1] · Dangxiong Wang[1]

Abstract A train–bridge coupling vibration model for low-speed maglev traffic was established to study the dynamic response of low-medium speed Maglev Train and simply supported beam interation system. In this model, the vehicle is assumed as a secondary suspension mass-spring-damper model with 12 degrees of freedom. The finite element model is supplied to simulate the bridge structure. The influence of the track irregularity is also taken into account. Then, the program for vertical coupling vibration simulation of low-speed maglev traffic is developed. Finally, a field test was carried out on a 25-m simply supported beam in a low-speed maglev test line. The results show that the fundamental frequency, the vibration mode, the mid-span dynamic deflection and the acceleration of the track beam measured values are slightly smaller than the simulation results. When the maglev train passes the simply supported beam, the measured acceleration values of the vehicle are slightly less than the simulation ones. These research results show that the vertical coupling vibration model of low-speed maglev train–track beam is reasonable and the simulation results are credible. With the increase of speed, the accelerations of the bridge and car body are increased in the overall trend. But, the acclerations of the bridge and the maglev car body are both small versus the speed, wich indicates that the dynamic performanceof the bridge and maglev train are good.

✉ Dangxiong Wang
Dangxiong333@163.com

[1] Department of Bridge Engineering, School of Civil Engineering, Southwest Jiaotong University, Chengdu 610031, China

Editor: Eryu Zhu

Keywords Low-speed maglev · PD control · Coupled vibration · Track beam · Dynamic responses

1 Introduction

Since the twenty-first century, China's urban rail transit has experienced rapid development. Low-speed maglev transportation has become one of the most competitive green ground transportation systems by the virtue of low noise, excellent climbing ability, small turning radius and many other advantages. The maglev system is different from the wheel–rail system, and its coupling effect is more complex. Therefore, it is necessary to study the coupling vibration of the low-speed maglev train–track beam system.

At present, the coupling vibration research of the track beam of maglev train at home and abroad is mostly conducted for high-speed maglev, such as Yang [1] used moving uniform mass model to analyze the vibration of track beam; Zeng [2] analyzed coupling vibration of vehicle–track beam by simplifying the maglev vehicle as a 7 rigid body vibration system connected by a spring damping element; Teng [3], based on the prototype of the German maglev vehicle, established the coupling vibration analysis model of high-speed maglev vehicle–track beam which includes 10 DOF vehicle model, electromagnet model, track beam model and suspension control model.

This paper, based on the existing research results, introduced of a feedback mechanism with the active control of the suspension controller and used finite element method to build the bridge structure model. The maglev vehicle is simplified as a vertical model with 12 degrees of freedom and secondary suspension. Track irregularity is also considered. The vertical coupling vibration analysis model of the maglev car bridge is established.

The dynamic load test was carried out on the 25-m simply supported beam of a low-speed maglev test line. The contrast between the simulated values and the measured values of the field test shows that this model is reasonable.

2 Theoretical Model

In this paper, the train–bridge dynamic simulation analysis model of low-speed maglev vehicle system is composed of maglev vehicle model, bridge model, electromagnet system model and suspension controller model, with consideration of the track irregularity, according to a certain track relationship (as shown in Fig. 1). The numerical simulation analysis method is used to solve the dynamic response of maglev train–bridge system.

2.1 Vertical Dynamic Model of Maglev Vehicle Model

The vertical model of the maglev vehicle consists of six rigid bodies, includinga car body and five suspension frames. Each rigid body vertically considers two degrees of freedom of nods, ups and downs. The vertical dynamic model of maglev vehicle is shown in Figs. 2 and 3.

In Fig. 2, M_{ci}, J_{ci}, y_{ci} and α_i denote the mass, moment of inertia, displacement and nod angle at the center of gravity (or geometrical center) of the ith maglev vehicle. f_{kim} and f_{sim}, respectively, represent the maglev vehicle department of suspension point of the spring force and damping force:

$$f_{\text{kim}} = K_p\left(y_{\text{im}}^T - y_{\text{im}}^B\right) \tag{1}$$

$$f_{\text{sim}} = C_p\left(\dot{y}_{\text{im}}^T - \dot{y}_{\text{im}}^B\right) \tag{2}$$

where K_p and C_p represent the two-line suspension stiffness and damping, and y_{im}^T and y_{im}^B represent the displacement of the top and bottom points of the vehicle body to the corresponding secondary suspension points.

Using the d'Alembert principle, the vertical dynamic equation of a maglev vehicle system can be written.

Considering that the maglev vehicle is composed of three carriages, the whole vehicle system has 36 degrees of

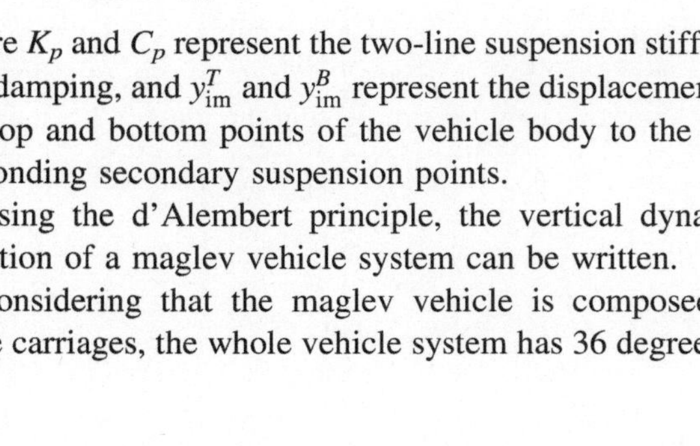

Fig. 1 Dynamic simulation analysis model

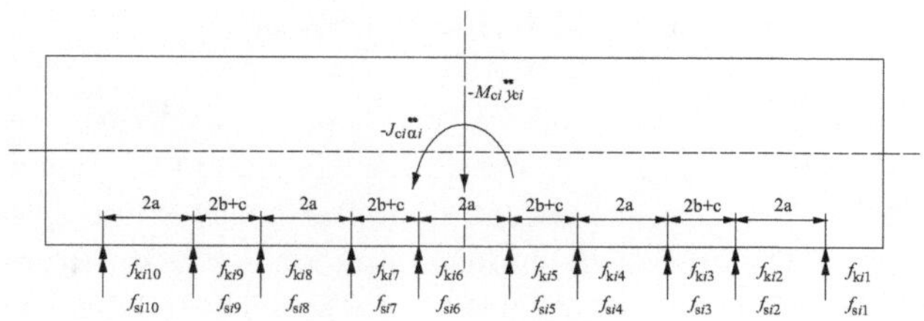

Fig. 2 Vertical model of vehicle

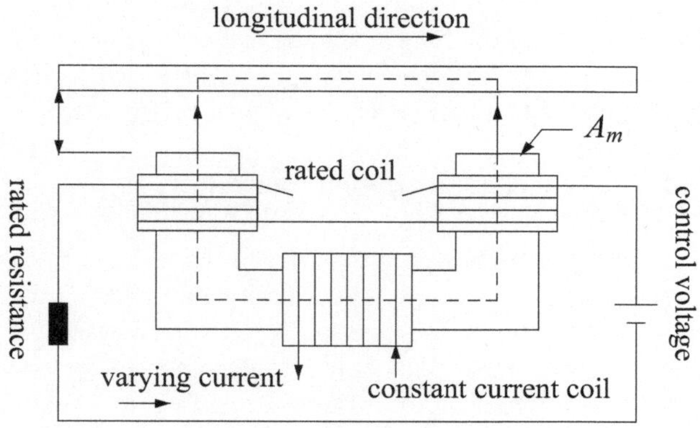

Fig. 3 Electromagnet system model

freedom in the vertical direction. In the simulation analysis, the numerical method is used for solving.

2.2 The Model of Bridge

When finite element method is used to analyze the dynamic behavior of the bridge structure, two parts are analyzed including the characteristics of bridge vibration and dynamic response. The dynamic response analysis is mainly used to solve the structural response (such as displacement, velocity and acceleration) with the time course of change.

The finite element method is used to analyze the dynamic behavior of the bridge structure. The derivation of the motion equation of the bridge structure is shown as follows:

$$M\ddot{X} + C\dot{X} + KX = F \tag{3}$$

where M, C, and K are the mass matrix, the damping matrix and the stiffness matrix for the global coordinate system, respectively. X is the total column vector of the system displacement, and F is the external load column vector.

The vibration superposition method and step-by-step integration method [4] are two methods which are used to solve system motion equation. In this paper, the Newmark-β integration method is used for solving.

2.3 The Model of Electromagnet System

The electromagnet model is based on the EMS-type SC magnets in Ref. [5], including guideway iron rail, iron-core magnet, SC coil and normal coil. The model is shown in Fig. 3. The current I_{SC} in the constant current coil is provided by a constant current source in which the total weight of the maglev vehicle is balanced by the electromagnetic force generated by the constant current.

The time-varying current i in the rated coil, which is driven by a control voltage source, is used to maintain the suspension gap h in the vicinity of its rated value. Electromagnet system parameters are shown in Table 1.

2.4 The Model of Suspension Controller

Considering that maglev system itself is an unstable system [2, 6, 7], to achieve suspension stability, the introduction of active control of the suspension controller is necessary. The suspension controller function is to generate the appropriate time-varying control voltage according to a certain algorithm generated by the feedback signal of the sensor installed at the bottom of the suspension. Then, the time-varying current is generated to adjust the maglev vehicle and maintain the suspension gap in the rated suspension gap.

Considering all factors, the PD control method, which is commonly used in industrial control, is induced in this project. The model is shown in Fig. 4. h_O is the rated floating clearance; k_p and k_d are the two key control parameters of the suspension controller, respectively; u denotes the time-varying control voltage.

3 The Equations of Vertical Coupled Vibration

The vertical motion equations the vehicle and bridge structure can be written as the following matrix form.

$$M_{v+b}\ddot{u}_{v+b} + C_{v+b}\dot{u}_{v+b} + K_{v+b}u_{v+b} = P_{v+b} \quad (4)$$

Among them, M_{v+b}, C_{v+b}, K_{v+b} denote the mass, damping and stiffness matrices of the system, respectively. u_{v+b} represents the displacement column vector of the system, and P_{v+b} represents a load column vector.

Table 1 Electromagnetic system parameters

Parameter	Unit	Value
Constant current coil turns	—	1020
Control coil turns	—	96
Magnetic pole area	m^2	0.04
Control coil resistance	Ω	1
Air permeability	w/Am^{-1}	$4\pi E^{-7}$

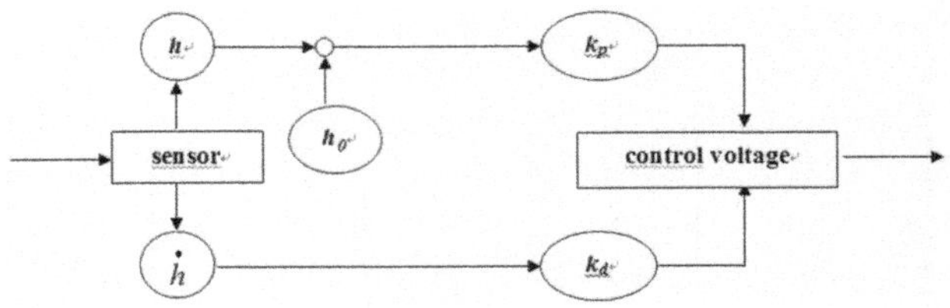

Fig. 4 Suspension controller model

Formula 4 form a large-scale complex time-varying nonlinear dynamic system. To solve such a dynamic system, only the time of numerical integration method can be used. The Newmark-β method is used to solve the dynamic responses of the maglev vehicle and the bridge structure. In the calculation, the time step is 0.001 s.

The main idea is to calculate the suspension clearance h and the change rate of floating gap at the current time based on the previous maglev vehicle and bridge response. It can be substituted into the suspension controller equation for the current control voltage u and solved for the time-varying current i according to the solenoid system loop equation. The electromagnetic force f at the current moment can be calculated. Finally, the current maglev vehicle and bridge response can be obtained by solving the vibration equation. Repeating the above process, the entire maglev vehicle bridge vibration response of the whole process can be solved.

4 Field Test

In order to verify the rationality of the model and the reliability of the simulated program, a field test of a 25-m simply supported beam on a maglev test line was carried out.

Figure 5 shows the mid-span section of the concrete simply supported track beam. Two sides of the beam are, respectively, single track beam. The bridge span is 24.94 m. The height is 2.1 m, and the width is 1.3 m. The material of the beam is C55 concrete, and the second phase of the dead load is 10 kN/m. The design speed is 100 km/h.

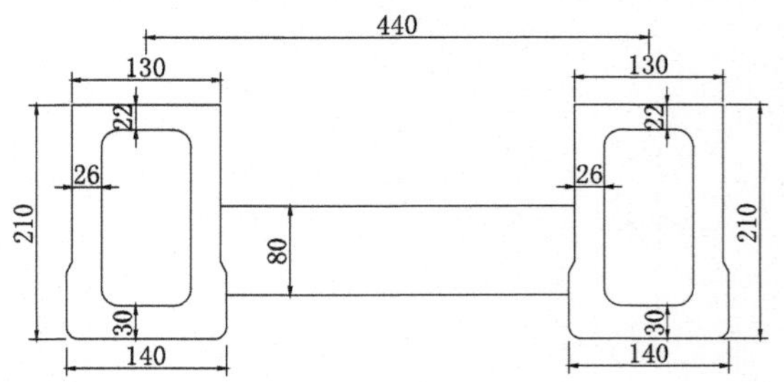

Fig. 5 Mid-span section of the track beam (unit: cm)

Round-shaped pier is used with height of 11.25 m. Figure 6 shows the field test photos.

During the test, a typical span was selected according to the site conditions. The vertical displacement, vibration and acceleration sensors were arranged at the mid-span to measure the dynamic displacement and acceleration of the bridge.

Loading conditions include car body loads of 24, 30 and 33 t. The speed rating is 50–100 km/h. Due to the random response of the dynamic factors and in order to ensure the test accuracy, each test condition includes no less than 6 groups.

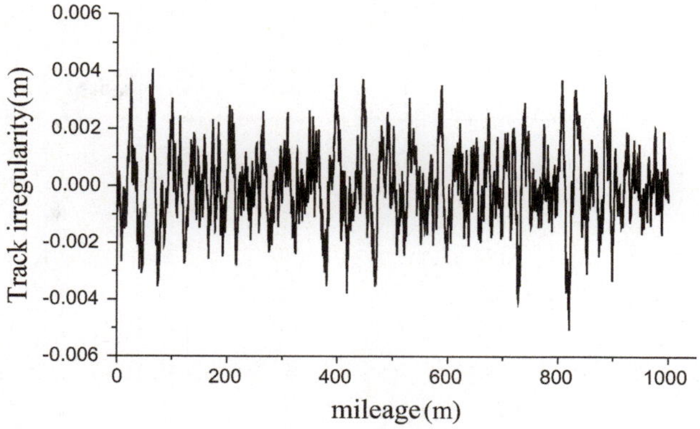

Fig. 7 Track irregularity

5 Comparison and Analysis of Simulation and Measurement

Using the program developed in this paper, the vehicle–bridge coupling vibration of the 25-m simply supported beam on the maglev test line is analyzed and compared with the measured values in the field test. The effect of track irregularities is also considered. Track irregularity is shown in Fig. 7.

The maximum displacement and acceleration of the bridge structure are obtained by peak analysis of the recorded acceleration and displacement time history curves. The measured value of the vertical base frequency of the bridge is 6.768 Hz, slightly larger than the calculated value 6.030 Hz.

The measured results show that the measured base frequency of the bridge is higher than the simulation value, due to the impact of the track structure. The measured simulation mode is the same as the simulation mode.

When the test maglev vehicles pass the 25-m simply supported beam in the speed of 50–100 km/h, the contrast of the beam in the dynamic deflection and vertical acceleration between the simulation value and the measured values is shown in Tables 2 and 3. The contrast of vertical acceleration of the maglev train body between the simulation values and the measured values is shown in Table 4.

It can be seen from Tables 2, 3, 4, when maglev vehicles pass through the 25-m simply supported beam:

1. Through the theoretical simulation, it could be seen that values of the mid-span deflection and acceleration of the track beam vary with the speed of the vehicle.
2. The theoretical simulation value of the mid-span deflection and acceleration of the track beam is at the same order of magnitude of the test value however it is slightly bigger than the measure value. The change rule is consistent with the measured value.
3. The measured values of the vertical acceleration of the maglev vehicle are close to the theoretical simulation values, and the measured values are basically smaller than the theoretical simulation values.

Fig. 6 Photographs of the field test

Table 2 Contrast of maximum deflection of bridge between measured and predicted values (Unit: mm)

Condition		33 t	30 t	24 t
50 km/h	Predicted	2.270	2.049	1.640
	Measured	1.587	1.608	1.428
60 km/h	Predicted	2.238	2.053	1.636
	Measured	1.620	1.642	1.768
70 km/h	Predicted	2.296	2.047	1.640
	Measured	1.885	1.612	1.443
80 km/h	Predicted	2.376	2.074	1.670
	Measured	1.732	1.629	1.482
90 km/h	Predicted	2.419	2.103	1.681
	Measured	1.730	1.651	1.457
100 km/h	Predicted	2.300	2.101	1.701
	Measured	1.726	1.648	1.443

Table 3 Contrast of maximum acceleration of bridge between measured and predicted values (Unit: m/s^2)

Condition		33 t	30 t	24 t
50 km/h	Predicted	0.589	0.537	0.433
	Measured	0.351	0.31	0.268
60 km/h	Predicted	0.624	0.562	0.435
	Measured	0.464	0.351	0.284
70 km/h	Predicted	0.614	0.547	0.448
	Measured	0.379	0.37	0.456
80 km/h	Predicted	0.596	0.547	0.447
	Measured	0.54	0.365	0.383
90 km/h	Predicted	0.607	0.556	0.456
	Measured	0.568	0.378	0.328
100 km/h	Predicted	0.603	0.55	0.445
	Measured	0.584	0.342	0.368

Table 4 Contrast of maximum acceleration of vehicle between measured and predicted values (Unit: m/s^2)

Condition		33 t	30 t	24 t
50 km/h	Measured	0.351	0.310	0.268
	Predicted	0.589	0.537	0.433
60 km/h	Measured	0.464	0.351	0.284
	Predicted	0.624	0.562	0.435
70 km/h	Measured	0.379	0.370	0.456
	Predicted	0.614	0.547	0.448
80 km/h	Measured	0.540	0.365	0.383
	Predicted	0.596	0.547	0.447
90 km/h	Measured	0.568	0.378	0.328
	Predicted	0.607	0.556	0.456
100 km/h	Measured	0.584	0.342	0.368
	Predicted	0.603	0.550	0.445

4. The heavier the maglev train, the greater deflection of the bridge and the acclerations of the bridge and car body are. The regular of simulation values between the weight of the cay body and the dynamic responses is the same as that of the measured.

6 Conclusions

In this paper, based on the existing research results, the vertical vibration analysis model of the maglev car bridge is established including a feedback mechanism with active control of the suspension controller, the finite element method to build the bridge structure model, the maglev vehicle for 12 degrees of freedom of the two-line vertical model and the track irregularities. Relevant program is worked out based on this. A field test is carried out.

The result of field test shows that:

1. The measured base frequency of the bridge span structure is slightly higher than the simulation calculation frequency, but vibration mode is the same.
2. The measured values of the vertical acceleration in the beam are close to and slightly smaller than the simulation values. The theoretical simulation value of vehicle acceleration is consistent with the measured law. The variation trend is the same. The analysis result of the compiled software is credible.
3. With the increase of speed, the accelerations of the bridge and car body increases in the overall trend. With the increase of the maglev weight, the accelerations of the bridge and car body increases.
4. The acclerations of the bridge and the maglev car body are both small versus the speed and the maglev train weight, wich indicates that the dynamic performance of the bridge and maglev train are good.

Acknowledgements This work was supported by Chinese National Natural Science Foundation (51378429, U1434205) and the Key Project of Chinese National Programs for Fundamental Research and Development (973 Program) (2013CB036301).

References

1. Yang Y, Teng NG, Huang XC, Teng YF (2005) Vibration analysis of a simply supported beam traversed by uniform distributed moving mass. J Vib Shock 24(3):19–22
2. Zeng YW, Wang SH, Zhang KL (1999) A study of vertical coupling dynamics of EMS maglev train and guideway systems. J China Railw Soc 4(2):21–26
3. Teng YF (2008) The study of vibration of high speed maglev guideway interacting with vehicle. Shanghai Jiao Tong University Doctor Degree Dissertation, pp 12–13
4. Yuan XB, Zhao X, Fang DH (2010) A new step-by-step integration method based on 3-order hermite interpolation by double-parameter for dynamic response. Eng Mech 10:42–46
5. Wang SK (1995) Levitation and guidance of a maglev vehicle using optimal preview control. Carnegie Mellon University, Pittsburgh, pp 20–25

Recent Trend of Regenerative Energy Utilization in Traction Power Supply System in Japan

Hitoshi Hayashiya[1]

Abstract In 2006, the first Lithium-ion battery in Japan was installed in traction power supply system by the West Japan Railway Company and now more than 20 energy storage systems have already been installed in traction power supply system in Japan. In this article, the recent Japanese trends of regenerative energy utilization are summarized not only in DC railway but also in AC railway. We, East Japan Railway Company, have installed three energy storage systems for DC railway and one railway static power conditioner (RPC), not energy storage system, for AC railway for regenerative energy utilization. The effects of these systems are also shown quantitatively in this paper.

Keywords Regenerative energy · Electric railway · DC railway · AC traction power supply · Energy storage system · Railway static power converter · Regenerative inverter

1 Introduction

Railway transportation is recognized as one of the most eco-friendly transportation modes. For example, CO_2 emissions per passenger per kilometer by mode of transportation in Japan are shown in Fig. 1 which was reported by the Japanese Ministry of Land, Infrastructure, Transport and Tourism. As shown in this figure, CO_2 emission from railway transportation in carrying one passenger was one-fifth of the amount emitted from airplane and one-seventh of the amount emitted from car in 2013 [1]. The total annual electric energy consumption in railway transportation was, however, 17.7 TW in 2013, and it corresponded to about 1.9% of total electric energy supply in Japan [2]. It was not a small value, and a lot of railway companies are making efforts to reduce their energy consumptions as small as possible to contribute to solving global warming problem. Such efforts are also important to accomplish our national target to reduce CO_2 emission about 26% of 2013 by 2030 which is mentioned in "Submission of Japan's Intended Nationally Determined Contribution" in UNFCCC COP21 (the United Nations Framework Convention on Climate Change, the Conference of the Parties), Paris [3].

Figure 2 shows the changes in energy consumption in East Japan Railway Company. As shown in this figure, the energy consumption in 2014 was about 10% smaller than that in 1990. Reduction of energy consumption is eminent in traction power supply for conventional railway, more than 20% of that in 1990, by virtue of introduction of energy-saving trains with light bodies, regenerative brakes and VVVF inverters. For station power supply, on the other hand, introductions of LED lightings and high-efficiency air conditioners contributed to the reduction of energy consumption at railway station. For example, the office building of Tokyo Branch Office of East Japan Railway Company accomplished 33% reduction in electric energy consumption compared with that in 2010, before the huge earthquake in 2011 and nuclear crisis after that, as shown in Fig. 3 [4]. Regarding renewable energy utilization, there are a lot of examples of rooftop photovoltaic system installations at railway stations in Japan [5]. Among these solutions to improve energy efficiency in railway system,

✉ Hitoshi Hayashiya
hayashiya@jreast.co.jp

[1] Tokyo Branch, East Japan Railway Company, 2-20-68, Higashi-Tabata, Kita-Ku, Tokyo 114-8550, Japan

Editor: Dr. Xuesong Zhou

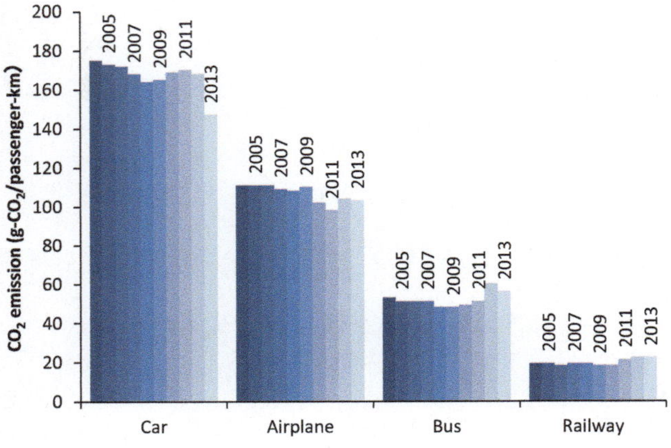

Fig. 1 CO_2 emissions per passenger per kilometer

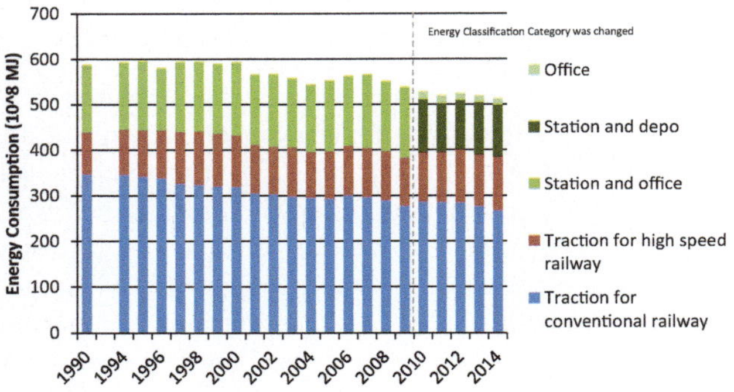

Fig. 2 Change in energy consumption in East Japan Railway Company (JR East)

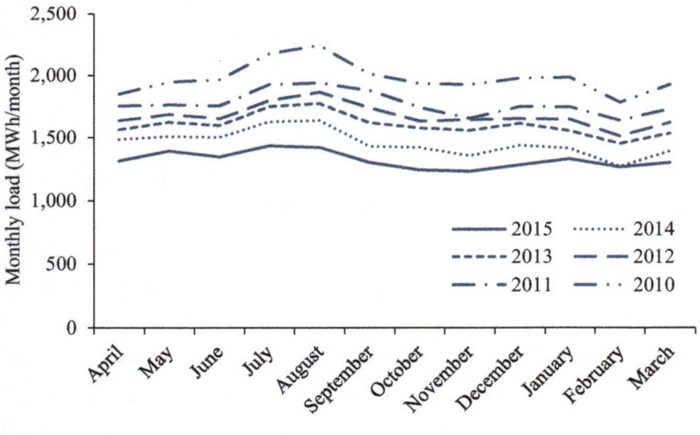

Fig. 3 Changes in electric energy consumption at Tokyo Branch Office building of JR East

the enhancement of regenerative energy utilization is one of the most actively discussed technical issues, and the recent Japanese trends in this topic will be summarized in this article.

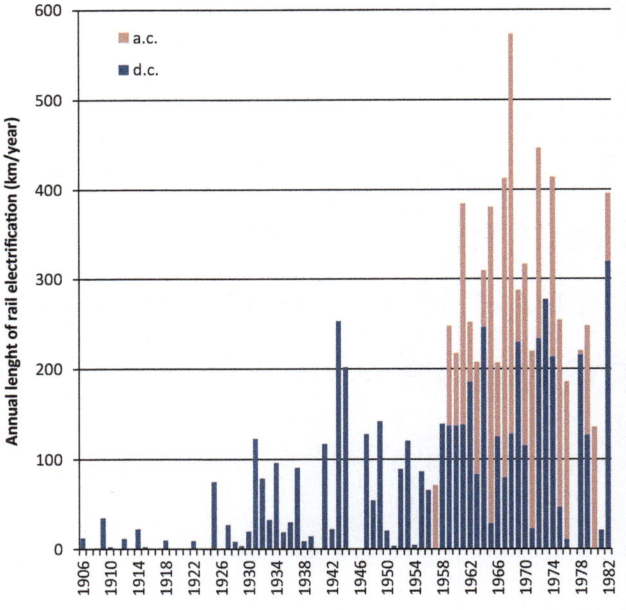

Fig. 4 Annual length of rail electrification in Japan National Railway

2 Development of Electric Railway in Japan

Figure 4 shows the history of rail electrification in former Japan National Railway. The first demonstration of electric railway was performed by Dr. Ichisuke Fujioka at the third domestic industrial exposition at Ueno in Tokyo in 1890. Five years later, the first electric tram started business operation in Kyoto and the first electrification of full-scale railway was realized in Tokyo in 1904 which was merged with Japan National Railway in 1906. As shown in this figure, rail electrification was strongly promoted during the period of rapid economic growth around 1960s. As a result, the electric railway network all over Japan was formed as shown in Fig. 5. In this figure, only the mass rail transit lines operated by JR group, former Japan National Railway, are described.

3 Regenerative Energy Utilization

In DC traction power supply system around large cities, regenerative energy utilization is actively discussed during the last decade to improve the energy efficiency of DC electric railway in Japan. To make use of the residual regenerative power efficiently, there have been some solutions conventionally. Figure 6 summarizes conventional possible solutions mainly for avoiding the cancelation of regeneration which have already been realized in Japan [6, 7].

As shown in Fig. 6a, regenerative resistance is one of the most effective solutions but the residual power is not utilized in this solution. Tie feeding shown in Fig. 6b is a low-cost solution, and it was reported that it contributed to realize about 2% of energy saving by some railway companies. If there are enough electric loads on the 6 kV distribution lines to feed residual regenerative power to stations and signaling systems along the railroad trackside, regenerative inverter shown in Fig. 6c will be able to become one of the most effective solutions for regenerative power utilization. In a DC traction power supply system for subway, it is the most feasible countermeasure against regenerative brake cancelation because the load of railway station for subway is large enough for simultaneous utilization of residual regenerative power. PWM converter shown in Fig. 6d is almost the same as regenerative inverter, but the regenerative power is reversed from DC power system for railway to the utility grid directly.

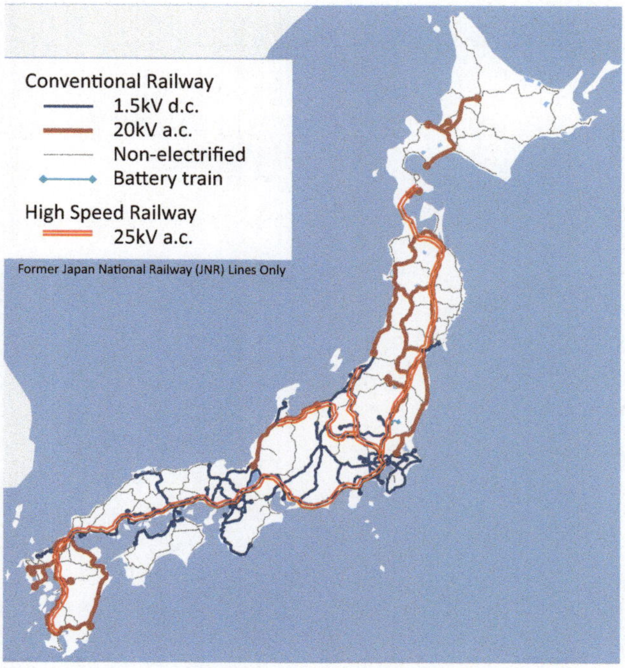

Fig. 5 Electrification of Japanese railway network

Compared with these existing solutions, the installation of energy storage system can be effective and is independent to other electric loads and operation schedules of trains. Furthermore, it is getting feasible more and more because of the cost reduction of energy storage system, such as lithium-ion battery, in Japan. Figure 7 shows the simplified concept of the function of energy storage system installation into DC traction power supply system. As shown in Fig. 7a, if there is another powering train adjacent to the braking train with regenerative brakes, the regenerative energy from braking train is consumed by the other powering train simultaneously. Because of the voltage drop of the catenaries system, on the other hand, it is impossible to transfer the regenerative energy for a long distance, and the regenerative energy is canceled if there is no powering train near the braking train as shown in Fig. 7b. To avoid such a regenerative brake cancelation, the energy storage system is installed along the railroad trackside and the residual regenerative power is charged to energy storage system during the braking period and will be utilized for the powering later.

Figure 8 shows the existing energy storage systems in traction power supply system, 1.5 kV or 750 V, in Japan. As shown in this figure, Li-ion battery is most popular for the energy storage system in traction power supply system because of its cost-effectiveness. As other storage mediums, nickel–metal hydride batteries and electric double-layer capacitors, called "Super-capacitor" in China and Europe, are selected in some railway companies. Generally, the required capacity to charge all regenerative energy for one braking opportunity is more than 10 kWh in

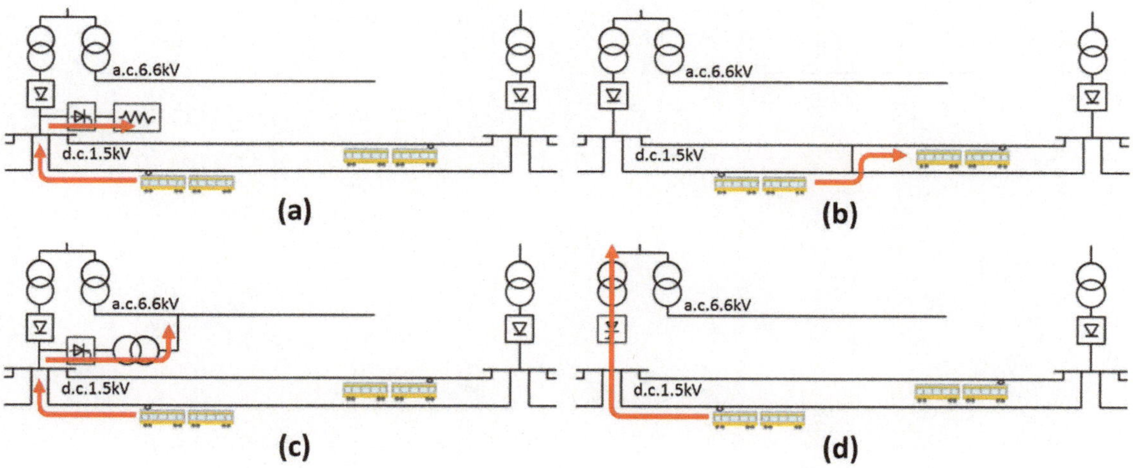

Fig. 6 Conventional solutions against residual regenerative energy. **a** Regenerative resistance, **b** tie feeding, **c** regenerative inverter, **d** PWM converter

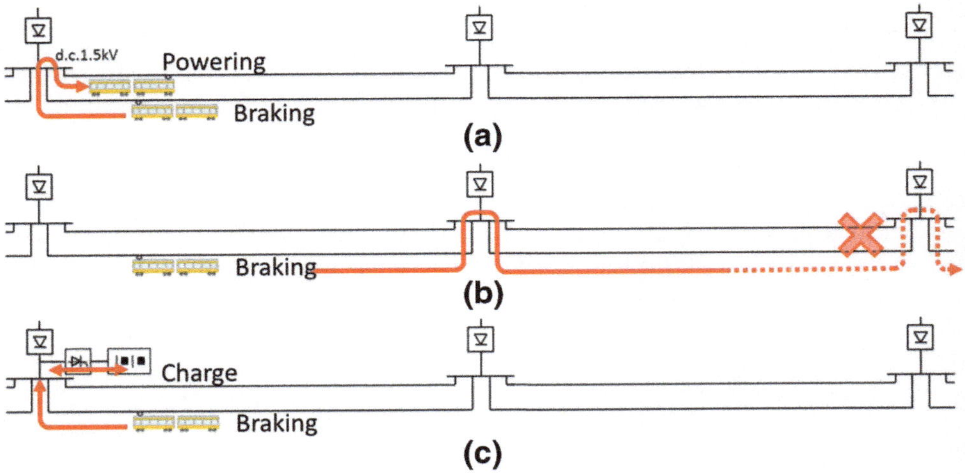

Fig. 7 Concept of energy storage system in DC traction power supply system. **a** Braking with powering train, **b** braking without powering train, **c** braking without powering train, but with energy storage system

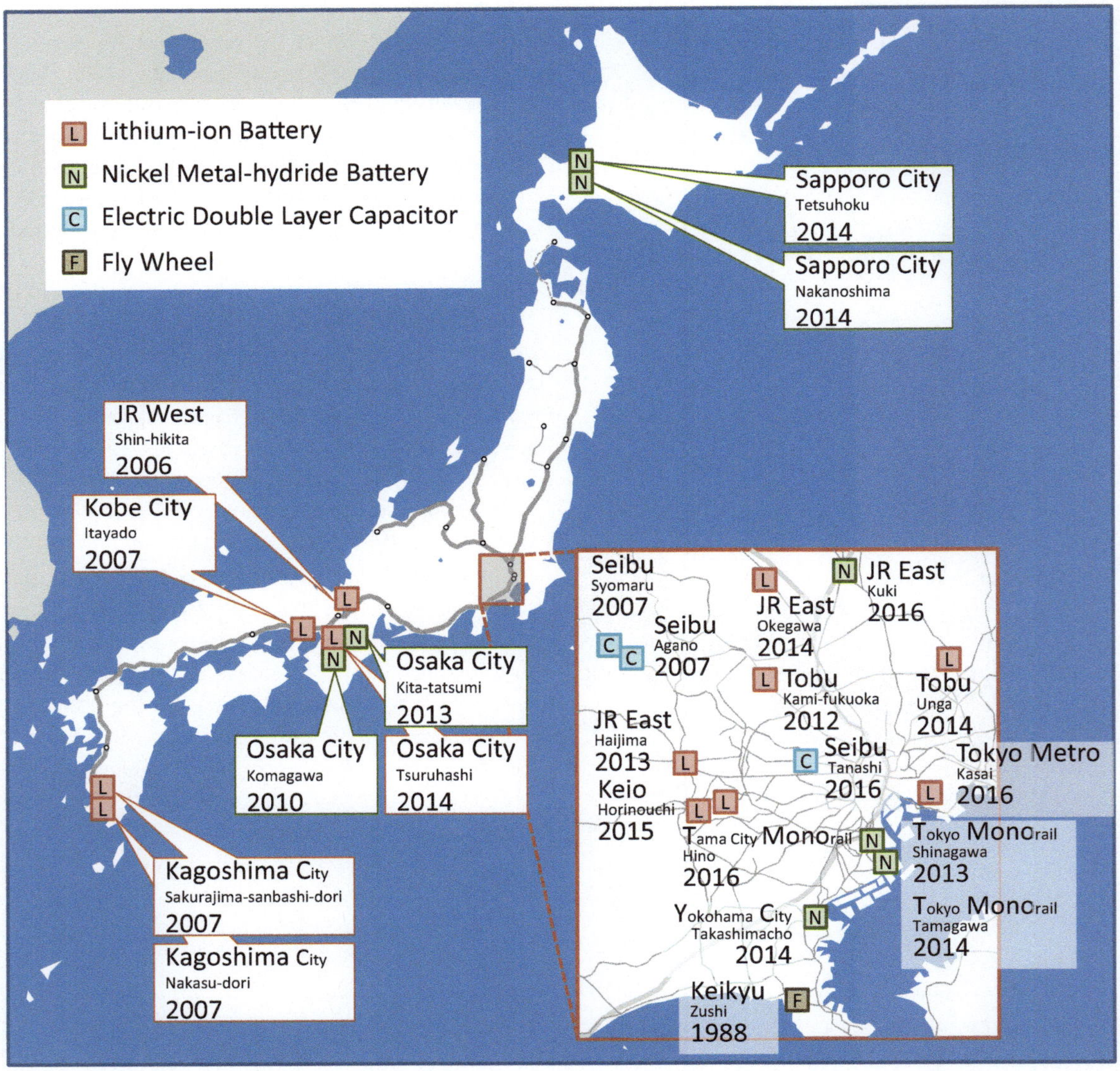

Fig. 8 Existing energy storage systems in DC traction power supply system in Japan

Japanese commuting trains with more than 10 cars, and the battery system is more appropriate to realize such a large capacity than a capacitor system. As a special case, fly wheel system was installed more than 25 years ago on Keihin Express Line for voltage drop compensation at the end of the DC traction power supply system. As shown in Fig. 8, more than 20 energy storage systems have already been installed in DC traction power supply system in Japan since 2006. The main purpose of installation of energy storage system is as follows:

- Compensation for voltage drop
- Avoidance of regenerative braking cancelation
- Utilization of regenerative energy
- Emergency power supply

Figure 9 shows the change in the number of traction substations where energy storage systems were newly installed. The installation of energy storage system is strongly promoted recently not only because of the support in relation to the subsidy from government, but also because of the cost reduction of energy storage system. Such a trend has been strongly supported by the governmental promotion to utilize renewable energy after the huge earthquake of 2011 in the eastern part of Japan and nuclear crisis immediately after the earthquake.

4 Experiences in East Japan Railway Company

4.1 Regenerative Energy Utilization in DC Traction Power Supply System

We, East Japan Railway Company, have already installed three energy storage systems for the utilization of regenerative energy around the metropolitan area of Tokyo as shown in Fig. 10. They were installed at Haijima SS, Okegawa SS and Kuki SS, respectively. These substations are located about 40 km away from Tokyo central station,

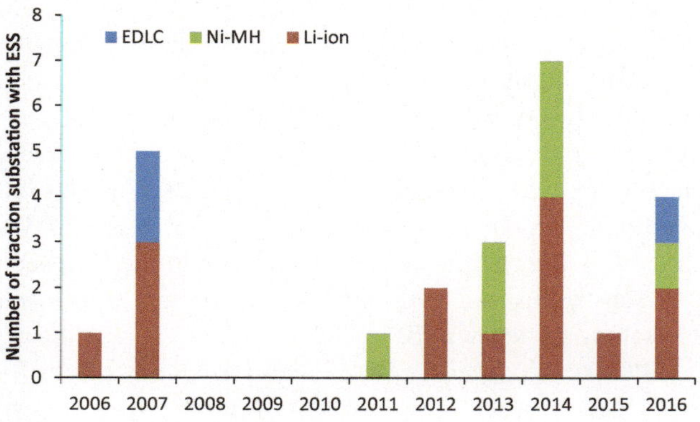

Fig. 9 Change in the number of traction substation where ESS is newly installed and the installed mediums

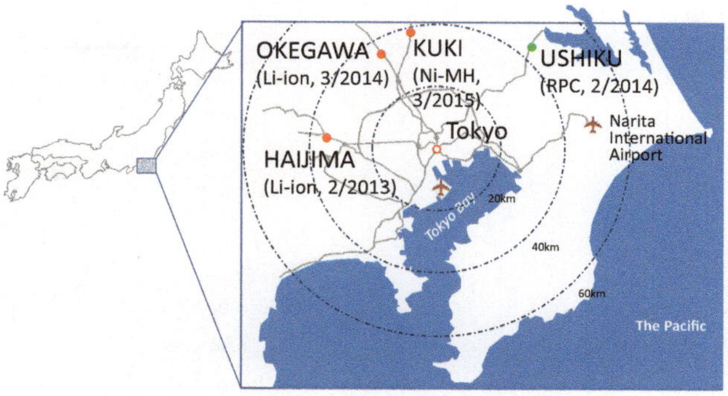

Fig. 10 Location of three energy storage systems in East Japan Railway Company

and according to our previous experience, such a suburban area is appropriate for energy storage system installation for the utilization of regenerative energy. In the central part of the metropolitan area of Tokyo where the time interval between train operations is less than 5 min, the operation frequency of railway is so high that all regenerative energy is utilized by the other powering trains simultaneously and it is not necessary to install energy storage system. On the other hand, in the countryside where the time interval between train operations is more than 10 min, the total amount of residual regenerative energy is not enough for the investment of energy storage system installation. Based on the detail measurements of consumed and residual regenerative power around the Osaka area which was carried out by West Japan Railway Company 5 years ago, they concluded that 7 min interval length of train operation was the most appropriate for energy storage system installation [8]. It is important to choose appropriate location in order for installations of energy storage system to be cost effective.

Table 1 shows the specifications of energy storage systems at Haijima SS and Okegawa SS. The energy-saving effects based on the first one-year operation are also shown in this table. Figure 11 shows the comparison of the daily energy supply from Okegawa SS before and after the installation of energy storage system in March 2014, and the ratio of discharged energy from battery to total energy supply from Okegawa SS is shown in Fig. 12. As shown in these figures, the installation of energy storage system contributes to the reduction of supply energy from traction substation of about 10% and it becomes maximum values during the middle seasons, in spring and in autumn. In summer and in winter, onboard air-conditioning loads are large and some part of the residual regenerative energy is utilized on board. The amount of residual regenerative energy becomes small in these seasons compared with the middle seasons.

As shown in Table 1, annual energy-saving effect is about 700 MWh/year at Okegawa SS and the payback year

Table 1 Specifications of energy storage system installed at Haijima SS and Okegawa SS

	Haijima SS	Okegawa SS
System		
Rated capacity	2000 kW	
Rated voltage	1650 V	
Rated current	1210 A	
Battery		
Devices	Lithium-ion	
Module capacity	173 V, 5.5 Ah	
Configuration	4 series, 20 parallel	4 series, 36 parallel
Maximum current	3000 A	3000 A
Rated capacity	76.12 kWh	137.02 kWh
DC/DC converter		
Devices	IGBT	
Parallel number	6	
Carrier frequency	600 Hz	
Operation start	February, 2013	March, 2014
Energy-saving effect of 1st 1 year	400 MW/year	700 MWh/year
Manufacturer	Hitachi Co., Ltd.	

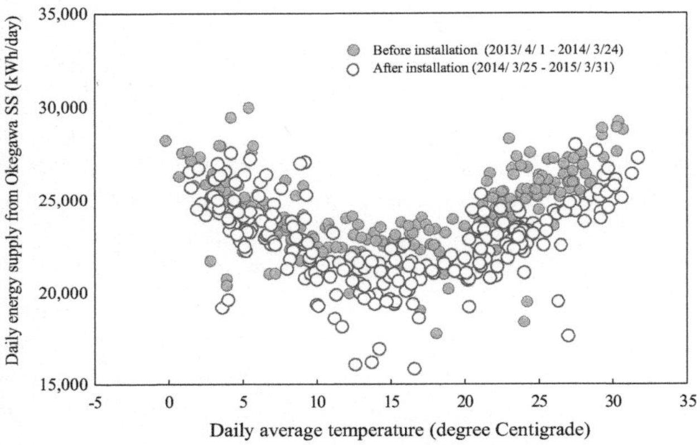

Fig. 11 Effect of energy storage system at Okegawa SS

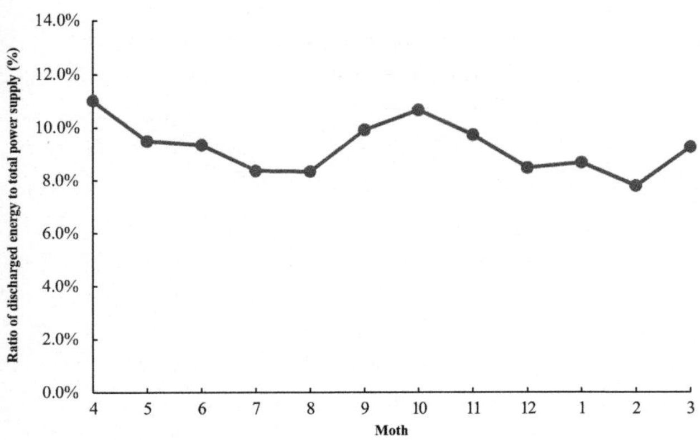

Fig. 12 Ratio of discharged energy to total energy supply from SS

for investment is about 30 years considering maintenance cost and battery replacement cost. Here, it is supposed that the lifetime of battery will be 15 years because only 25% of battery capacity is used for daily operation now to make battery lifetime longer. In the normal operation, state of charge (SOC) of the battery is controlled to be from 30% to 50% and such a narrow range operation is expected to make the lifetime of the battery system last about 15 years.

4.2 Regenerative Energy Utilization in AC Traction Power Supply System

Generally speaking, AC traction power supply system is applied for rural areas in Japan as shown in Fig. 5, and the total amount of residual regenerative energy tends to be small in AC traction power supply system because of its low operation frequency. The Joban Line was, exceptionally, electrified by AC 20 kV near Tokyo area because there is Kakioka Magnetic Observatory of Japan Meteorological Agency and so the DC electrification of railway was restricted by the low frequency of operations to avoid the influence of DC stray current on the survey of geomagnetic intensity. In this area, the number of operated electric trains is moderate enough because Ushiku SP (Sectioning Post) on Joban Line is located near Tokyo, about 40 km from city center. Here, SP is located between two substations and two single-phase circuits from each substation are faced at SP.

RPC, railway static power conditioner, was originally developed to solve the unbalanced three-phase voltage problem at the receiving point of traction substation for high-speed railway caused by single-phase train loads. Two

RPCs were installed at Shin-Numakunai SS and Shin-Hachinohe SS on Tohoku Shinkansen Line in 2002. Because RPC can realize power interchange between two different electric circuits with phase difference via AC/DC/AC conversion, it can increase the opportunity of simultaneous utilization of residual regenerative power by interchanging the residuals to adjacent electric circuit. Figure 13 shows the circuit configuration around RPC at Ushiku SP [10]. Ushiku SP is located between Fujishiro SS and Tsuchiura SS on Joban Line, and RPC can interchange residual regenerative power from one feeder circuit to another. Conventionally, residual regenerative energy is reversed from traction power supply system to the grid via traction transformers at substations, and it is not reflected to the electric charge for railway operation. Furthermore, there is no confirmation that such a back-flown single-phase power is utilized appropriately in the grid. Installation of RPC at SP can realize certain utilization of regenerative energy within traction power supply system. Figure 14 shows the concept of regenerative power flow without and with RPC at SP.

The change in interchanged energy per day from April to November in 2015 is shown in Fig. 15. Here, the interchanged power is controlled by referring the power supply from adjacent traction substations, Fujishiro SS and Tsuchiura SS. Hence, the interchanged power is always utilized in the adjacent circuit and it is possible to recognize the amount of interchanged energy as the effect of RPC. It is confirmed that the converted regenerative power is not back-flown from the railway power supply system to the grid via traction transformer at another traction substation. The effect of RPC depends on the seasons as shown in Fig. 15, similar to the tendency of energy storage system mentioned in the previous section. The annual effect of RPC at Ushiku SP was estimated to be about 2500 MWh/year or more according to the first 8-month operation results. This effect is estimated according to the interchanged energy through RPC. If it is possible to estimate the effect according to the history of total supplied energy, it will be easier for gaining understanding. However, it is difficult to distinguish the effect of RPC from other factors, such as weather, operation diagrams and types of rolling stocks, and such an estimation from interchanged energy is applied here.

4.3 Future Prospects

Now, the designing of one energy storage system is underway to be installed at Kita-Senju SS on Joban Line

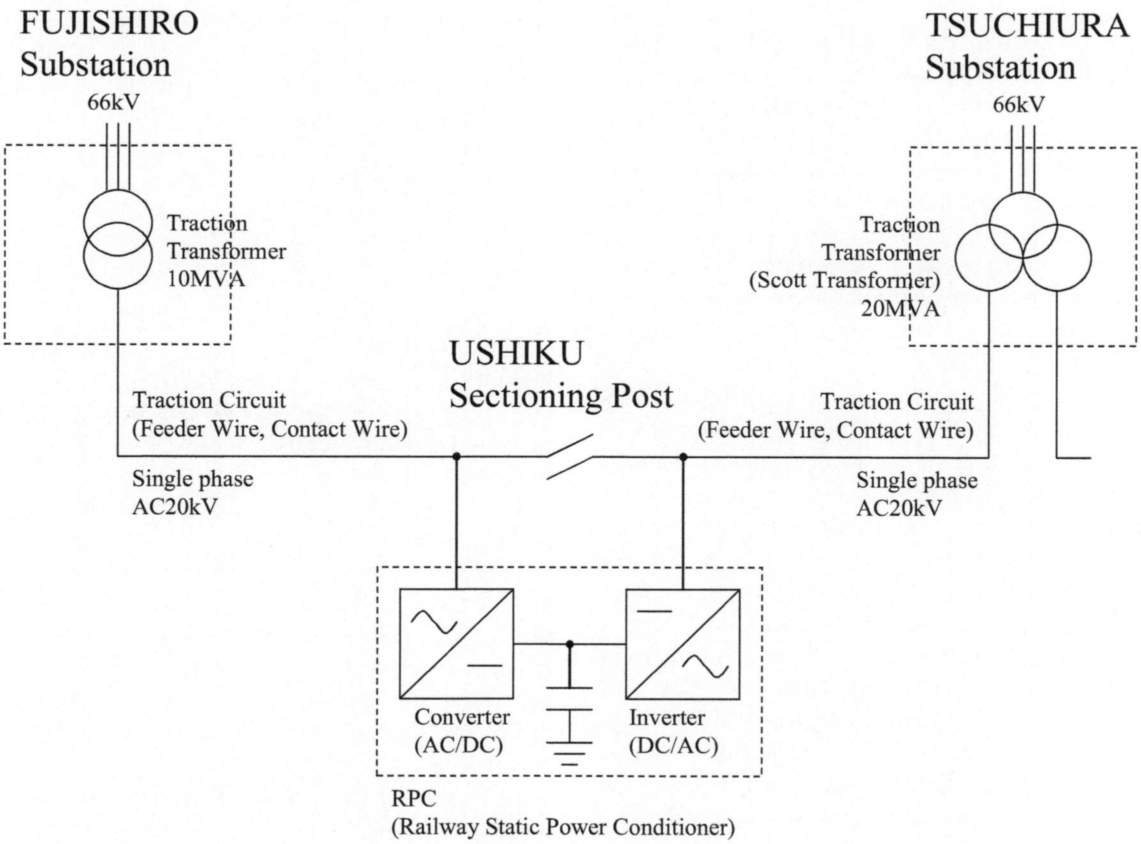

Fig. 13 Circuit configuration around RPC at Ushiku SP

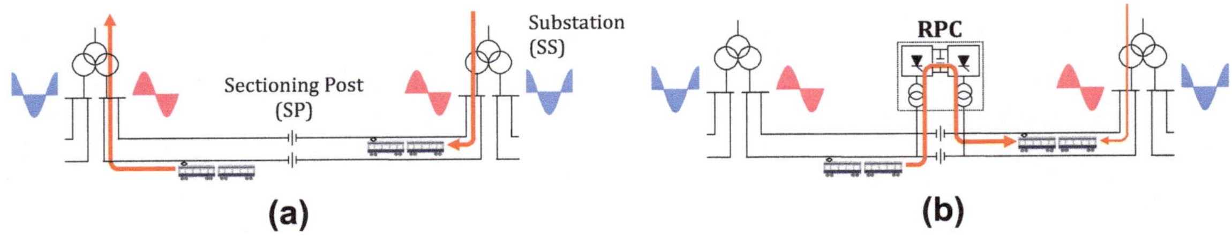

Fig. 14 Regenerative power flow without and with RPC at SP. **a** Without RPC, **b** with RPC

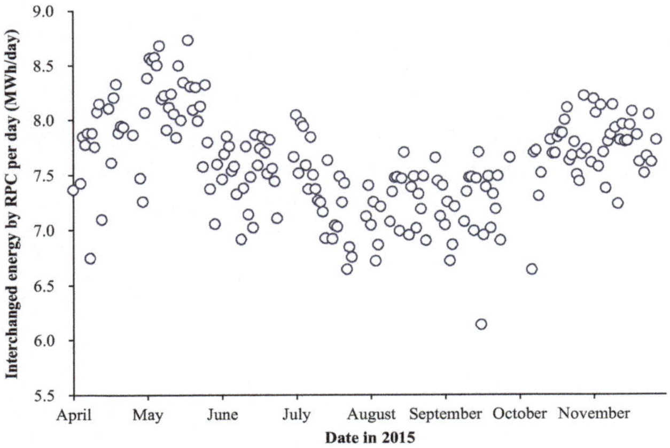

Fig. 15 Daily effect of RPC at Ushiku SP during the first 8 months operation

near Tokyo. This will be the fourth energy storage system installed by East Japan Railway Company. As a storage medium, lithium-ion battery will be applied in this system. In these days, the total cost of energy storage system installation is about 200 million Japanese Yen including the construction cost and the payback year will be around 30 years. Due to the fact that the substations where large effect of energy storage system is expected have already been selected for the previous installation, additional cost reduction is required for more installations of energy storage systems around Tokyo in the future.

Two regenerative inverters are under construction in JR East at Kajibashi SS at Tokyo Statin and Fukiage SS on Takasaki Line. Since the station load is smaller than the regenerative power at average stations, we did not choose regenerative inverter as a solution for regenerative energy utilization previously. Regenerative inverter is generally a solution for subway system whose station load is large enough. At Kajibashi SS, however, there is a huge load of Tokyo Station and the regenerative inverter is selected as a solution for regenerative energy utilization for the first time in JR East. At Fukiage SS, there is a dedicated high-voltage transmission line to connect several traction substations on Takasaki Line, and the converted regenerative power can be consumed only for the railway system along the line simultaneously.

Regarding regenerative energy utilization in high-speed railway, there is a large potential but the installation costs is not feasible under the actual conditions [9].

5 Conclusions

In this paper, the recent trend of regenerative energy utilization in traction power supply system in Japan is summarized. As a conclusion, I would like to mention some topics which should be discussed and considered more in the future.

Firstly, evaluation methods of the effects of the energy storage system and regenerative inverter should be established and standardized. The installation of such solutions can improve the level of energy efficiency at the installed substation but it can deteriorate those at adjacent substations in some cases. Also, power supply is influenced by weather, operation diagram, train characteristics, and so on. The effect of solutions has to be distinguished from the influences by other factors. Proposal of international standard to evaluate the effects of energy storage system or regenerative inverter system should be required in the future.

Secondly, the control scheme of charge and discharge of battery system and the design scheme on deciding the capacity of converter and energy storage system should be established. We are now making efforts by repeating trials and errors. The designing methods of energy storage system for regenerative energy utilization will be gradually matured in the future.

Lastly, but not least, optimal system design between on-ground and onboard energy storage system will be required in the future. Nowadays, energy storage system tends to be installed on ground because of its volume and weight problems. In the future, if the energy density of the storage medium is improved and the rapid charging technology is developed, the optimal solution will be changed. A comprehensive design scheme for regenerative energy utilization will be required in the future.

In November 2016, I was invited to the Chinese symposium of regenerative energy utilization in railway system at Zhenjiang held by Beijing Jiaotong University. I came to know that there are a lot of discussions in China about almost the same issues related to the regenerative energy utilization as in Japan. As you know, subway systems have already been constructed in 25 cities in China, and thus has a lot of experiences not only in AC high-speed railway [11] but also in DC urban railway [12]. I hope our collaborations between China and Japan in this technical field will be promoted more in the future and result in the realization of a more efficient, reasonable and advanced traction power supply system, and we are able to continue to contribute to the sustainable development of our future society together.

References

1. Web page of the Ministry of Land, Infrastructure, Transport and Tourism, Japan. http://www.mlit.go.jp/common/001117515.pdf (**in Japanese**)
2. Web page of the Ministry of Land, Infrastructure, Transport and Tourism, Japan. http://www.mlit.go.jp/tetudo/tetudo_tk2_000027.html (**in Japanese**)
3. Web page of the Ministry of Foreign Affairs, Japan. http://www.mofa.go.jp/mofaj/files/000090898.pdf
4. Hayashiya H (2016) Energy consumption in railway company. In: 2016 IEE-Japan industry application society conference, No.5-S1-1 (**in Japanese**)
5. Hayashiya H, Furukawa T, Itagaki H, Kuraoka T, Morita Y, Fukasawa Y, Mitoma Y, Oikawa T (2012) Potentials, peculiarities and prospects of solar power generation on the railway premises. In: 1st international conference on renewable energy research and applications (ICRERA2012)
6. Hayashiya H, Suzuki T, Kawahara K, Yamanoi T (2014) Comparative study of investment and efficiency to reduce energy consumption in traction power supply. In: 16th international conference on power electronics and motion control (PEMC2014), No.229
7. Hayashiya H, Suzuki T, Hino M, Hara D, Tojo M, Shimada S, Kudo K (2015) Effect evaluation of Li-ion battery for regenerative energy utilization in traction power supply system. In: 17th European conference on power electronics and applications (EPE2015), No.0035
8. Hayashiya H, Suzuki T, Kawahara K, Yamanoi T (2014) Comparative study of investment and efficiency to reduce energy consumption in traction power supply. In: 16th international conference on power electronics and motion control (PEMC2014), No.229
9. Hayashiya H, Yokokawa S, Iino Y, Kikuchi S, Suzuki T, Uematsu S, Sato N, Usui T (2016) Regenerative energy utilization in a.c. traction power supply system. In: Proceedings of the 2016 IEEE international power electronics and motion control conference (PEMC)
10. Hayashiya H, Kikuchi S, Matsuura K, Hino M, Tojo M, Kato T, Ando M, Oikawa T, Kamata M, Munakata H (2013) Possibility of energy saving by introducing energy conversion and energy storage technologies in traction power supply system. In: 15th European conference on power electronics and applications (EPE2013)
11. Li Q, Liu W, Shu Z, Xie S, Zhou F (2014) Co-phase power supply system for HSR. In: International power electronics conference (IPEC-Horoshima 2014-ECCE Asia), No.19J1-2
12. Yang Z, Xia H, Wang B, Lin F (2014) An overview on braking energy regeneration technologies in Chinese urban railway transportation. In: International power electronics conference (IPEC-Horoshima 2014-ECCE Asia), No.20G4-4

20

The Analysis of Tram Tracks Geometric Layout Based on Mobile Satellite Measurements

Cezary Specht[2] · Wladyslaw Koc[1] · Piotr Chrostowski[1] · Jacek Szmaglinski[1]

Abstract In this article, the results of the research in a field of which uses active global navigation satellite system (GNSS) geodetic networks for the inventory of geodetic geometric tram tracks are presented. The applied measurement technique has been adapted for the designing of the geometric layout of tram tracks. Several configurations of receivers and settings of an active GNSS networks with the objective to increase the accuracy of positioning and the availability of accurate localization are investigated. The measurement methods are optimized in order to increase the accuracy of determining positions from 3 mm up to 6 mm. Thus, the study of deformations in geometric layouts—according to the authors—is already possible. The implementation techniques of the mobile satellite measurements in a field of tram track inventory process are presented in this article. The course of the measurements and the results of the inventory of the tram system in Gdansk, Poland have been discussed. The results have turned out to be extremely useful for the geometric track layout evaluation. It has been proved that the applied method allows a comprehensive tram network inventory to be performed based on satellite measurements. The presented method is fast and cost-efficient.

Keywords Urban transit · Tramways · Satellite measurements · Geometric layout

List of symbols

a_h	The slope of a horizontal line
b_h	The vertical intercept of a horizontal line (m)
c	The value of speed of light (m/s)
i	GPS satellite number
j	Moment of time number
k	Overtone number
l	Fixes number
$L_{o,i}(t_j)$	Approximate length between satellite and receiver (m)
N	Number of samples
N_i	Number of unspecified phase cycles
R	Arc radius (m)
w	Regression weight
x_0	Approximate x position of receiver (m)
$x_{s,i}(t_j)$	x position of i number satellite in j moment
X	Precise x position of receiver (fix) (m)
X_0	x position of circle center (m)
X_k	Fourier transform
y_0	Approximate y position of receiver (m)
$y_{s,i}(t_j)$	y position of i number satellite in j moment
Y	Precise y position of receiver (fix) (m)
Y_0	y position of circle center (m)
z_0	Approximate x position of receiver (m)
$z_{s,i}(t_j)$	z position of i number satellite in j moment

✉ Jacek Szmaglinski
jacszmag@pg.gda.pl

Cezary Specht
c.specht@geodezja.pl

Wladyslaw Koc
kocwl@pg.gda.pl

Piotr Chrostowski
piotrchrost@gmail.com

[1] Department of Rail Transportation and Bridges Structures, Faculty of Civil and Environmental Engineering, Gdansk University of Technology, ul. Narutowicza 11/12, 80-233 Gdańsk, Poland

[2] Department of Geodesy and Oceanography, Faculty of Navigation, Gdynia Maritime University, Aleja Jana Pawła II 3, 81-345 Gdynia, Poland

Editor: Dr. Xuesong Zhou

Z	Precise z position of receiver (fix) (m)
$\delta_0(t_j)$	Receiver clock error in j moment (s)
$\delta_{s,i}(t_j)$	i satellite clock error in j moment (s)
λ	GPS wavelength (m)
$\rho_i(t_j)$	Pseudorange between i satellite and receiver in j moment (m)
σ_x	x position determination error (m)
σ_y	y position determination error (m)
σ_z	z position determination error (m)
σ_t	Time determination error
χ^2	Chi-squared distribution (m^2)

1 Introduction

For several years, the tram systems have been rapidly developed and upgraded all over the world. The undoubted leader in this field is France. Since the late 1980s, tram networks have been opened in 25 cities of France. French trams and technical track-designing solutions are used in all the continents. Along with the revival of trams, the rolling stock has been changed. The vehicles are being adapted to the needs of people with limited mobility. Consequently, the approach to the designing and maintenance of track structure is changing. In order to eliminate common operational problems, there are attempts being made to modify the classical tramways. An example of such solution is the rubber-tired tramway (Translohr), shown in Fig. 1.

Precise defining of the track layout is one of the most important issues enabling to draw up track inventory. Accordingly, the detection of defects in a track shape provides proper design works. However, the process of setting out the tram route and determining the geometric layout of operating track is based on traditional surveying methods. The main disadvantage of the described methods is that the railway needs to be divided to the various sections which differ in the accuracy of positioning.

Fig. 1 Rubber-tired tramway on T6 line in Paris (photo Jacek Szmaglinski)

The simplest methods are based on the direct measurements of the horizontal rise of arc and determining the geometric layout, created from the curvature graph. However, this procedure is performed on the local reference system with transformation, which can lead to serious errors. The most common mistakes are connected with the inaccurate definition of the returning route angle. As a result, during the track lining some unintentional curvature smoothing might occur.

Within the measurements of track axis coordinates (optionally stretch of rails), the problem resulting from the transformation can be eliminated. Frequently, the track axis is measured with a tachymeter referring to the surveying grid-fixed points. This procedure is performed locally and involves the individual parts of a route. Thus, the coordinates of the prism (set by the crew in the track axis) are determined by a stationary tachymeter (placed on the fixed point). Alternatively, the prisms are stationary (on fixed points) while the tachymeter is moved along the track, at this time the coordinates of the total station are defined [2]. However, the locally measured coordinates require adjustments. As a result, additional errors may appear and the assessment of the geometrical layout of the track becomes problematic, especially for the long sections of the railway route.

2 Modern Measuring Methods

The classical methods of specifying the track geometrical layout are time-consuming and require accurate maintenance of surveying grids (located along the track). For this reason, the work on developing a comprehensive method is being continued for a long time. It should be faster and more accurate in determining the values of basic track geometrical parameters. One of the currently used solutions is to measure the acceleration occurring on the rail vehicle (i.e., accelerometry). The measurement is performed with the use of track-recording cars moving with relatively high velocity (sometimes exceeding 100 kph). This allows researchers to determine the irregularities of the track and its impact on the appearing acceleration (especially vertical) resulting in the decrease in comfort. However, this - method of defining the track geometrical layout is facing a number of difficulties related with the exponential nature of errors rising together with the distance [20]. Simultaneously, due to the high velocity of ride, the identification of the characteristic elements of track (such as turnouts, semaphores, etc.) may be difficult.

Estimating the track geometrical layout with laser scanning method has been developed recently [1]. This technique involves mobile laser scanning of the track and its surroundings. Subsequently, the geometrical layout is determined due to so-called point cloud often containing huge collection of points (with their coordinates). It is possible to install the scanner on either moving rail vehicle or on plane (helicopter, airplane, etc.). Route scanning performed from above uses photogrammetry measuring methods [19]. The accuracy of the point measurement is approximately 2–3 cm and allows researchers to determine the position of turnouts and track equipment. However, using laser scanning method with the moving base is connected with a rapid increase in errors when defining the position.

The accuracy of the positioning might be increased by combining different measurement methods. For example, the *UFM-120* track-recording car determines the track geometrical layout using the accelerometer and gyroscope. Simultaneously, the error correction is carried out by using global positioning system (GPS) method [20, 21]. The measurement of the position with GPS system plays supplementary role in the described case. For this reason, using inexpensive single-frequency GPS receivers without differential correction is well-grounded.

Ceaseless development of the satellite radio navigation measurement techniques creates an opportunity to search for a way to determine the track geometrical layout. Since the beginning of the twenty-first century, the dual-frequency receivers have been used for precise measurements in the USA (High Accuracy Differential GPS technology) [7, 14]. In order to determine the coordinates of the track, the GPS antenna (rover) was installed on the Nolans *TS-1* hand trolley (Fig. 2). Differential corrections were transmitted via radio from the second receiver (reference) set on the point with known coordinates. The coordinates of the track were measured by real-time kinematics (RTK) method. During the research, the receivers were installed on rail–road vehicles. The repeatability of the measurements was tested with different velocities (from 2.23 to 6.70 m/s). The results were very promising. The obtained horizontal accuracy of position determination was 12 mm while the vertical was 22 mm. However, the measuring distance was relatively short (425 m long). Therefore, the decrease in the accuracy connected with the rising distance from the base station was not specified. Using the described method for longer sections (several kilometers), results in the decrease in accuracy of approximately 10 mm for each additional kilometer. An alternative solution is to shift the reference station position continuously with the change of the part of a measured length. Nevertheless, the additional errors related to the movement of the geodetic fixed points should be taken into consideration.

3 Mobile Satellite Measurements

In 2009, a research team consisting of the Gdansk University of Technology, the Gdynia Maritime Academy employees and the Polish Naval Academy was created. The purpose of the team was to investigate the possibility of using GNSS measurements to determine the shape of the rail track. The first surveys were carried out in February 2009 in relation to about 50-km section of railway line from Koscierzyna to Kartuzy in northern Poland. In the following years, the research included both railway and tram lines. The first measurements were taken using four Leica *1200* receivers placed above the wheels of the two-axle platform wagon. The measuring set was pulled at a constant speed by a *WM-15* diesel motor car. Observations were performed in the distance mode with 300-mm interval, and communication with ASG-SERVICE [9] was provided by a general packet radio service (GPRS) modem. The coordinates or the track were recorded and calculated

Fig. 2 Lightweight Nolan's Tool and Supply Cart *TS-1*, used as a basis for the GPS receiver [17]

Fig. 3 The measuring set used in the 2010 campaign

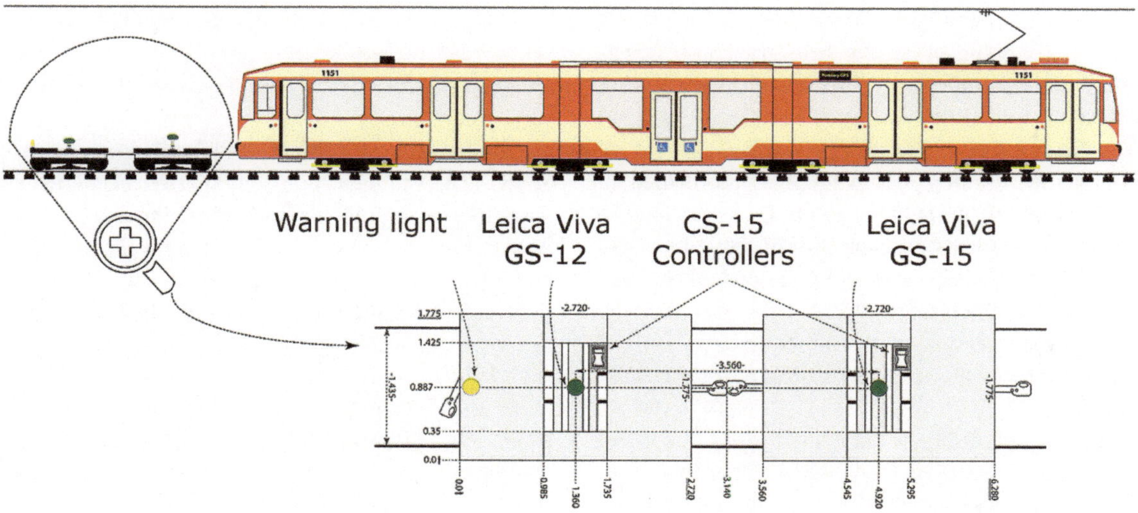

Fig. 4 The tram track measuring set

in real-time mode. Studies have confirmed the inventory of long sections of railway lines with respect to GNSS measurements [4, 13].

In subsequent measurements, performed in the following year, the settings of receivers were modified. Due to the fact that the location of the power car driver's cab was near to the two antennas, partial horizon obscuration and increase in geometrical factor dilution of precision (DOP) occurred. Dilution of precision factor is responsible for the obtained accuracy. The additional platform with three antennas mounted above the axles was attached to the measurement set [5] as shown in Fig. 3.

The experience and technology developed during the tram tracks research in 2012–2014 were used in the 2015 measurement campaign on railway tracks. It proved that tram boogies prepared by the scientific team could be easily used on the railroad tracks measurements.

4 Satellite Measurements of Tram Tracks

After a thorough analysis of the results obtained from the campaigns carried out in the years 2009–2010, significant changes in the methodology of research were introduced [11]:

- Resignation from implementation of the real-time calculations due to a momentary interruptions in differential corrections packet transmitting, caused by a large number of users in specific times of the day,
- Performing the post-processing calculations which allowed greater flexibility in the use of signals from different reference stations,
- The single-system (GPS) Leica *1200* receiver was changed to dual-system [GPS/GLONASS (Globalnaja Nawigacionnaja Sputnikowaja Sistiema)] Leica Viva receiver because the measurement accuracy depends directly on the number of available satellites,
- An alternative solution to *ASG-EUPOS* corrections (in 2012 it did not support *GLONASS*) was used.

Based on the presented assumptions, in February 2012 the first measurement campaign on the tram track in the city of Gdansk in northern Poland was carried out. The boogies from pre-war tramway Danziger Wagon Fabrik (DWF) *300* series were used as a measuring set. The GNSS receivers were mounted in the geometrical center of each boogie. Position of antennas was determined using the Leica *TCRA 1103* precise tachymeter. The *GS-15* antenna was mounted on the first boogie, and *GS-12* on the last one. Antennas transmitted data to *CS-15* controllers using the Bluetooth radio communication. Boogies were attached to *N8C-MF01 #1151* tramway power car. The choice of such a specific car was caused by its universality (it was suited to operate on entire tram network in Gdansk) and the ability to drive at a constant speed of 4.17 m/s without a risk of damage to the vehicle. The measuring train is shown in Fig. 4.

Due to the required low speed of the ride, a need to organize the measurements at night was decided so as not to disturb the functioning of regular tram transport. It was planned that the inventory of 54 km of single track throughout the city would be executed between 23:15 and 04:39 (about 5.5 h) [11].

The second measurement campaign was performed in September 2013 during night. This time, it was planned to focus on the measurements repeatability analysis by making a triple ride in the section between the city center and tram loop Łostowice Świetokrzyska. During this test, the short (about 250 m) section of a straight track in one of the tracks has been chosen to be measured more precisely by several rides [12].

The control measurement on the described section above was performed in April 2014. Coordinates of the track axis were designated every 30 cm using tachymeter. Subsequently, several rides were carried out using hand-operated *GRAW iTEC* troley with mounted Leica *Viva GS-15* receiver. The resulting image of tram lines, measured in 2012–2014, is shown in Fig. 5.

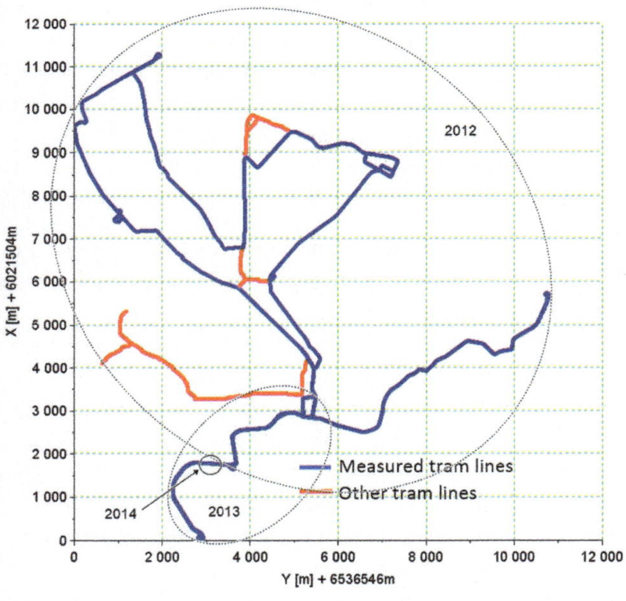

Fig. 5 Measured coordinates of tram tracks in Gdansk (Gauss conformal projection)

5 The Accuracy Analysis

During the first measurement campaign (in 2012), 15,854 coordinates were determined by the Leica *Viva GS15* receiver and 16,015 by Leica *Viva GS12* receiver. It turned out that the distance between the points was not meeting assumptions of the method. As a matter of fact, the achieved density of the coordinates was 3.6 m, while assumed level was less than 0.3 m. The wrong way of setting the recording frequency of the receivers was the cause of failure. Still, the achieved results of the inventory were interesting.

The second measurement campaign was performed in 2013. The studies considered the need to meet the assumptions of the method. Simultaneously, the analysis was limited to multiple passage between the City Center and Lostowice tram loop due to the data obtained in the previous measurements of the tram line. In total, 240,620 designations were obtained with their average density of 0.13 m.

In 2014, 70-m-long track section was investigated. This time, two different references Active Geodetic Network (Leica *SmartNet* and *ASG-EUPOS*) were used. Number of position for each networks was 29,318 and 12,102 fixes. The average density of the measurements was 0.05 m. A total of 219 coordinates of reference points were determined with the precise tachymeter (density of 0.3 m).

The measurements have been conducted in three different modes:

- Code—collects only GNSS code, no phase solution, meter accuracy,
- *xRTK*—phases GNSS measurement, decimeter accuracy,
- Phase—phases GNSS measurement, centimeter accuracy.

Only the phase measurement allows for track inventory to be with the necessary accuracy. It consists of measuring the differences in phase of waves with different frequencies (due to Doppler effect related to the very high speed of satellites). In addition to the unknown three-dimensional coordinates *XYZ*, there are additional unknowns in the equation system associated with unspecified phase cycles and GNSS receiver clock errors. For this reason, it is necessary to measure the pseudoranges to four satellites for three times, which results in a system of twelve equations with ten unknowns. This system can be written as $L = G \times X$, where L is the matrix of measurements, G is the matrix of coefficients and X is the matrix of unknowns [8]. Equations 1–4 represent the matrix coefficients *L*, *G i X*.

$$\begin{vmatrix} \lambda \cdot \rho_1(t_0) - L_{o,1}(t_0) + c \cdot \delta_{s,1}(t_0) \\ \lambda \cdot \rho_2(t_0) - L_{o,2}(t_0) + c \cdot \delta_{s,2}(t_0) \\ \lambda \cdot \rho_3(t_0) - L_{o,3}(t_0) + c \cdot \delta_{s,3}(t_0) \\ \lambda \cdot \rho_4(t_0) - L_{o,4}(t_0) + c \cdot \delta_{s,4}(t_0) \\ \lambda \cdot \rho_1(t_1) - L_{o,1}(t_1) + c \cdot \delta_{s,1}(t_1) \\ \lambda \cdot \rho_2(t_1) - L_{o,2}(t_1) + c \cdot \delta_{s,2}(t_1) \\ \lambda \cdot \rho_3(t_1) - L_{o,3}(t_1) + c \cdot \delta_{s,3}(t_1) \\ \lambda \cdot \rho_4(t_1) - L_{o,4}(t_1) + c \cdot \delta_{s,4}(t_1) \\ \lambda \cdot \rho_1(t_2) - L_{o,1}(t_2) + c \cdot \delta_{s,1}(t_2) \\ \lambda \cdot \rho_2(t_2) - L_{o,2}(t_2) + c \cdot \delta_{s,2}(t_2) \\ \lambda \cdot \rho_3(t_2) - L_{o,3}(t_2) + c \cdot \delta_{s,3}(t_2) \\ \lambda \cdot \rho_4(t_2) - L_{o,4}(t_2) + c \cdot \delta_{s,4}(t_2) \end{vmatrix} = \begin{vmatrix} u_{x,1}(t_0) & u_{y,1}(t_0) & u_{z,1}(t_0) & \lambda & 0 & 0 & 0 & c & 0 & 0 \\ u_{x,2}(t_0) & u_{y,2}(t_0) & u_{z,2}(t_0) & 0 & \lambda & 0 & 0 & c & 0 & 0 \\ u_{x,3}(t_0) & u_{y,3}(t_0) & u_{z,3}(t_0) & 0 & 0 & \lambda & 0 & c & 0 & 0 \\ u_{x,4}(t_0) & u_{y,4}(t_0) & u_{z,4}(t_0) & 0 & 0 & 0 & \lambda & c & 0 & 0 \\ u_{x,1}(t_1) & u_{y,1}(t_1) & u_{z,1}(t_1) & \lambda & 0 & 0 & 0 & 0 & c & 0 \\ u_{x,2}(t_1) & u_{y,2}(t_1) & u_{z,2}(t_1) & 0 & \lambda & 0 & 0 & 0 & c & 0 \\ u_{x,3}(t_1) & u_{y,3}(t_1) & u_{z,3}(t_1) & 0 & 0 & \lambda & 0 & 0 & c & 0 \\ u_{x,4}(t_1) & u_{y,4}(t_1) & u_{z,4}(t_1) & 0 & 0 & 0 & \lambda & 0 & c & 0 \\ u_{x,1}(t_2) & u_{y,1}(t_2) & u_{z,1}(t_2) & \lambda & 0 & 0 & 0 & 0 & 0 & c \\ u_{x,2}(t_2) & u_{y,2}(t_2) & u_{z,2}(t_2) & 0 & \lambda & 0 & 0 & 0 & 0 & c \\ u_{x,3}(t_2) & u_{y,3}(t_2) & u_{z,3}(t_2) & 0 & 0 & \lambda & 0 & 0 & 0 & c \\ u_{x,4}(t_2) & u_{y,4}(t_2) & u_{z,4}(t_2) & 0 & 0 & 0 & \lambda & 0 & 0 & c \end{vmatrix} \times \begin{vmatrix} X \\ Y \\ Z \\ N_1 \\ N_2 \\ N_3 \\ N_4 \\ \delta_0(t_0) \\ \delta_0(t_1) \\ \delta_0(t_2) \end{vmatrix} \tag{1}$$

$$u_{x,i}(t_j) = -\frac{x_{s,i}(t_j) - x_0}{L_{0,i}(t_j)} \tag{2}$$

$$u_{y,i}(t_j) = -\frac{y_{s,i}(t_j) - y_0}{L_{0,i}(t_j)} \tag{3}$$

$$u_{z,i}(t_j) = -\frac{z_{s,i}(t_j) - z_0}{L_{0,i}(t_j)} \tag{4}$$

where c is the value of speed of light (m/s), i GPS satellite number, j moment of time number, $L_{o,i}(t_j)$ approximate length between the satellite and receiver (m), N_i number of unspecified phase cycles, x_0 approximate x position of receiver (m), $x_{s,i}(t_j)$ x position of i number satellite in j moment, X precise x position of receiver (fix) (m), y_0 approximate y position of receiver (m), $y_{s,i}(t_j)$ y position of i number satellite in j moment, Y precise y position of receiver (fix) (m), z_0 approximate x position of receiver (m), $z_{s,i}(t_j)$ z position of i number satellite in j moment, Z precise z position of receiver (fix) (m), $\delta_0(t_j)$ receiver clock error in j moment (s), $\delta_{s,i}(t_j)$ i satellite clock error in j moment (s), λ GPS wavelength (m) and $\rho_i(t_j)$ pseudorange between i satellite and receiver in j moment (m).

Table 1 Availability of GNSS measurements methods during measurement campaigns [10]

Campaign	2012		2013	2014
Receiver	GS-12	GS-15	GS-15	GS-15
Code (%)	27.817	17.276	6.303	0.681
xRTK (%)	2.222	1.230	–	–
Phase (%)	69.959	81.493	93.697	99.319
Mean error: $\overline{\text{GDOP}}$ (mm)	25.02	22.65	19.35	18.79
Error STD: σ_{GDOP} (mm)	13.00	25.14	15.54	24.28

Depending on the numerous conditions, the receiver is able to select one of the available methods of coordinate calculations. As a percentage factor, the availability, i.e., the share of determined positions of a particular method in the total number of measurements has been defined. Table 1 shows the availability of particular measurement techniques in the subsequent years. Interesting results were obtained in 2012. Receiver GS-15, due to its location (near the tram), should indicate lower availability of phase solution. However, as indicated in Table 1, the availability of phase method is 11.5% more with respect to the further receiver (GS-12). This fact undoubtedly proves the increasing accuracy offered by successive generations of GNSS receivers (in 2012 the GS-15 receiver was the most modern GNSS receiver offered by Leica).

The accuracy of each position is determined by geometric dilution of precision (GDOP) factor which is defined as:

$$\text{GDOP} = \sqrt{\sigma_x^2 + \sigma_y^2 + \sigma_z^2 + \sigma_t^2} \tag{5}$$

where σ_x x position determination error (m), σ_y y position determination error (m), σ_z z position determination error (m) and σ_t time determination error.

In order to assess the accuracy of the measurement in constant conditions, in 2014 the track section without any field obstacles was chosen. This resulted in almost perfect coverage of phase GNSS solution.

Therefore, the average accuracy of points 3D determination (calculated automatically by the receiver) is given only for phase measurements (Table 1). Figure 6 shows the probability density function of position error.

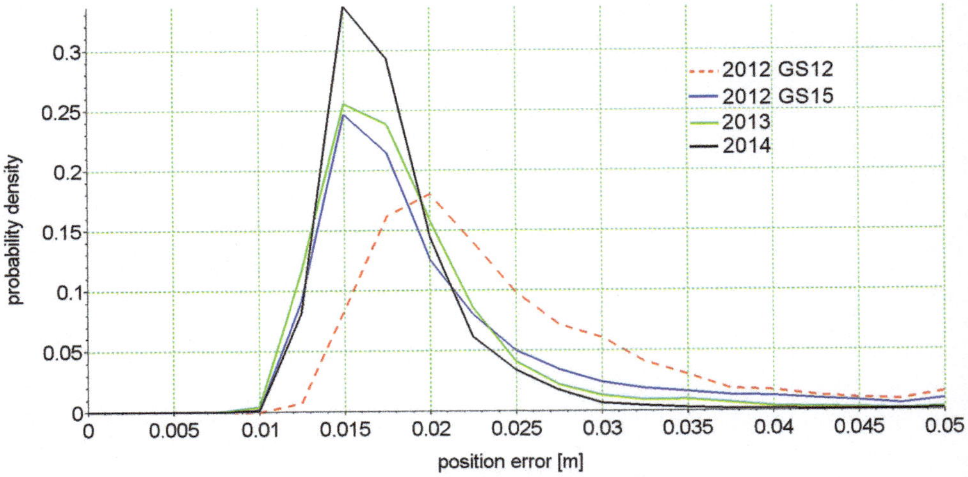

Fig. 6 The probability density function of 3D position determination error

6 Assessment of Tram Track Geometric Layout

The geometric layout of railway in horizontal plane consists of straight sections and arches with a constant or variable curvature. Straight sections are defined in the geometrical point of view. That is why, the assessment of the layout determined from the satellite measurements can be the basis of obtained accuracy. Any deviation from the planned course would be treated as a measurement error if the track is assumed to be a perfectly straight section of the track. However, during the building process and operation some distortions may appear in the track. The stiffness of the surface structure limits the course of distortions. Thus, it can be assumed that the coordinates will contain both the components with the distortions of the track and errors resulting from the adopted measurement technique.

To determine the accuracy obtained in the field conditions, the algorithm described in work [3] was used. All rides made on the test sections has been extracted and transformed. The transformation was based on the coordinate rotation and translation.

The equation of the theoretical straight line is determined by the weighted regression:

$$\chi^2(a_h, b_h) = \sum_{l=1}^{n} w_l \cdot (X_l - a_h \cdot Y_l - b_h)^2 \tag{6}$$

The values of coefficients of equation were determined under the condition of necessity of the existence of functions minima:

$$\frac{d\chi^2}{da_h} = -2\sum_{l=1}^{n} w_l \cdot Y_l \cdot (X_l - a_h \cdot Y_l - b_h) = 0 \tag{7}$$

$$\frac{d\chi^2}{db_h} = -2\sum_{l=1}^{n} w_l \cdot (X_l - a_h \cdot Y_l - b_h) = 0 \tag{8}$$

The weight of each element is equal to the inverse of the square of the measurement uncertainty:

$$w_l = \frac{1}{\sigma_l^2} \tag{9}$$

Then, the measured coordinates are subjected to matrix operations to describe them as a function in a local coordinate system:

$$\mathrm{FIX_{rot}} = \mathrm{FIX_{glob}} \cdot \mathrm{TRAN_1} \cdot \mathrm{ROT} \tag{10}$$

$$\mathrm{FIX_{loc}} = \mathrm{FIX_{rot}} \cdot \mathrm{TRAN_2} \tag{11}$$

Matrix of fixes in a global coordinate system:

$$\mathrm{FIX_{glob}} = \begin{vmatrix} Y_1 & X_1 & Z_1 & 1 \\ Y_2 & X_2 & Z_2 & 1 \\ \cdot & \cdot & \cdot & \cdot \\ \cdot & \cdot & \cdot & \cdot \\ \cdot & \cdot & \cdot & \cdot \\ Y_n & X_n & Z_n & 1 \end{vmatrix} \tag{12}$$

Translation and rotation matrices:

$$\mathrm{TRAN_1} = \begin{vmatrix} 1 & 0 & 0 & 0 \\ 0 & 1 & 0 & -b_h \\ 0 & 0 & 1 & 0 \\ 0 & 0 & 0 & 1 \end{vmatrix} \tag{13}$$

$$\mathrm{ROT} = \begin{vmatrix} \cos(atg(a_h)) & -\sin(atg(a_h)) & 0 & 0 \\ \sin(atg(a_h)) & \cos(atg(a_h)) & 0 & 0 \\ 0 & 0 & 1 & 0 \\ 0 & 0 & 0 & 1 \end{vmatrix} \tag{14}$$

$$\mathrm{TRAN_2} = \begin{vmatrix} 1 & 0 & 0 & -\mathrm{FIX_{rot}}(1,1) \\ 0 & 1 & 0 & 0 \\ 0 & 0 & 1 & 0 \\ 0 & 0 & 0 & 1 \end{vmatrix} \tag{15}$$

a_h the slope of a horizontal line, b_h the vertical intercept of a horizontal line (m), l fixes number, N number of samples, w regression weight and χ^2 Chi-squared distribution (m^2).

Fig. 7 Graphical interpretation of XTE idea [6]

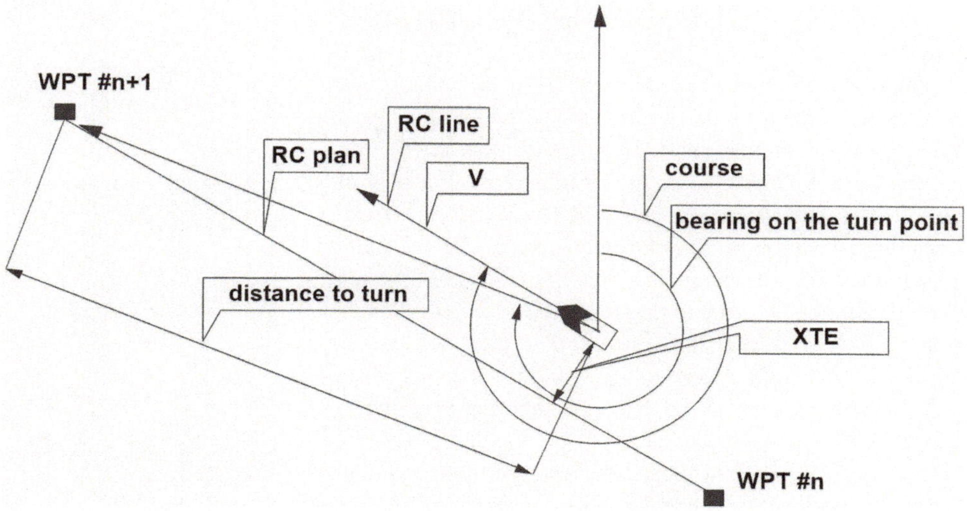

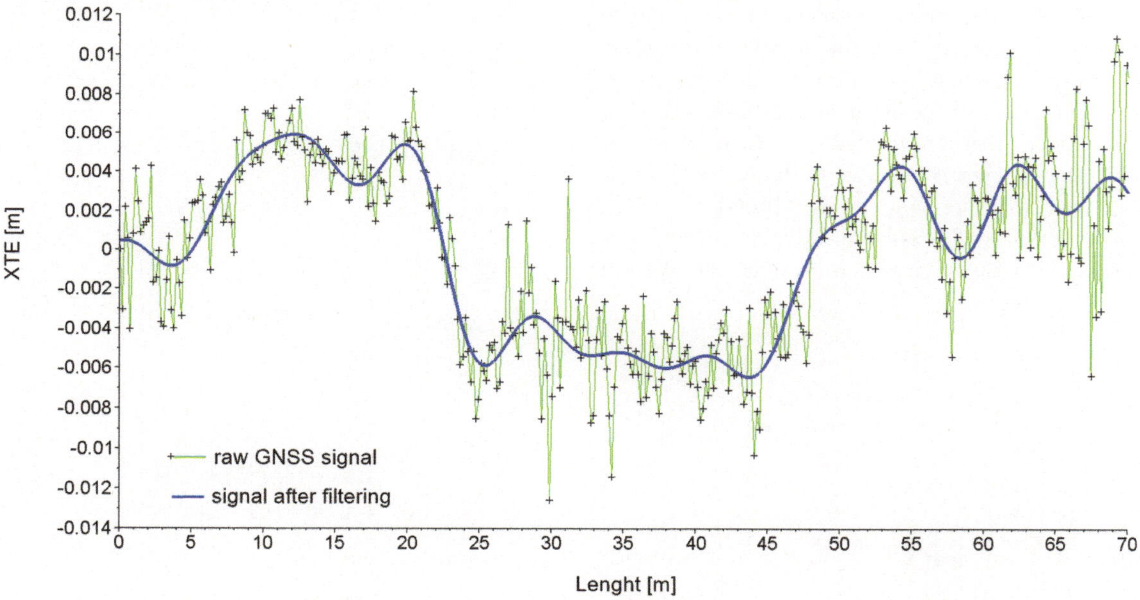

Fig. 8 Comparison of XTE signals before and after filtration (filter cutoff 0.15 1/m)

In such way in local coordinate system, the direction of the straight coincided the abscissa and the ordinates were defining the deviation from that direction. In the terminology used in navigation, it is described as the cross-track error (XTE). This means the distance from the designated course and the measure of the error of the moving object position (Fig. 7).

In the case of analyzing a straight section transformed into a local coordinate system, the XTE value is determined by the relationship:

$$\mathrm{XTE}_l = \mathrm{FIX}_{\mathrm{loc}}(l, 2) \tag{16}$$

The course of XTE function (signals) analysis confirmed that the registered shape of track (characterized by relatively smooth curvatures due to track stiffness) is possible to be assessed after removing the uncertainty connected with measurement technique. The fast Fourier transformation (FFT) algorithm in Scilab [16] environment was used to perform Fourier transformation in order to analyze the XTE signal (frequency) by the relationship:

$$X_k = \sum_{l=0}^{N-1} \mathrm{XTE}_l \cdot \mathrm{e}^{\left(\frac{-2\cdot\pi\cdot i}{N}\cdot k\cdot l\right)} \tag{17}$$

where i imaginary unit, k overtone number and X_k Fourier transform.

After implementation (extension the matrix to the size of 2^k), the transformation followed with the display of resulting transformations. Then, the low-pass filter parameter was determined and the transformation was filtered. After that, the inverse transformation was performed. Under the assumption that filtration makes the measured shape of the track to be approximately the same as the real one, it might be considered that the differences between the filtered and unfiltered signal are measurement errors. The cut-off filter for the measured section was set to 0.15 1/m, and the achieved course of the XTE function is shown in Fig. 8.

For a single series of measurements, the mean standard error of horizontal plane position determination was $\sigma = 2.3$ mm. After the filtering, the expected position of the track axis XTE_{mean} was determined for all measurement rides. Based on the reference measurements, it can be concluded that the incorrections of straight track shape were determined sufficiently accurate. In further analysis, XTE_{mean} was a reference position of the track axis. The ΔXTE differences between filtered signal of individual rides and the expected value were interpreted as the accuracy of each measurement series. Absolute values of ΔXTE can be considered as uncertainty radius of positions determination in the global coordinate system. The Weibull distribution was adopted for further statistical analysis (Fig. 9). The expected value of $|\Delta XTE|$ was 2.6 mm, and the measurement error with a 95% probability does not exceed 6.5 mm (Table 2).

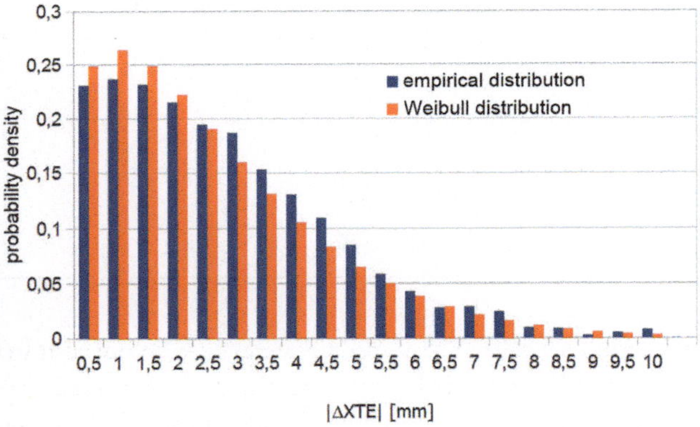

Fig. 9 The probability density function of Weibull distribution $W(\delta = 1.3, \lambda = 2.8)$ against empirical distribution $|\Delta XTE|$ calculated for the five measurement series

Fig. 10 Orthofotomap of the analyzed geometrical layout (with an indications of driving directions). Own study based on [18]

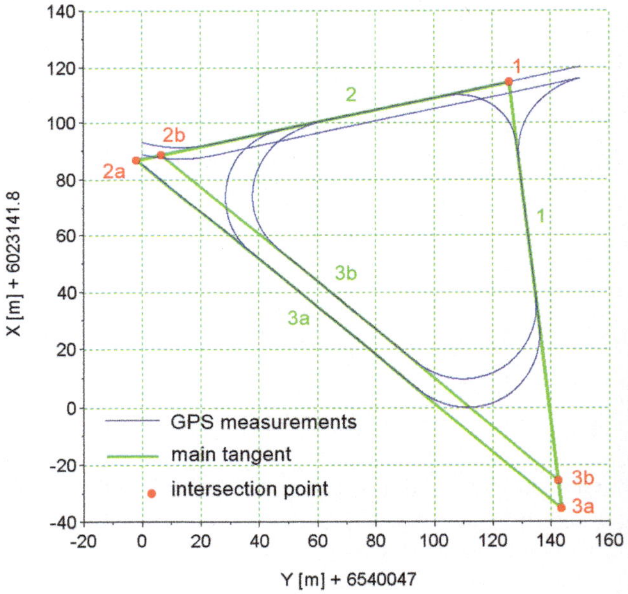

Fig. 11 Main tangents determined on the basis of satellite measurements (Gauss conformal projection)

Table 2 The statistics of distribution of satellite measurements uncertainties

Plane	Parameters of Weibull distribution	Expected value (mm)	Variance (mm^2)	Standard deviation (mm)	Probability level 95% (mm)
Horizontal	$\delta = 1.3, \lambda = 2.8$	2.6	3.9	2.0	< 6.5
Vertical	$\delta = 1.4, \lambda = 4.7$	4.3	10.1	3.2	< 10.5

Table 3 Calculated values of directional azimuth of straights on the tram loop

Main tangent	Measurement	Azimuth Az (°)	Expected value of azimuth $\overline{Az}$ (°)	Azimuth standard deviation σ_{Az} (°)	Azimuth delta Δ_{Az} (°)	Mean azimuth delta $\overline{\Delta_{Az}}$ (°)
1	1	353.2239263	353.2214942	0.0069377	0.0024320	0.0054753
	2	353.2098389			0.0116553	
	3	353.2243640			0.0028698	
	4	353.2283042			0.0068100	
	5	353.2273846			0.0058904	
	6	353.2226554			0.0011612	
	7	353.2139860			0.0075082	
2	1	257.6019966	257.6029893	0.0097036	0.0009927	0.0072064
	2	257.6072277			0.0042384	
	3	257.6197245			0.0167352	
	4	257.6007888			0.0022005	
	5	257.5924230			0.0105662	
	6	257.5915263			0.0114629	
	7	257.6072380			0.0042487	
3a	1	130.0287088	130.0264066	0.0047961	0.0023021	0.0035827
	2	130.0192412			0.0071654	
	3	130.0283197			0.0019131	
	4	130.0293568			0.0029502	
3b	1	130.0803666	130.0822637	0.0040273	0.0018971	0.0030837
	2	130.0795353			0.0027284	
	3	130.0868892			0.0046255	

Table 4 The list of geometrical parameters of the tram loop arcs

Arc	Radius (m)	Compound curve 1 (m)	Compound curve 2 (m)	Mean delta $\bar{\Delta}_{poss}$ (mm)
1	20.0	$R = 50$ $L = 5.5$	$R = 50$ $L = 5.5$	18.134
2a	24.8	$R = 50$ $L = 5.5$	$R = 50$ $L = 5.5$	102.082
2b	25.0	$R = 50$ $L = 5.5$	$R = 50$ $L = 5.5$	26.948
3a	24.8	$R_1 = 100$ $L_1 = 5.236$	$R = 50$ $L = 5.5$	35.395
		$R_2 = 50$ $L_2 = 5.236$		
3b	24.8	$R_1 = 100$ $L_1 = 5.236$	$R = 50$ $L = 5.5$	51.829

The straight sections with uniform inclination were selected to assess the accuracy in vertical plane. The accuracy analysis was performed similarly to the horizontal plane. Its results are shown in Table 2.

7 Evaluation of Other Geometrical Elements Based on Mobile Satellite Measurements

As the satellite measurement allows researchers to determine in a very precise way the track axis coordinates, there is a possibility to use the collected data not only for the inventory of tram lines, but also in geometrical layout designing. The computer program SATTRACK [3] which operates in Scilab [16] environment was developed for the designing purposes. The sample geometrical layout of tram line in Gdansk was analyzed using the described tool.

The inventory of the tram loop Chelm Witosa in Gdansk was made in order to illustrate the possibilities offered by the presented method. Tracks were measured twice, in 2012 and 2013. Track system of the loop had been put into operation in 2007, and then, it was extended in 2012. It consists of two tracks of the Lostowice–Downtown direction, two tracks of reception siding and one track of departure siding (Fig. 10). Eight turnouts were installed on the loop. Coordinates of 26,868 points were determined during the measurements. Each of the tracks was measured at least three times.

At the beginning, the straight sections were separated from the arches. Cut down straight sections were

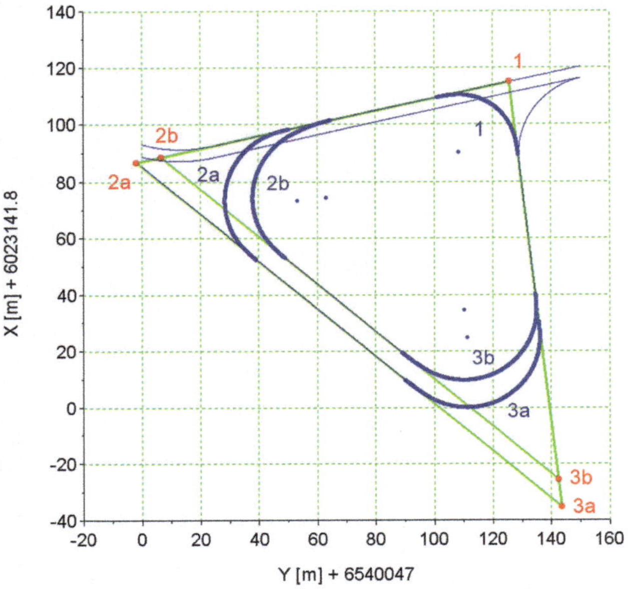

Fig. 12 The recreated arc section filed in the measured geometrical layout (Gauss conformal projection)

subsequently processed. The equation of the straight section that fits best to the data obtained from the measurements was determined using the least squares method. The vertices of the polygon system (Fig. 11) were calculated from the coefficients of straights. In the next step, the radii of the arcs and the length of transition curves were determined. The curved sections were located between the straight sections of the track. The straights 1 and 2 (marked on Fig. 11) were measured seven times, straight 3a was measured four times, and 3b three times. The azimuths (angle between the direction of the north and the direction of the straight) and the coordinates of the vertices were the basic parameters describing these straights. Table 3 shows the accuracy of the designation of the different straight azimuths. The expected value was calculated as the average value for subsequent test rides, and the delta is a measure of the difference between the azimuth of the measurement signal and the expected value.

Next, an attempt to fit arcs consisting of a circular part and the transition curves was made for the measured track. The average difference between the design and location of existing (measured) track axis was a critical parameter in determining the quality of the fit, which was defined as $\bar{\Delta}_{poss}$—mean delta (Table 4). The Δ_{poss} values along the straight section is equal to the XTE while along the arc is calculated as follows:

$$\Delta_{\text{poss}} = \sqrt{(X_0 - X_l)^2 + (Y_0 - Y_l)^2} - R \tag{18}$$

where R arc radius (m), X_0 x position of circle center (m) and Y_0 y position of circle center (m).

Due to the fact that all the circular arcs, at least in its initial or final area, were a part of the turnout's diverging tracks, two kinds of transition curves might have been used. The compound curve consisting of transition arc with 50 m radius and length of 5.236 m was used in diverging track of turnout. When the arc was combined with the straight section, two kinds of transitions can be applied: cubic parabola or compound arc consisting of two transition arcs (5.236 m length) with the 100 and 50 m radius each with 5.236 m length [15].

The increased mean delta of fitting of the arc #2a might be the result from the construction process of the tram loop, which was carried out in several stages. During the expansion of the tram loop, the arc #2a has been rebuilt. Two turnouts and track crossing were added on its length, which caused deformation of the shape and resulted in the

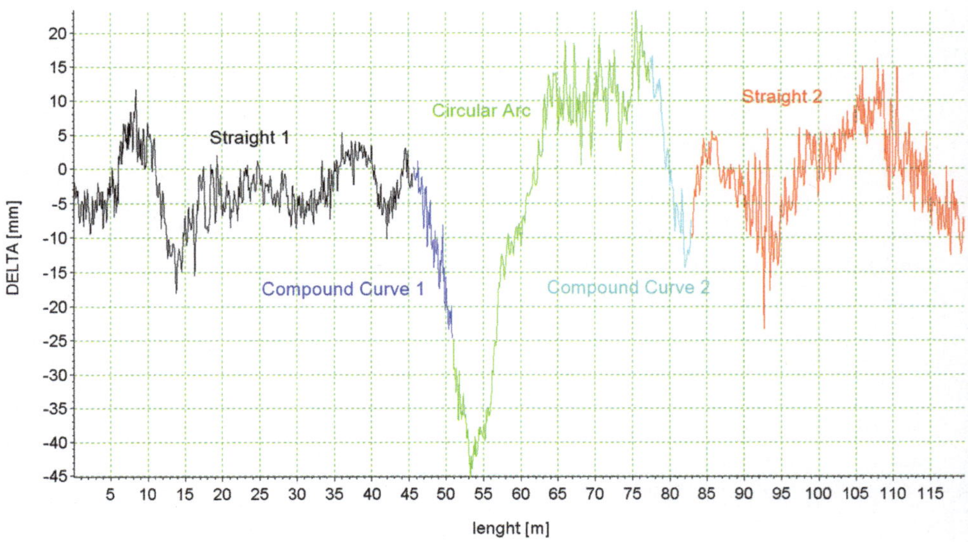

Fig. 13 The horizontal differences chart for the single measurement on arc #1 (with the adjacent straights)

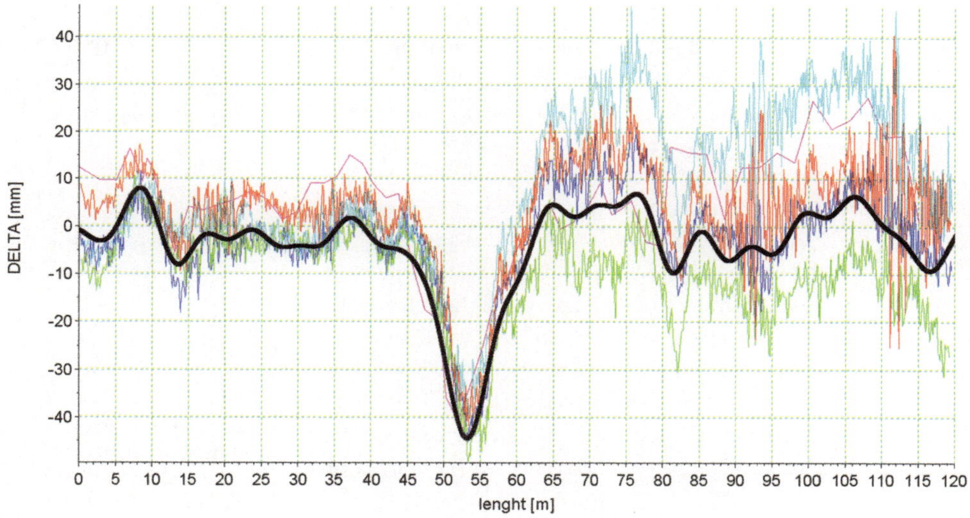

Fig. 14 The horizontal differences chart for all measurements on arc #1 against the expected value of track centerline deformation

Table 5 Statistic estimators of delta parameters

Measurement	Mean value $\bar{\Delta}_{poss}$ (mm)	Standard dev. $\sigma_{\Delta_{poss}}$ (mm)
1	5.51	6.31
2	1.93	2.26
3	6.14	4.71
4	2.81	2.80
5	3.05	2.65

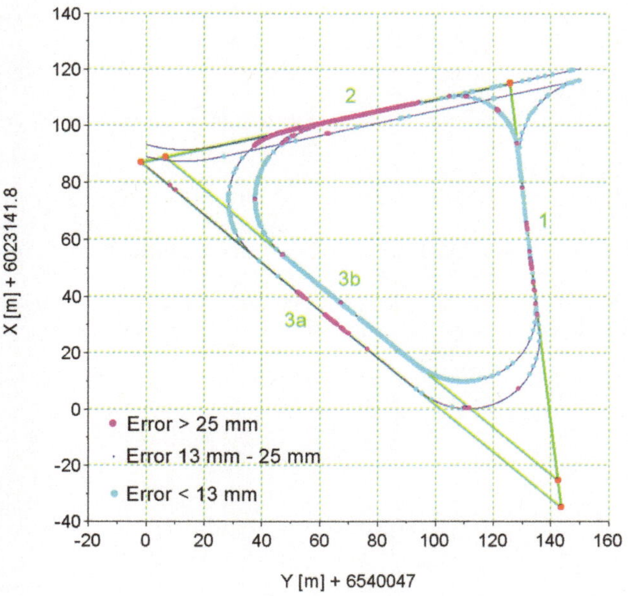

Fig. 15 Determination error distribution. Chelm Witosa tram loop

decrease of passenger comfort. The reconstructed geometric layout of the loop, against the measured layout, is shown in Fig. 12.

Most of the measurements were conducted on the arc #1 and—as it turned out—its shape was the closest to the project. For this reason, it was expected that the results of further analysis of this arc would be the most reliable. The main tangent polygon system, best suited to the actual shape of the tram tracks, was created on the basis of the seven measurement rides. It was calculated based on the expected azimuth and coordinates of the vertices. Then, the difference between the measuring signal and designed geometrical layout was calculated for each ride (Fig. 13). The next step was to determine—by filtering the signal—the expected value of track shape (Fig. 14). This made it possible to determine the actual error of position determination for each of the rides (Table 5).

The graphs in Figs. 13 and 14 clearly show that the error of the positions determination on the straight section #1 is considerably smaller than that in the straight section #2. From the 26 868 coordinates recorded on the tram loop, the coordinate positions error of 1637 (6.1%) was in the range of less than 13 mm. With the distribution of designations accuracy, shown in Fig. 15, the biggest number of these points is located on the straights sections of #1 and #3b. On the other hand, 712 (2.6%) coordinates are determined with an error of more than 25 mm, and most of them are located on the straight #2. The distribution of errors might be the result of multi-storey buildings, located on the north side of the tram loop, as it is a significant obscuration of horizon.

8 Conclusions

The development of urban rail transport is the cause of exploration (by designers and managers of infrastructure) of innovative methods to facilitate the process of upgrading and expanding the tram network. Particularly, an important course of action is to find the most efficient way to define the shape of the track and determine its deformation.

The technique of Mobile Satellite Surveying provides fast and precise determination of the track axis coordinates. The experiences of the research team allows the researchers to formulate the thesis that *with the development of measurement techniques the accuracy of methods of satellite positioning will be continuously growing.*

Based on the specified coordinates, it can be very easy to create a precise reproduction of the track layout: main tangents, angles of return, curve radii and the length of transition curves. The computer program SATTRACK is developed for adapting it to the specific design of tram routes (small curve radii, large values of cant, short transition curves, large variability of velocity on the length of the route).

Precise planning of the measurement campaigns becomes particularly important when measurements were performed within the highly urbanized area due to the lack of available satellites. It is important so that the number of available satellites should be maximized. This allows for a reduction in the impact of horizon obscuration on the accuracy of position determination. The combination of satellite measurements and modern design methods will significantly improve the quality of track designing, both for tram and rail.

References

1. Elberink SO, Khoshelham K (2015) Automatic extraction of railroad centerlines from mobile laser scanning data. Remote Sens 7(5):5565–5583
2. Izvoltova J, Villim A, Kozak P (2014) Determination of geometrical track position by Robotic Total Station. Proc Eng 91:322–327
3. Koc W, Chrostowski P (2014) Computer-aided design of railroad horizontal arc areas in adapting to satellite measurements. J Transp Eng 140(3):04013017-1–04013017-8
4. Koc W, Specht C (2010) Application of the Polish active GNSS geodetic network for surveying and design of the railroad. In: First international conference on road and rail infrastructure—CETRA 2010, Opatija, Croatia, pp 757–762
5. Koc W, Specht C (2011) Selected problems of determining the course of railway routes by use of GPS network solution. Arch Transp 23(3):303–320
6. Koc W, Specht C, Chrostowski P, Palikowska K (2012) The accuracy assessment of determining the axis of railway track basing on the satellite surveying. Arch Transp 34(3):307–320
7. Munson DC (2004) High-precision GPS for continuous monitoring of rail. Final report for High-Speed Rail IDEA Project 26, Washington
8. Specht C (2007) GPS system. Pelplin **(in Polish)**
9. Specht C et al (2008) Infrastructure and services test execution of ASG-EUPOS precise satellite positioning system. A research project commissioned by the Polish Office of Geodesy and Cartography (in Polish), Scientific Consortium: Gdansk University of Technology, University of Warmia and Mazury in Olsztyn, and Polish Naval Academy in Gdynia
10. Specht C, Koc W, Chrostowski P, Szmagliński J (2014) Assessment of the mobile satellite surveying accuracy in horizontal and vertical planes (in Polish). In: VII Sci.-Tech. conference “Designing, building and maintenance of infrastructure in rail transportation INFRASZYN 2014”, Zakopane, Poland, pp 264–278
11. Specht C, Koc W, Chrostowski P, Szmagliński J (2015) Satellite inventory of tram track geometrical layout. In: 13th International conference and exhibition railway engineering 2015. Engineering Technics Press, Edinburgh
12. Specht C, Koc W, Smolarek L, Grządziela A, Szmagliński J, Specht M (2014) Diagnostics of the tram track shape with the use of the global positioning satellite systems (GPS/Glonass) measurements with a 20 Hz frequency sampling. J VibroEng 16(6):3076–3085
13. Specht C, Nowak A, Koc W, Jurkowska A (2011) Application of the Polish Active Geodetic Network for railway track determination, transport systems and processes—marine navigation and safety of sea transportation. CRC Press, London, pp 77–81
14. Szwilski TB (2003) Determining rail track movement trajectories and alignment using HADGPS. In: AREMA conference, Chicago
15. Technical guidelines for designing, constructing and maintenance of tram tracks (in Polish 1983), The Polish Ministry of Administration, Land Development and Environmental Protection, The Department of Public Transport and Roads, Warsaw
16. Web page: www.scilab.org (the Free Platform for Numerical Computation)
17. Web page: www.westernsafety.com, Nolan page 1—tool and supply carts
18. Web page: www.google.pl/maps
19. Wienia R (2015) Combined aerial and train mounted lidar system provide a fast and innovative approach to surveying railway infrastructure and track geometry. In: International conference railway engineering 2015, Edinburg
20. Wildi T, Glaus R (1999) A multisensor platform for kinematic track surveying. In: International workshop on mobile mapping technology, Bangkok
21. Zywiel J, Oberlechner G (2001) Innovative measuring system unveiled. Int Railw J 41(9):31–33

21

Dynamic Property Analysis of At-Ground Structure Under Moving Maglev Train

Dangxiong Wang[1] · Xiaozhen Li[1] · Jie Geng[1] · Yanlong Ge[1]

Abstract In order to study the vertical coupling vibration of low–medium-speed Maglev train and at-ground structure system, the vertical coupling vibration model of low–medium-speed Maglev train-bridge system is established firstly based on the co-simulation of SIMPACK and ANSYS. The method of co-simulation is verified through the test experiments on a 20-m simply supported beam dynamic load experiment test line. Later, the vertical coupling vibration dynamics simulation model of low–medium-speed Maglev train and at-ground structure system is built and the dynamic simulation analysis is performed. According to the analysis, the vibration frequency of the at-ground structure is relatively high and the first vertical vibration frequency is 32.9 Hz; the vertical displacement and acceleration of the frame's center of the at-ground structure are bigger than the bottom's center; the vibration of the frame is the high-frequency vibration (comparing with the bottom), and the acceleration of the frame's center is obviously greater than the bottom's between 50 and 100 Hz.

Keywords Low–medium-speed Maglev train · At-ground structure · Coupling vibration · High-frequency vibration · Co-simulation

✉ Dangxiong Wang
Dangxiong333@163.com

[1] Department of Bridge Engineering, School of Civil Engineering, Southwest Jiaotong University, Chengdu 610031, China

Editor: Eryu Zhu

1 Introduction

Nowadays, with the rapid growth of China's economic and urban population, a fast, economic land occupation, environment protection way of transport is eagerly needed. Maglev transport has been born with many advantages, including well climbing capacity, small turning radius, low noise, low pollution, and good environmental compatibility. As a representative of maglev transportation, low–medium-speed maglev transportation is particularly suitable for medium–short-distance transport in cities and has good development in the future urban transportation [1].

The low–medium-speed maglev line consists of bridges and subgrade structure. Because the Maglev train embraces the rail, in subgrade section, the maglev guideway has to be propped up by a special structure and ensure the Maglev train having enough space to embracing the rail in vertical. The structure is called at-ground structure [2]. The at-ground structure is a concrete beam which is placed on the processing subgrade in advance and supported by uniform elastic subgrade, forming the combination system of Maglev train-at-ground structure subgrade. Low–medium-speed Maglev train stabilizes near the rated suspension gap through adjusting the active control electromagnetic force and keeps running smoothly. Therefore, both the bridge and at-ground structure will produce deformation under the action of Maglev train and affect the suspension gap, then affecting the smooth running of the Maglev train. Lots of scholars are all committed to Maglev train-bridge system coupling vibration: Ref. [3] analyzed the bridge dynamic responses when the high-speed Maglev train runs on the bridge at the resonance speed; Ref. [4] considering the PI control and establishing two degrees of freedom Maglev train-simply supported beam (simplified as Euler–Bernoulli beam) vertical coupling vibration model, discussed the resonance effect of coupling model and the influences of the suspension control parameters on the coupling vibration; Ref. [5] building 10 degrees of freedom TR06 Maglev train model and simplifying the suspension control as linear spring-damping system, discussed the Maglev train-bridge vertical coupling vibration under random irregularity excitation; Refs. [6–9] establishing Maglev train-bridge vertical coupling vibration model, studied the effect of bridge stiffness, material and structure form to the dynamic responses of Maglev train-bridge system; Refs. [10, 11] analyzed dynamic characters of suspension control system and bridge and also studied the changing rules of suspension gaps at different spans and bridge stiffness; Refs. [12] considering 12 degrees of freedom Maglev train model and PD control, discussed the impact effect of Maglev train to bridge; Refs. [13, 14] analyzed the influence of wind load on the coupling vibration of Maglev train-bridge system.

The at-ground structure is supported by the uniform elastic subgrade. Under the train load, the stress mechanism is obviously different from the bridge. Under the long time operation, some uncontrollable factors will affect the Maglev train-at-ground structure vertical coupling vibration, including the supported stiffness change of subgrade, uneven settlement and the supported void. The at-ground structure is generally used frame structure, and ordinary Euler–Bernoulli beam model cannot predict the dynamic performance accurately. So, it is necessary to research the Maglev train-at-ground structure system vertical coupling vibration and the related articles is little. On the basis of predecessors' research, this paper considered the PID active suspension control and used the delicate FEM model of bridge and at-ground structure based on co-simulation of SIMPACK and ANSYS. Firstly, the Maglev train-bridge vertical system coupling vibration model was validated based on one 20-m simply supported beam load test of a low–medium-speed maglev test line. Then, the Maglev train-at-ground structure vertical system coupling vibration model was established to be analyzed.

2 Theoretical Model

2.1 The Model of Maglev Train

In this paper, the five modules low–medium-speed Maglev train was considered. Each car is supported by five pairs of suspension frames. The vehicle and suspension frame are connected through air spring, slide, etc. Each suspension frame has four independent suspension electromagnets, which produced levitation force and act on the suspension frame and bridge to achieve the maglev car suspension steady. The vehicle has 40 suspension electromagnets totally, and the neighboring electromagnets longitudinal distance is 0.7 m. Figure 1 shows the Maglev train suspension.

Car body and suspension frame are both considered six degrees of freedom. The air spring is a treated spring-damping element. The vehicle's degree of freedom is shown in Table 1, and the total number is 86. According to d'Alembent, the movement differential equations of car body and suspension frames are established [1]. Table 2 shows the main calculation parameters of Maglev train.

2.2 Suspension Control Model

The relationship among electromagnetic force, suspension gap and current is [15]:

$$F(c, i(t)) = \frac{\mu_0 A n^2}{4}\left[\frac{i_0 + \Delta i(t)}{c_0 + \Delta c(t)}\right]^2 \tag{1}$$

This paper considers PID control based on displacement–speed–acceleration feedback. Because the speed signal cannot be measured directly, it should be reconstructed by displacement and acceleration signals in the state observer. The state of construction is as follows:

$$\Delta\dot{\hat{c}}(t) = \Delta\dot{\hat{z}}(t) + 2\xi_0\omega_0[\Delta c(t) - \Delta\hat{c}(t)] \tag{2}$$

$$\Delta\ddot{\hat{z}}(t) = \Delta\ddot{z}(t) + \omega_0^2[\Delta c(t) - \Delta\hat{c}(t)] \tag{3}$$

After using state reconstruction, feedback current can be expressed as the function of displacement speed, acceleration feedback signal, as follows:

$$\Delta i(t) = K_p\Delta c(t) + K_v\Delta\dot{\hat{c}}(t) + K_a\Delta\ddot{z}(t) \tag{4}$$

So, the electromagnetic levitation force of control point is shown as:

$$F(c, i(t)) = \frac{\mu_0 A n^2}{4}\left[\frac{i_0 + K_p\Delta c(t) + K_v\Delta\dot{\hat{z}}(t) + K_a\Delta\ddot{z}(t)}{c_0 + \Delta c(t)}\right]^2 \tag{5}$$

The levitation force is applied to the electromagnet, and the active suspension control of Maglev train can be realized.

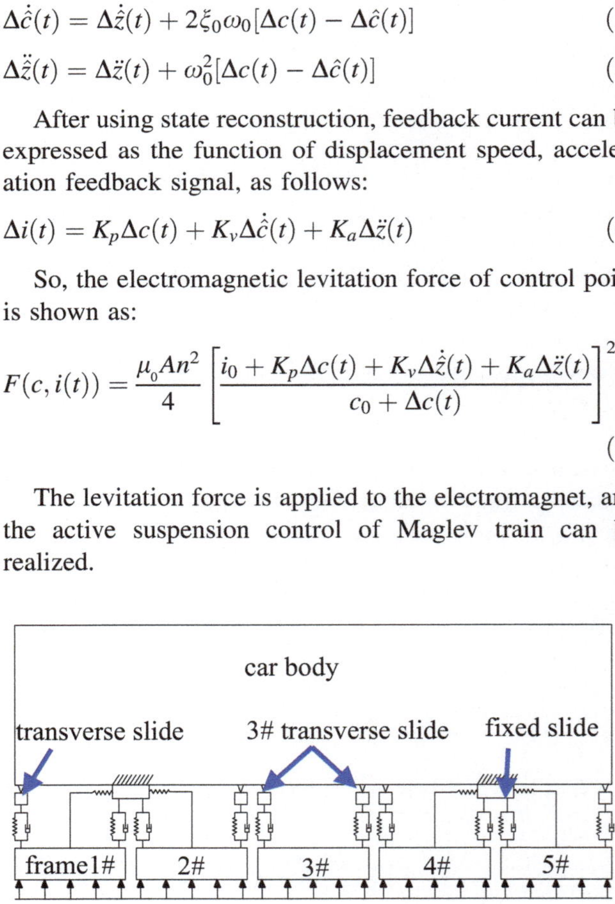

Fig. 1 Suspension of five modules *L_M*-speed Maglev train

Table 1 Dofs of the vehicle model

Component	x	y	z	α	β	γ	Remark
Suspension frame	x_{fi}	y_{fi}	z_{fi}	α_{fi}	β_{fi}	γ_{fi}	$i = 1$–10
3# sliding table	–	y_{smi}	–	–	–	–	$i = 1$–4
Other sliding tables	x_{si}	y_{si}	–	–	–	–	$i = 1$–8
Vehicle	x_{ci}	y_{ci}	z_{ci}	α_{ci}	β_{ci}	γ_{ci}	$i = 1$

Table 2 Calculation parameters of the Maglev train (30*t*)

Sign	Name	Unit	Parameters
M_c	Vehicle mass	kg	18.77×10^3
M_f	Frame mass	kg	1000
I_{cx}	Vehicle around *x*-axis	kg m^2	2.21×10^3
I_{cy}	Vehicle around *y*-axis	kg m^2	3.85×10^5
I_{cz}	Vehicle around *z*-axis	kg m^2	3.88×10^5
I_{fx}	Frame around *x*-axis	kg m^2	460
I_{fy}	Frame around *y*-axis	kg m^2	1150
I_{fz}	Frame around *z*-axis	kg m^2	1200
K_{sz}	Air spring stiffness	kN/m	80
C_{sz}	Air spring damping	kN/s/m	5

2.3 The Model of At-Ground Structure

The at-ground structure is a concrete beam which is placed on the processing subgrade in advance and supported by uniform elastic subgrade. It consists of five spans continuous rigid bridge (on top) and bottom. Subgrade consists of graded crushed stone and AB packing and the function of subgrade likes stiffness. As shown in Fig. 2, the subgrade is simplified uniform spring for theoretical model. Figure 3 shows the FE model of at-ground structure.

Because of using the modal superposition, SIMPACK eventually recognizes the modal information of at-ground structure. ANSYS has the powerful flexible body modal analysis function and can build any exquisite flexible structure model. So, for any structure (including bridge), this co-simulation method can be adopted and the theory is the same.

3 Model Validation

This co-simulation method was verified based on the 20-m simply supported beam dynamic load experiment of a test line. Using the same theory, the Maglev train-bridge vertical system coupling vibration model was established. Figure 4 shows the middle span section of the simply supported beam, using C50 concrete.

Using three Maglev train marshalings, the weight is $25t + 30t + 30t$ and the velocity of train is 80 km/h. The track irregularity spectrum recommended by Ref. [16] is adopted. Figure 5 shows the comparison of the simulation and measurement time–history curves of vertical dynamic deformation and acceleration in middle span of the simply supported beam. Figure 6 shows the comparison of the simulation and measurement frequency–history curves of acceleration in middle span of the simply supported beam.

According to the comparison, in both time–history and frequency–history, the simulation values are fit with the measured values. That is to say, the Maglev train-bridge vertical system coupling vibration model based on co-simulation method is reliable. Because of the same simulation theory, the Maglev train-at-ground structure vertical system coupling vibration model is also reliable.

4 Analysis of Maglev Train-At-Ground Structure Vertical System Coupling Vibration

4.1 Natural Frequency of At-Ground Structure

The at-ground structure FE model was established in ANSYS using solid95-type element and C40 concrete. The combin14-type element was used to simulate the uniform stiffness, and the stiffness value is 200 MPa/m. The section size is shown in Fig. 2.

Table 3 shows the nature frequency of the at-ground structure within 100 Hz. Figure 7 shows the typical vibration modes. As shown in Table 3 and Fig. 7, the nature frequencies of the at-ground structure are obviously high. There are only seven modes within 100 Hz, and the first vertical nature frequency is 32.9 Hz, and before the 32.9 Hz, there have no vertical vibration modes. The reason is the at-ground structure is supported by the strong uniform elastic subgrade and its stiffness is big. Before 80 Hz, the vibration is mainly the overall vertical modes. In the second overall vertical modes, there also has the frame's local bend vibration.

4.2 Dynamic Response Analysis of the At-Ground Structure

Using three Maglev train marshalings, the weight is $25t + 30t + 30t$ and the velocity of train is 80 km/h.

Figure 8 shows the vertical dynamic deformation in frame's center (as shown in Fig. 2, point B) and bottom's center (as shown in Fig. 2, point A) of at-ground structure. As shown in Fig. 8, the dynamic deformation time–history curves frame's center and bottom's center both have three peaks because of the three Maglev train marshalings. The maximum value is 0.466 mm (frame's center) and

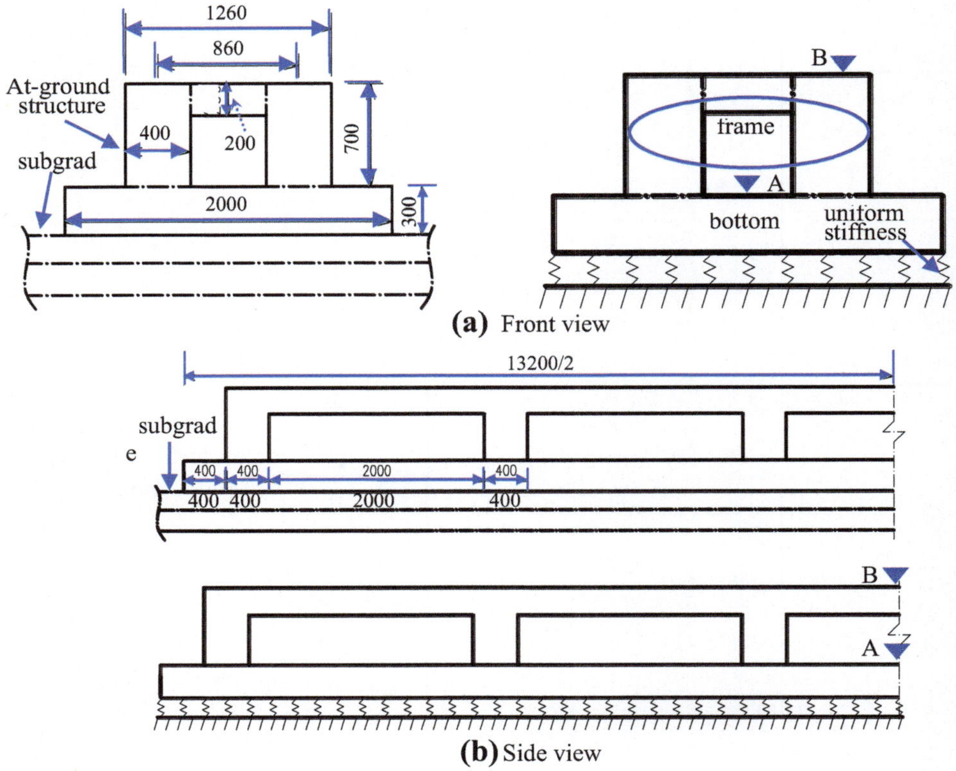

Fig. 2 At-ground structure simplified model (unit: mm). **a** Front view. **b** Side view

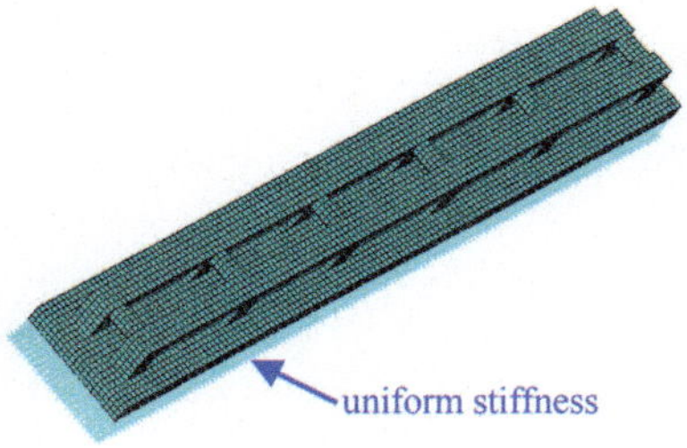

Fig. 3 At-ground structure FE model

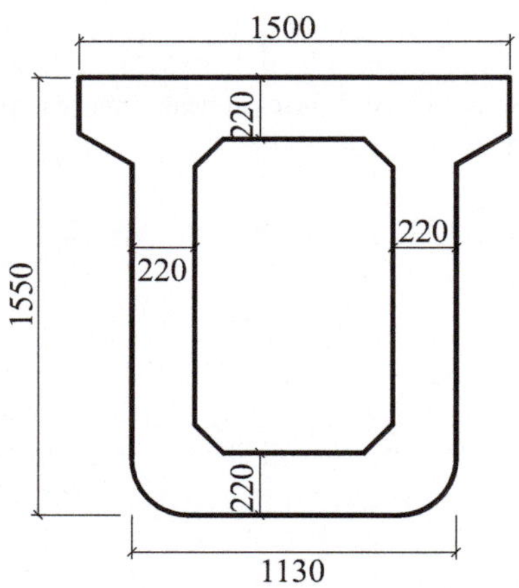

Fig. 4 Cross section of 20-m simply supported beam (unit: mm)

0.378 mm (bottom's center), respectively. Because the total dynamic deformation of frame's center is composed of the total bottom's center dynamic deformation and the local frame's center dynamic deformation, the dynamic deformation of frame's center is bigger than the bottom's center dynamic deformation.

Figure 9 shows the vertical acceleration in frame's center and bottom's center of at-ground structure. As shown in Fig. 9, the peak values of vertical acceleration in frame's center and bottom's center are 6.017 and 1.428 m/s^2. The value of frame's center is much larger than the value of bottom's center. In order to analysis the vibration characteristics of frame and bottom's center in-depth, Fig. 10 shows the acceleration frequency–history curves of frame's center and bottom's center. As shown in Fig. 10, there are obviously acceleration peaks of frame's center and bottom's center within 100 Hz and the frequency is frame's center: 31.2, 62.5, 87.8, 95.7 Hz; bottom's center: 31.2, 62.5, 95.7 Hz. Before the 50 Hz, the acceleration of frame's center is little different with bottom's center acceleration and after the 50 Hz, the acceleration of frame's center is much bigger than bottom's center acceleration.

In order to analyze the vibration, due to the natural vibration analysis of at-ground structure, the first and second frequency of vertical overall bend vibration is 32.9 and 62.4 Hz separately and at 88.2, 91.4, 97.2 Hz, there is

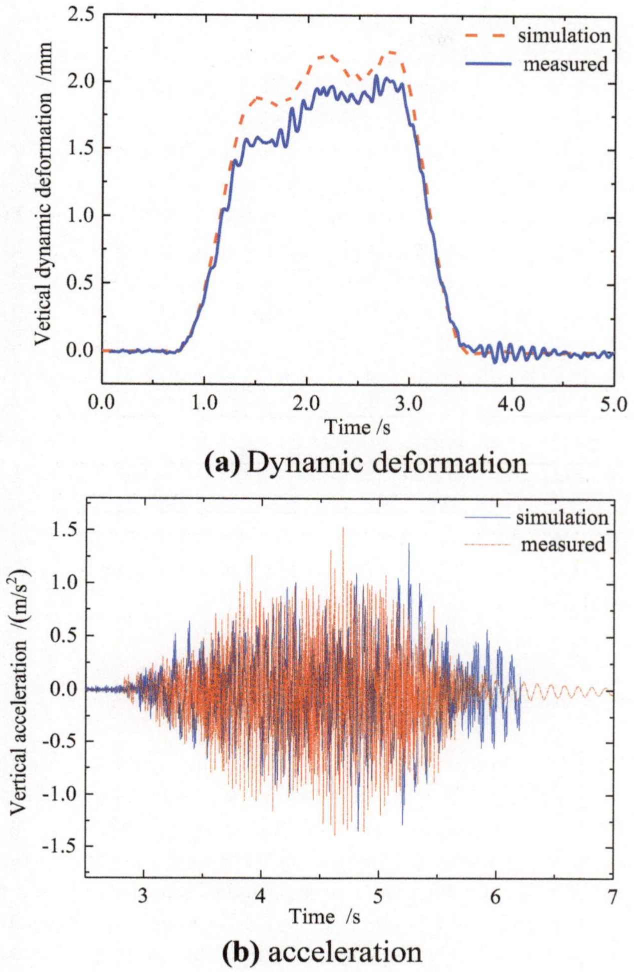

Fig. 5 Comparison of the simulation and measurement time–history curves. **a** Dynamic deformation. **b** Acceleration

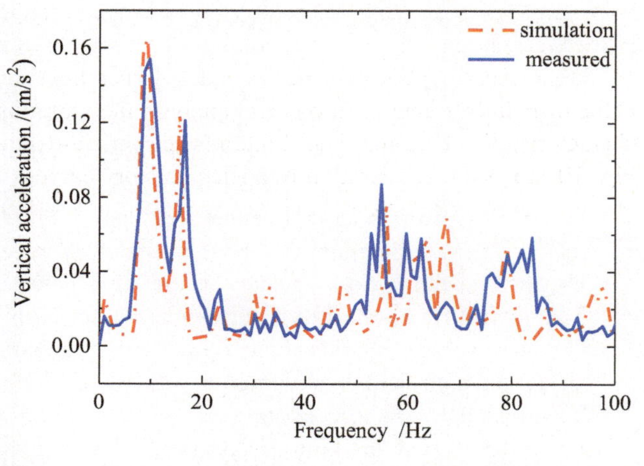

Fig. 6 Comparison of the simulation and measurement frequency–history curves

frame's local vertical vibration. So, the vibration peaks of frame's center and bottom's center are existed at this

Table 3 The modal and nature frequency of at-ground structure

Frequency	Modal
32.9	1st symmetric overall vibration
45.6	1st antisymmetric overall vibration
62.4	2nd symmetric overall vibration
77.5	2nd antisymmetric overall vibration
88.2	Frame local vibration
91.4	Frame local vibration
97.2	Frame local vibration

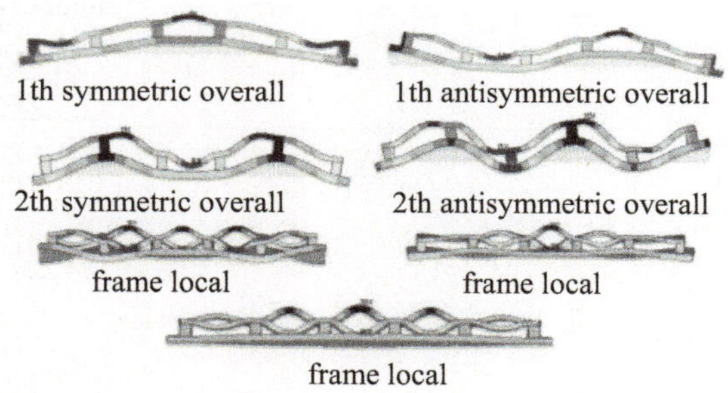

Fig. 7 The typical modal of at-ground structure

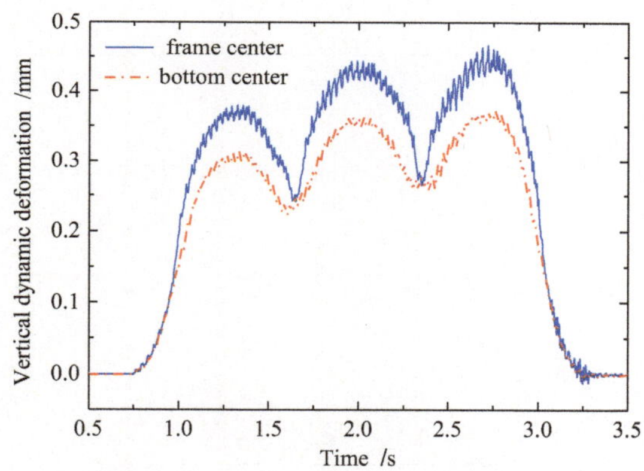

Fig. 8 The vertical dynamic displacement time–history curves of at-ground structure

frequency. From the second vertical bend modal (62.4 Hz), the frame's local vertical vibration modal appeared. So, before about 50 Hz, because of the overall vertical vibration of at-ground structure, the acceleration values of frame's center and bottom's center are nearly equal and between 50 and 100 Hz, and the acceleration of frame's center is much bigger than bottom's center acceleration based on the frame's local vertical vibration. Therefore, the frame's vertical vibration is high frequency (compared with the bottom's vibration).

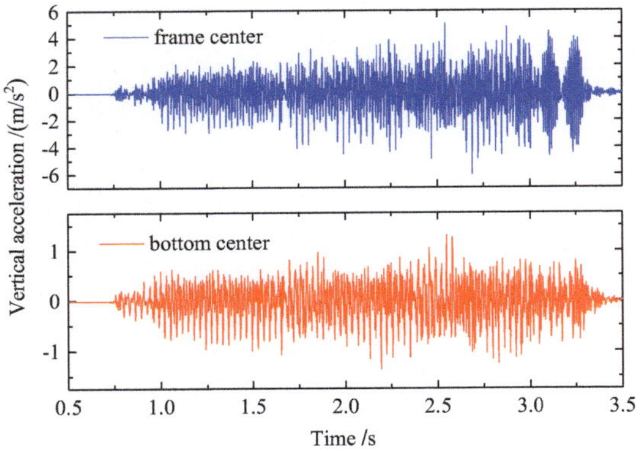

Fig. 9 The vertical acceleration time–history curves of at-ground structure

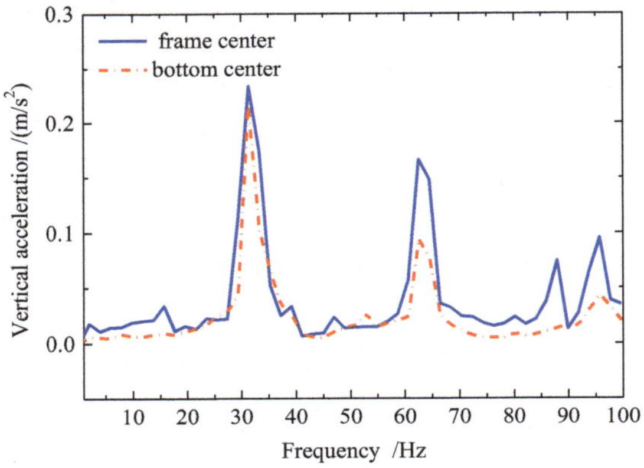

Fig. 10 The vertical acceleration frequency–history curves of at-ground structure

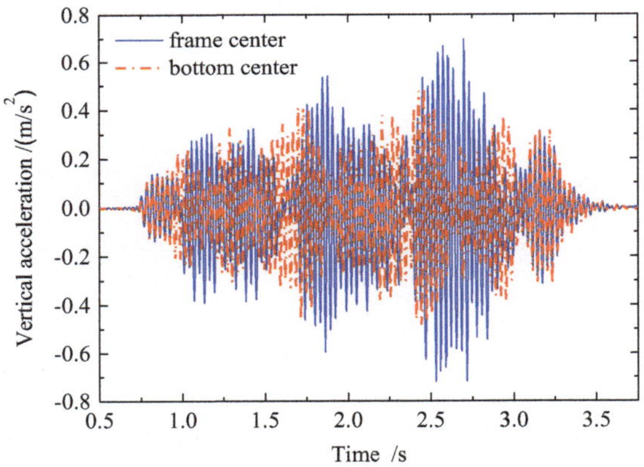

Fig. 11 The vertical acceleration of at-ground structure (50-Hz low-pass filter)

Figure 11 shows the vertical acceleration time–history curves of frame's center and bottom's center (50-Hz low-pass filter).

As shown in Fig. 11, comparing with the unfiltered curves, the 50-Hz low-pass filter acceleration curves of frame's center are little different with bottom's center acceleration curves and both the curves (50-Hz low-pass filter) have obviously three peaks (due to three Maglev train marshalings). So, although the frame's center acceleration value is much bigger than bottom's center acceleration value (unfiltered curves and time–history), it is caused by the frame local vertical vibration and the vibration is high frequency (more than 50 Hz). The different low-frequency vibration is little caused by the overall vertical vibration (less than 50 Hz). Due to the China's <High-Speed Railway Design Standard>, the bridge's limited value of vertical acceleration at ballast less track is 0.5 g considering the 20-Hz low-pass filter. The limited filter value references the European Standard. Because the vibration of frame and bottom belongs to two different frequency domains, using the filter limited value (20-Hz low-pass filter) to filter the vertical acceleration curves based on the China's <High-Speed Railway Design Standard> is slightly unsuitable.

5 Conclusions

This paper considered the PID active suspension control and used the delicate FEM model of bridge and at-ground structure based on co-simulation of SIMPACK and ANSYS. Firstly, the Maglev train-bridge vertical system coupling vibration model was validated based on one 20-m simply supported beam load test of a low–medium-speed maglev test line. Then, the Maglev train-at-ground structure vertical system coupling vibration model was established to be analyzed. The main conclusions are shown:

1. The Maglev train-at-ground structure vertical system coupling vibration model based on the co-simulation method is reliable.
2. Being supported by the strong uniform elastic subgrade, the first-order nature frequency of the at-ground structure is 32.9 Hz. Due to the frame local vibration, the acceleration of frame's center is 6.017 m/s^2, which is bigger than the bottom's center acceleration (1.428 m/s^2). The acceleration of frame's center is much bigger than the bottom's center acceleration between 50 and 100 Hz and has little difference within 50 Hz. Comparing with the vibration of bottom, the frame shows the high frequency vibration (more than 50 Hz), which lead to the acceleration of frame's center is much bigger than the bottom's center in time–history. In view of the obvious difference between at-ground structure and the conventional bridge, the vibration evaluate standard should be researched in further study.

Acknowledgements This research was supported by National Natural Science Foundation of China (NSFC, Grant Nos. 51308469, 51378429).

References

1. Deng XX (2009) The study of middle-low speed maglev vehicle system dynamic. Southwest Jiaotong University, Chengdu **(in Chinese)**
2. Wang LD (2014) Research on key technology of low and medium speed maglev guideway girder. Southwest Jiaotong University, Chengdu **(in Chinese)**
3. Shi J, Yau JD, Wang YJ (2012) Dynamic response of guideway girders due to high-speed maglev trains moving at resonant speeds. Eng Mech 29(12):196–203 **(in Chinese)**
4. Yau JD (2009) Vibration control of maglev vehicles traveling over a flexible guideway. J Sound Vib 321:184–200
5. Zhao CF, Zhai WM (2002) Maglev vehicle/guideway vertical random response and ride quality. Veh Syst Dyn 38(3):185–210
6. Cai CB (2001) Maglev vehicle/elevated-beam guideway vertical coupling dynamics. J China Railw Soc 23(5):27–33 **(in Chinese)**
7. Shi J, Wei QC, Wu FY (2003) Study on vibration of the beam of magnetic levitation express railway and its control. China Saf Sci J 10:76–80 **(in Chinese)**
8. Jiang WL, Gao MM (2004) Study of the effect of track beam parameters on vertical coupled dynamic response of maglev vehicle-viaduct. China Railw Sci 25(3):71–75 **(in Chinese)**
9. Lee JS, Kwon SD, Kim MY et al (2009) A parametric study on the dynamics of urban transit maglev vehicle running on flexible guideway bridges. J Sound Vib 328(3):301–317
10. Zhai WM, Zhao CF (2005) Dynamics of maglev vehicle/guideway systems (I)—magnet/rail interaction and system stability. Chin J Mech Eng 41(7):1–10 **(in Chinese)**
11. Zhai WM, Zhao CF (2005) Dynamics of maglev vehicle/guideway systems (II)—modeling and simulation. Chin J Mech Eng 41(8):163–175 **(in Chinese)**
12. Shan CS (2014) Study on vertical coupling vibration of low-medium speed maglev train-bridge system. Southwest Jiaotong University, Chengdu **(in Chinese)**
13. Yau JD (2009) Aerodynamic vibrations of a maglev vehicle running on flexible guideways under oncoming wind actions. J Sound Vib 329:1743–1759
14. Kwon SD, Lee JS, Moon JW et al (2008) Dynamic interaction analysis of urban transit maglev vehicle and guideway suspension bridge subjected to gusty wind. Eng Struct 30:3445–3456
15. Shi J, Wei QC, Zhao Y (2007) Analysis of dynamic response of the high-speed EMS maglev vehicle/guideway coupling system with random irregularity. Veh Syst Dyn 45(12):1077–1095
16. Liang X (2015) Study on maglev vehicle/guideway coupled vibration and experiment on test rig for a levitation stock. Southwest Jiaotong University, Chengdu **(in Chinese)**

Permissions

All chapters in this book were first published in URT, by Springer International Publishing AG.; hereby published with permission under the Creative Commons Attribution License or equivalent. Every chapter published in this book has been scrutinized by our experts. Their significance has been extensively debated. The topics covered herein carry significant findings which will fuel the growth of the discipline. They may even be implemented as practical applications or may be referred to as a beginning point for another development.

The contributors of this book come from diverse backgrounds, making this book a truly international effort. This book will bring forth new frontiers with its revolutionizing research information and detailed analysis of the nascent developments around the world.

We would like to thank all the contributing authors for lending their expertise to make the book truly unique. They have played a crucial role in the development of this book. Without their invaluable contributions this book wouldn't have been possible. They have made vital efforts to compile up to date information on the varied aspects of this subject to make this book a valuable addition to the collection of many professionals and students.

This book was conceptualized with the vision of imparting up-to-date information and advanced data in this field. To ensure the same, a matchless editorial board was set up. Every individual on the board went through rigorous rounds of assessment to prove their worth. After which they invested a large part of their time researching and compiling the most relevant data for our readers.

The editorial board has been involved in producing this book since its inception. They have spent rigorous hours researching and exploring the diverse topics which have resulted in the successful publishing of this book. They have passed on their knowledge of decades through this book. To expedite this challenging task, the publisher supported the team at every step. A small team of assistant editors was also appointed to further simplify the editing procedure and attain best results for the readers.

Apart from the editorial board, the designing team has also invested a significant amount of their time in understanding the subject and creating the most relevant covers. They scrutinized every image to scout for the most suitable representation of the subject and create an appropriate cover for the book.

The publishing team has been an ardent support to the editorial, designing and production team. Their endless efforts to recruit the best for this project, has resulted in the accomplishment of this book. They are a veteran in the field of academics and their pool of knowledge is as vast as their experience in printing. Their expertise and guidance has proved useful at every step. Their uncompromising quality standards have made this book an exceptional effort. Their encouragement from time to time has been an inspiration for everyone.

The publisher and the editorial board hope that this book will prove to be a valuable piece of knowledge for researchers, students, practitioners and scholars across the globe.

List of Contributors

Wenshan Fang and Stefano Bruni
Mechanical Engineering Department, Politecnico di Milano, 20156 Milan, Italy

Jose´ Martinez-Casas
Centro de Investigacio´n en Ingenierı´a Meca´nica, Universitat
Polite`cnica de Vale`ncia, Camino de Vera s/n, 46022 Valencia, Spain

Anna Fraszczyk
Mahidol University, Nakhon Pathom, Thailand

Corinne Mulley
The University of Sydney, Sydney, Australia

Ni Yuancheng and Zhu Eryu
School of Civil Engineering, Beijing Jiaotong University, No. 3, Shangyuan Village, Haidian District, Beijing, China

Chen Zhuo
China Railway Fifth Survey and Design Institute Group CO
LTD, No. 9, Kangzhuang Road, Daxing District, Beijing, China

Chun Zhang and Haishan Xia
School of Architecture and Design, Beijing Jiaotong University, Beijing, China
UC-BJTU Future City Co-Lab, Beijing, China

Yan Song
Department of City and Regional Planning, University of North Carolina at Chapel Hill, Chapel Hill, NC, USA

Fei Yan
National Engineering Research Center for Rail Traffic
Operation and Control, Beijing, China

Chunhai Gao
Beijing Traffic Control Technology Co., Ltd., Beijing, China

Tao Tang
State Key Laboratory of Rail Traffic Control and Safety, Beijing Jiaotong University, Beijing, China

Yao Zhou
School of Electronic and Information Engineering, Beijing Jiaotong University, Beijing, China

Daizong Zhang and Haishan Xia
School of Architecture and Design, Beijing Jiaotong University, Beijing, China

Chen Shen
School of Civil Engineering, Beijing Jiaotong University, Beijing, China

Christopher Auffrey, Xin Fu, Xinhao Wang and Alyssa Wissel McClearnon
School of Planning, University of Cincinnati,USA
UC-BJTU Future City Co-Lab, Beijing, China
4 School of Design, Architecture, Art and Planning, University of Cincinnati, Cincinnati, OH, USA

Hongliang Wang and Eryu Zhu
School of Civil Engineering, Beijing Jiaotong University, No. 3, Shangyuan Village, Haidian District, Beijing, China

Zhuo Chen
China Railway Fifth Survey and Design Institute Group CO, LTD., No. 4, Kangzhuang Road, Huangcun Town, Daxing District, Beijing, China

Xiaoshu Cao
School of Geography and Planning, Sun Yat-Sen University, No. 135, Xingang West Road, Gangzhou 510275, Gangdong Province, People's Republic of China

Linna Li
Department of Geography, Beijing Normal University, Room 9719, Jingshi Building, No. 19, Xin Jie Kou Wai Street, Hai Dian District, 100875 Beijing, China

Samuel A. McKinley
Price Hill Will, Sustainable Places Studio, LLC, Cincinnati Area, KY, USA

Heng Wei
Beijing University of Technology, Beijing, China

Chang'an University, Xian, China 4 Department of Civil and Architectural Engineering and Construction Management, The University of Cincinnati, 792 Rhodes Hall, 2850 Campus Way, Cincinnati, OH 45221-0071, USA

Hesam Soleimani and Majid Moavenian
Department of Mechanical Engineering, Ferdowsi University of Mashhad, Mashhad, Iran

James Kelly and Marin Marinov
NewRail, Mechanical and Systems Engineering School, Newcastle University, Newcastle upon Tyne, UK

Afaq Khattak and Jiang Yangsheng and Zhu Juanxiu
School of Transportation and Logistics, Southwest Jiaotong
University, Chengdu 610031, Sichuan, China

Hu Lu
Department of Industrial and Systems Engineering, National
University of Singapore, Singapore 119077, Singapore

Zhouzhou Xu Zhen Yang and Liang Chen
Chongqing Jiaotong University, Chongqing, China

Zhen Yang, Zixue Du, Cheng Chen, Xiaoxia Wen, Zhouzhou Xu and Chongqing Jiaotong
University, No. 66 Xuefu Road, Nan'an District, Chongqing, China
School of Mechanical and Vehicular Engineering, Chongqing Jiaotong University, Chongqing, China

Rang Zhang, Yuanjin Ji, Lihui Ren and Gang Shen
Institute of Rail Transit of Tongji University, No. 4800. Cao an gong Road, Jia ding District, Shanghai, China

Luca D'Acierno, Marilisa Botte, Chiara Caropreso and Bruno Montella
Department of Civil, Architectural and Environmental Engineering, Federico II University of Naples, via Claudio 21, 80125 Naples, Italy

Antonio Placido
D'Appolonia S.p.A., via del Fiumicello 7, 80142 Naples, Italy

Xiaozhen Li, Jie Geng and Dangxiong Wang
Department of Bridge Engineering, School of Civil Engineering, Southwest Jiaotong University, Chengdu 610031, China

Hitoshi Hayashiya
Tokyo Branch, East Japan Railway Company, 2-20-68, Higashi-Tabata, Kita-Ku, Tokyo 114-8550, Japan

Cezary Specht
Department of Geodesy and Oceanography, Faculty of Navigation, Gdynia Maritime University, Aleja Jana Pawła II 3, 81-345 Gdynia, Poland

Wladyslaw Koc, Piotr Chrostowski and Jacek Szmaglinski
Department of Rail Transportation and Bridges Structures, Faculty of Civil and Environmental Engineering, Gdansk University of Technology, ul. Narutowicza 11/12, 80-233 Gdan´sk, Poland

Dangxiong Wang, Xiaozhen Li, Jie Geng and Yanlong Ge
Department of Bridge Engineering, School of Civil Engineering, Southwest Jiaotong University,
Chengdu 610031, China

Index